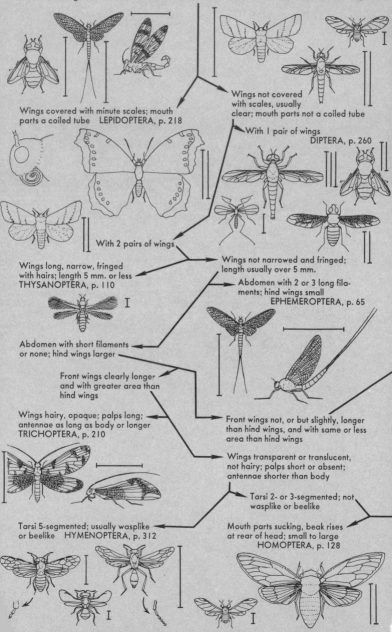

All wings obvious and membranous, but sometimes covered with scales or hairs

Wings covered with minute scales; mouth parts a coiled tube LEPIDOPTERA, p. 218

Wings not covered with scales, usually clear; mouth parts not a coiled tube

With 1 pair of wings DIPTERA, p. 260

With 2 pairs of wings

Wings long, narrow, fringed with hairs; length 5 mm. or less THYSANOPTERA, p. 110

Wings not narrowed and fringed; length usually over 5 mm.

Abdomen with 2 or 3 long filaments; hind wings small EPHEMEROPTERA, p. 65

Abdomen with short filaments or none; hind wings larger

Front wings clearly longer and with greater area than hind wings

Wings hairy, opaque; palps long; antennae as long as body or longer TRICHOPTERA, p. 210

Front wings not, or but slightly, longer than hind wings, and with same or less area than hind wings

Wings transparent or translucent, not hairy; palps short or absent; antennae shorter than body

Tarsi 2- or 3-segmented; not wasplike or beelike

Tarsi 5-segmented; usually wasplike or beelike HYMENOPTERA, p. 312

Mouth parts sucking, beak rises at rear of head; small to large HOMOPTERA, p. 128

A Field Guide
to the Insects

THE PETERSON FIELD GUIDE SERIES®
Edited by Roger Tory Peterson

Advanced Birding—*Kaufman*
Birds of Britain and Europe—*Peterson, Mountfort, Hollom*
Birds of Eastern and Central North America—*R.T. Peterson*
Birds of Texas and Adjacent States—*R.T. Peterson*
Birds of the West Indies—*Bond*
Eastern Birds' Nests—*Harrison*
Hawks—*Clark and Wheeler*
Hummingbirds—Williamson
Mexican Birds—*R.T. Peterson and Chalif*
Warblers—Dunn and Garrett
Western Birds—*R.T. Peterson*
Western Birds' Nests—*Harrison*
Backyard Bird Song—*Walton and Lawson*
Eastern Bird Songs —*Cornell Laboratory of Ornithology*
Eastern Birding by Ear—*Walton and Lawson*
More Birding by Ear: Eastern and Central—*Walton and Lawson*
Western Bird Songs—*Cornell Laboratory of Ornithology*
Western Birding by Ear—*Walton and Lawson*
Pacific Coast Fishes—*Eschmeyer, Herald, and Hammann*
Atlantic Coast Fishes—*Robins, Ray, and Douglass*
Freshwater Fishes (N. America north of Mexico)—*Page and Burr*
Insects (America north of Mexico)—*Borror and White*
Beetles—*White*
Eastern Butterflies—*Opler and Malikul*
Western Butterflies—*Opler and Wright*
Mammals—*Burt and Grossenheider*
Animal Tracks—*Murie*
Eastern Forests—*Kricher and Morrison*
California and Pacific Northwest Forests—Kricher and Morrison
Rocky Mountain and Southwest Forests—Kricher and Morrison
Venomous Animals and Poisonous Plants—Foster and Caras
Edible Wild Plants (e. and cen. N. America)—*L. Peterson*
Eastern Medicinal Plants and Herbs—*Foster and Duke*
Eastern Trees—*Petrides*
Ferns (ne. and cen. N. America)—*Cobb*
Mushrooms—*McKnight and McKnight*
Pacific States Wildflowers—*Niehaus and Ripper*
Western Medicinal Plants and Herbs—*Foster and Hobbs*
Rocky Mt. Wildflowers—*Craighead, Craighead, and Davis*
Trees and Shrubs—*Petrides*
Western Trees—Petrides
Wildflowers (ne. and n.-cen. N. America)—*R.T. Peterson and McKenney*
Southwest and Texas Wildflowers—*Niehaus, Ripper, and Savage*
Geology (e. N. America)—*Roberts*
Rocks and Minerals—*Pough*
Stars and Planets—*Pasachoff*
Atmosphere—*Schaefer and Day*
Eastern Reptiles and Amphibians—*Conant and Collins*
Western Reptiles and Amphibians—*Stebbins*
Shells of the Atlantic and Gulf Coasts, W. Indies—*Morris*
Pacific Coast Shells (including Hawaii)—*Morris*
Atlantic Seashore—*Gosner*
Coral Reefs (Caribbean and Florida)—*Kaplan*
Southeastern and Caribbean Seashores—*Kaplan*

THE PETERSON FIELD GUIDE SERIES®

A Field Guide to
Insects

America north of Mexico

Donald J. Borror
and
Richard E. White

Color and shaded drawings by
Richard E. White
Line drawings by
the Authors

*Sponsored by the National Audubon Society,
the National Wildlife Federation,
and the Roger Tory Peterson Institute*

HOUGHTON MIFFLIN COMPANY   BOSTON   NEW YORK

ISBN 0-395-91171-0 (hc)

ISBN 0-395-91170-2 (pbk.)

Library of Congress Catalog Card Number: 70-80420

Printed in the United States of America

VB   44   43 42 41 40   39   38   37   36   35

Editor's Note

ENTOMOLOGISTS fall into two categories: those who find insects endlessly fascinating and those who would get rid of them. Those in the first group, likely as not, begin as obsessive butterfly collectors and never quite lose their sense of wonder about the six-legged world. Those who would get rid of insects are afflicted with an impulse to drop bricks on beetles and all other small crawly things. They may eventually wind up working for chemical companies, devising more sophisticated techniques of annihilation.

Seriously, insects, because of their astronomical number, are undeniably important in our lives. They cannot be ignored. Many are "beneficial"; others, in human terms, are obviously harmful. There is much talk of the damage they do. However, if we evaluate insects across the board according to the measure of our economy, we find that they fall on the credit side of the ledger. The authors of this book point out that whereas damage by insects in the United States has been estimated to run into huge sums annually, their pollinating services alone each year are probably worth considerably more than the damage costs.

So then, indiscriminate eradication is out. Some control is necessary, but we deplore the unscientific spraying that eliminates defoliator and pollinator alike; we resent the primitive methods that kill not only the noxious insects but also their natural controls, their predators and parasites. We regret the decline of attractive butterflies along the roadsides and in our gardens. Particularly upsetting is the widespread use of persistent chemicals, such as the chlorinated hydrocarbons, which poison the ecosystem and travel through the food chain until even bald eagles, ospreys, and peregrine falcons are lethally affected.

There is a strong case for less chemical control of insects and more biological control. This requires a more critical knowledge of insects and demands the ability to differentiate between our allies and our enemies. This *Field Guide* should be useful to the new breed of economic entomologists who have the responsibility of resolving the dilemma, but it is really written primarily for the larger audience that includes the general naturalist and the ecologist, as well as the aesthetically oriented citizen who finds pleasure in the psychedelic patterns of butterflies and moths and in the porcelainlike textures of beetles.

North of the United States–Mexican border the species of insects outnumber the birds by more than 100 to 1. To be precise, about 88,600 species have been catalogued. It would be hopeless to include even a tenth of them in a book of field guide size; however, on the family level comprehensive coverage is possible. In this book the authors go below the family level in a few groups. For a complete treatment of eastern butterflies we refer the reader to *A Field Guide to the Butterflies* by Alexander B. Klots. A western butterfly guide is in preparation.

The identification of insects is more like the identification of flowers than it is like the field recognition of birds. Tiny and catchable, they may be examined in the hand. Their recognition is still a visual process, nevertheless, but more comparable to the bird-in-hand technique of early ornithology than that of present-day fieldglass bird study. The approach is more technical and a fairly complex terminology is often unavoidable. Instead of the binocular, the hand lens becomes the most useful optical instrument.

The device of the arrow in the illustrations, first used in the eastern bird guide and later applied to other books in the Field Guide Series, is particularly useful when dealing with insects, because wing venation and structural detail may be more determinative than obvious patterns and marks.

A hand lens, a folding net, and this *Field Guide* will take up little space in your knapsack or coat pocket. Take them with you on your travels and tick off the many new things you find.

ROGER TORY PETERSON

Preface

INSECTS are a remarkable group of animals. They occur almost everywhere and make up more than half of all the living things on this planet, they play a significant role in the world of nature and affect man directly or indirectly in many ways, and they exhibit some unusual physiological and structural peculiarities.

There are several hundred thousand different kinds of insects (about 88,600 in the area covered by this book), and they occur in almost every type of habitat. The only habitat they have not invaded to any extent is the ocean. Most of them are small, and some are minute. Our species vary in length from less than a milli meter to about 6 inches, but more than half of them are less than $\frac{1}{4}$ inch long. This means that they can live in small situations, and a small but diversified area may contain many kinds — there may be more kinds of insects on an acre than species of birds in the entire United States, and their numbers may be as high as several million per acre.

From man's point of view the insects are extremely important animals; some are very destructive, and many are very beneficial. Insects may damage or kill cultivated plants, they may damage or contaminate stored foods and other products, and they may attack man or animals and bite, sting, or act as vectors of disease. Annual losses caused by insects in the United States have been estimated to be about $3\frac{1}{2}$ billion. On the other hand, insects do a great deal of good. Many are important agents in the pollination of plants, including most orchard trees and many vegetables and field crops; some provide products of commercial value (honey, beeswax, silk, and shellac); many are important items in the food of birds, fish, and other animals; those parasitic or predaceous on other insects help keep noxious species under control; many are valuable scavengers; some have been used in the treatment of disease; many have been used in studies of heredity, evolution, stream pollution, and other biological problems; and insects are interesting and often very beautiful animals. Most people look upon insects as undesirable pests, but we believe insects do more good than harm: insects' pollinating services alone are probably worth about $4\frac{1}{2}$ billion annually in this country.

Most insects have an enormous reproductive capacity, and if it were not for the many checks on their increase (enemies, adverse environmental conditions, and the like) we would soon be overrun by them. To cite an extreme example: a pair of pomace flies (*Drosophila*), in which the female can lay a hundred eggs, may have 25 generations a year and could (if there were no checks) increase in a year to about 10^{41} flies. This number of flies, packed a thousand to the cubic inch, would form a ball 96 million miles in diameter! Many insects can reproduce parthenogenetically (without a male fertilizing the eggs), and the eggs of some insects hatch into not just one young but into many — over a thousand in the case of some of the chalcids.

We are concerned in this book primarily with the problem of identification, which is a first step in getting acquainted with any group of animals, but we have included some information on the structure, habits, and importance of the various insect groups. We hope that those beginning the study of insects with this *Field Guide* will go beyond identification to further study of these animals.

We have been aided in the preparation of this book by a great many people. Some have read portions of the manuscript and have made helpful suggestions and criticisms, and others have provided information on specific points. Various specialists, particularly in the combined United States Department of Agriculture and Smithsonian staffs of the United States National Museum, have assisted by loaning specimens used in preparing the color plates and many of the drawings. We are particularly indebted to Donald M. Anderson, Barnard D. Burks, Kellie Burks, George W. Byers, Oscar L. Cartwright, Arthur D. Cushman, Donald R. Davis, W. Donald Duckworth, J. Gordon Edwards, William D. Field, Oliver S. Flint, Richard H. Foote, Paul H. Freytag, Richard C. Froeschner, Raymond J. Gagné, Ashley B. Gurney, Jon L. Herring, Ronald W. Hodges, John M. Kingsolver, Josef N. Knull, James P. Kramer, Karl V. Krombein, John D. Lattin, Paul M. Marsh, Frank W. Mead, Frank J. Moore, C. F. W. Muesebeck, Lois B. O'Brien, André D. Pizzini, Louise M. Russell, Curtis Sabrosky, David R. Smith, Thomas E. Snyder, Paul J. Spangler, Ted J. Spilman, George C. Steyskal, Alan Stone, Eileen R. Van Tassell, Edward Todd, Charles A. Triplehorn, George B. Vogt, Luella M. Walkley, Rose Ella Warner, Donald M. Weisman, Janice White, Willis W. Wirth, and David A. Young. We wish to thank those on the staff of Houghton Mifflin Company whose expert advice and assistance have made the production of this book possible.

Contents

Illustrations

How to Use This Book

Identification. The identification of insects is not fundamentally different from the identification of birds, mammals, ferns, or other forms of life. It is simply a matter of knowing what to look for, and being able to see it. Three things complicate the problem of insect identification: there are so many different kinds of insects (some 88,600 species in North America), many are small and the identifying characters often difficult to see, and many undergo rather radical changes in appearance and habits throughout their life cycle, with the result that one may learn to recognize an insect in one stage but be unable to recognize it in another.

It is impossible in a book of this size to include all the information necessary for identifying the huge numbers of insects, so we carry the identification only to the family level (further in a few families). This reduces the problem considerably, but adds a complicating factor: many families consist of species that vary greatly in size, shape, and color. Identification in such cases must be based on certain structural details rather than color and general appearance. We reduce the problem further here by dealing principally with adults and only incidentally with immature stages.

The part of this book dealing with identification is based primarily on an examination of the insect *in the hand*. After identifying an insect by a detailed examination of its structural details, one often can recognize it in the field without examining these details. Some means of magnification is needed for seeing many of the minute features of insect structure. A 10× hand lens is sufficient in many cases, but a microscope (usually a stereoscopic microscope) is necessary for identification of many of the smaller insects.

Identification of an insect may be based on one or more characteristics: its general appearance (size, shape, and color), the form or character of various body parts (antennae, legs, wings, bristles, or other parts), how it acts (if it is alive when examined), where it is found (the type of habitat, and the part of the country), and sometimes such characters as the sounds it produces, its odor, or the hardness of its body. Many insects can be recognized by their general appearance as belonging to a particular order or group of families, but further identification often requires an examination of individual parts of the insect.

Identification of an insect is a process of progressively narrowing down the group to which the insect belongs. We suggest the following steps:

1. Identify the order to which the insect belongs. This is done by using the pictured key on the front and back endpapers. A comparison of the specimen with the first pair of alternatives will lead to further alternatives, and eventually to the order; if you are not sure at any step which way to proceed, try both alternatives. Then check your identification by consulting the paragraphs on *Identification* and *Similar orders* in the general account of that order, the information in the table on pp. 57–59, and the illustrations.

2. Identify the group of families in the order (if the order contains only a few families, proceed to step 3). The larger orders are variously subdivided, but the major groups in the order (suborders, superfamilies, etc.) are indicated in the introductory account (under *Classification*), and the distinguishing characters of each group are given at the beginning of the discussion of a group. If such a group is further subdivided, information will be given on the subdivisions, so that eventually the specimen can be narrowed down to a small group of families.

3. Identify the family. This is done by checking the paragraphs on *Identification* and also the accompanying black and white illustrations for the families in the group.

If the beginner will spend a little time studying the illustrations in this book and the information in the table on pp. 57–59, he should soon be able to recognize the order of most insects he finds — and further identification will be by the steps 2 and 3 indicated above.

Illustrations. This book contains one or more illustrations for most North American insect families, with diagnostic arrows pointing to the important distinguishing features. The families for which there are no illustrations are small or rare, and not very likely to be encountered by users of this book. A few illustrations are designed to explain the characters used in identification; the majority are intended to show the characters of individual families, and consist of drawings of individual insects and isolated body parts.

Most of the illustrations have been made of a particular species, though the species (or genus) is not always indicated. These species were selected because they are more or less typical of the group, or are the ones most likely to be encountered. Some of the simpler drawings are rather generalized and intended to represent a group of species rather than any particular one. Some families are illustrated solely by drawings of isolated body parts; the insects in such families are similar in general appearance to those in related families illustrated by drawings of entire insects.

The arrows on the illustrations indicate major diagnostic characters; usually there are italics in the accompanying text or legend page to link with the pointed arrows. Arrows ending slightly off a figure are intended to call attention to general features (shape, number of segments, etc.) of the body part indicated. Two or more arrows from a common point refer to a single character. Arrows are generally omitted when they would not indicate clearly such easily observable features as general shape and color.

The actual size of most insects illustrated is shown by a line near the drawing; for some large insects this line is in 2 or more sections. These lines generally represent body length (from front of the head to tip of the abdomen, or to the wing tips if they extend beyond the abdomen); horizontal lines in some cases represent wingspread. If there is no size line, information on size will be found in the text.

Terms, Abbreviations, and Symbols. Many terms referring to body parts or areas are defined where they are used. For terms not so defined consult the Glossary (p. 363).

A few abbreviations and symbols are used in the descriptions: FW, front wing; HW, hind wing; ♂, male; ♀, female. Measurements, unless otherwise stated, refer to body length. Smaller measurements are given in millimeters. Comparison of millimeter and inch scales is shown below.

All venational characters, unless otherwise stated, refer to the front wing. Any reference to the number of cells in a specific part of the wing refers to the number of closed cells (those not reaching the wing margin) unless otherwise indicated.

Geographical Coverage. This book covers the families of insects occurring in North America north of Mexico. The characters given for each group apply to North American species, and may not apply to all the species occurring elsewhere. The terms "N. America" and "N. American" refer to that portion of the continent north of Mexico. Groups for which no information is given on geographic range are widely distributed in North America. Most of the figures given for the number of species in a group are conservative estimates.

CENTIMETERS (1 CM. = 10 MM.)

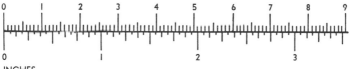

Comparison of millimeter and inch scales.

Collecting and Preserving Insects

COLLECTING and preserving insects is an ideal way to learn about them, and can be an interesting hobby. It provides the satisfaction of learning at first hand, and enables the collector to discover things about insects that he might not learn from books. It also provides material for the study of structural details that serve as identification characters. There are few restrictions against collecting insects compared to those for collecting other animals or plants. The average landowner has no objection to insect collecting on his property, and it is only in a few places such as parks where collecting may be restricted. Insects are so abundant that ordinary collecting has little or no effect on their numbers; the student of insects need not worry that his collecting activities will upset the balance of nature.

When and Where to Look for Insects

Insects occur almost everywhere, and the more places you examine the more kinds of insects you are likely to find. Insects are small and live in small habitats; different species may often be found only a few inches or a few feet apart. If you know where to look, you can find at least some insects virtually any time, but collecting will be more profitable from spring to fall than in the winter. Different insects are active at different times of the year, hence collecting throughout the year will yield a wider range of insects than collecting for only a part of the year.

You can begin your collecting at home, where you may find insects on pets, or infesting food, clothing, and other materials. On warm evenings outdoor insects are attracted to porch lights and windows of lighted rooms.

A good place outdoors to look for insects is on plants; each plant will attract certain species. Other insects occur in various concealed situations, such as in leaf litter or debris, under stones or other objects, in fungi, under bark, in dead logs, in decaying materials, and in the ground. Many insects live in water, either throughout their lives or during their immature stages, and may be found in different parts of specialized aquatic habitats. Adults of insects occurring in water only during their immature stages are usually found near water.

Collecting Equipment and How to Use It

The minimum equipment necessary to collect insects consists of your hands and a container for the insects collected. Certain tools are very helpful, nonetheless. For general collecting they include a net, killing jars, vials of preservative, envelopes, and small boxes; other tools are useful in some types of collecting. A pair of forceps is excellent for handling insects, as is a hand lens for examining them. Many of these items can be carried in a shoulder bag; the forceps and hand lens can be carried on a string around your neck.

The Net. A net for general collecting should be light and strong and have a fairly open mesh so that it can be swung easily and an insect can be seen through it. The size may depend on your personal preference, but most nets have a handle $2\frac{1}{2}$ to 3 ft. long and a rim about 1 ft in diameter; the bag should be about twice as long as the diameter of the rim, and rounded at the bottom. Marquisette, scrim, bobbinet, and bolting cloth are good materials for the bag, which should have a heavier material, such as muslin, around the rim. A fine mesh bolting cloth is probably the best material for the bag of a net that will be used primarily for sweeping.

Insect nets can be purchased from a supply house (prices ranging upward from a few dollars) or they can be homemade; homemade nets cost considerably less. An insect net can be made with a broom handle or similar stick, a wire for the rim, and a cloth bag, as shown in the illustration on the next page. If you prefer a net that can be taken apart and carried inconspicuously, use the frame of a fish-landing net, which you can buy from a sporting-goods store.

An insect net can be used in 2 general ways: you may look for an insect and swing at it or you may simply swing the net through vegetation. The first method requires a certain amount of speed and skill, especially for active or fast-flying insects. The second method, usually called "sweeping," can yield a considerable quantity and variety of small insects.

Insects caught in a net can be removed in various ways. Take care to prevent their escaping before you take them out of the net. Remove them with as little resultant damage as possible and — in the case of insects that bite or sting — without injury to yourself. You can keep an active insect from escaping by quickly turning the net handle to fold the bag over the rim.

Most insects can be removed from a net by grasping them through the net with the fingers. Small or fragile insects, easily damaged by this method, can be removed in these ways: (1) by inserting a box or bottle into the net and getting the insect directly into this container; (2) by working the insect into a fold of the net and placing this fold into a killing jar to stun the insect; or (3) by removing the insect with an aspirator (see p. 9). The first method is the one used by many collectors to remove butterflies or moths

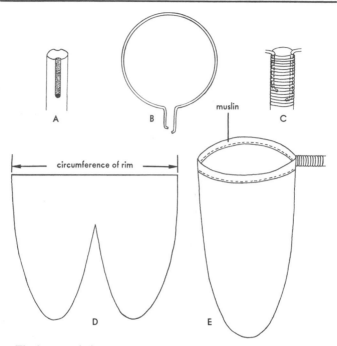

The homemade insect net. Cut grooves on opposite sides of one end of handle, as shown in A, 1 to about 2 in. from end and the other to about 3 in. from end; drill a hole about halfway through handle at end of each groove. Bend wire for rim (about No. 8 gauge) as shown in B, fit into holes and grooves, and fasten it there with fine wire or heavy cord (C). Cut material for bag as shown in D and sew. Completed bag, with muslin around rim, is shown in E.

from a net, since direct handling of the specimen often causes some of the scales to be rubbed off. If a butterfly or moth is grasped by the fingers through the net, grasp it by the body not the wings, and stun it by pinching the thorax before putting it into a killing jar (to reduce fluttering and wing damage in the jar).

Many beginners may be reluctant to grasp an insect for fear that it might bite or sting; such is much less likely than most people believe. An insect that bites does so by moving its jaws sideways and pinching or by piercing with the beak. Very few pinching insects are capable of causing pain or breaking the skin, and those that can are generally very large insects. Most biting insects are unable to bite if grasped firmly by the sides of the body.

An insect that stings does so with a structure at the posterior end of the body — usually quite readily. The only stinging insects are the bees, wasps, and some of the ants, and females alone can sting. Many flies and a few other insects strongly resemble bees or wasps but are quite harmless. If an insect in a net is one that might sting, remove it from the net in one of three ways: (1) put a fold of the net containing the insect into a killing jar until the insect is stunned (or pour an anesthetic such as chloroform over the fold of the net to stun the insect); (2) grasp the insect through the net with forceps and remove it; or (3) work the insect into a fold of the net, stun it by pinching the thorax, and then remove it. The third method is the simplest and quickest.

Insects caught by sweeping can be removed by shaking them to the bottom of the net and stunning them by the first method described above.

Killing Jars. Insects can be studied alive with considerable interest and profit, but it is well to kill and preserve a few. The characters that differentiate insects can be studied best in preserved specimens, and insect collections can be both attractive and instructive.

Killing jars are of various sizes and shapes, depending on the use they will receive. It is advisable to have 2 or more jars in the field, for insects of different types. Wide mouthed jars are preferable to narrow-mouthed ones, and glass jars should be reinforced with tape to reduce the hazards of breakage. Several materials can be used as killing agents, but the best are probably ethyl acetate and cyanide. Ethyl acetate is a much safer material to use than cyanide, but does not kill as quickly, and jars made with it must be recharged frequently; specimens killed by ethyl acetate are usually more relaxed than those killed by cyanide, and less likely to be discolored. Cyanide jars last much longer, and are quite safe to use if certain precautions are observed. Most collectors prefer to use killing jars made with cyanide.

Ethyl acetate (an ingredient of nail polish) is a clear liquid, and its fumes act as the killing agent. Jars made with it must contain something absorptive. Cotton or cloth can be used as the absorbent material, but plaster of paris is better. Pour a mixture of plaster of paris and water into the jar and allow it to set and dry; then place a few eyedroppers of the acetate on the plaster (which absorbs it), and the bottle is ready for use. Take care not to put in too much acetate, because wet insects make poor specimens. Add more ethyl acetate every few days.

Cyanide jars can be made with calcium, sodium, or potassium cyanide. Calcium cyanide is a dark gray powder often used as a fumigant and ordinarily is obtained from a store that sells insecticides. Sodium cyanide comes in the form of balls about an inch in diameter, which must be crushed before use in a killing jar; it is obtained from a company dealing in insecticides or chemicals.

Potassium cyanide (which looks like sugar) is usually available at a drugstore. The toxic agent in a cyanide jar is hydrogen cyanide, an extremely poisonous gas given off by the action of moisture on the cyanide. Calcium cyanide releases this gas very rapidly, and is perhaps the most dangerous type of cyanide to use; jars made with it ordinarily last only a month or two. Sodium and potassium cyanides give off the gas less rapidly and are less dangerous to use; jars made with them may last a year or more. Cyanide jars are easy to make but we recommend that the beginner not attempt to make his own. He can ask someone experienced in handling cyanide to make them for him or else buy them from a supply house.

The chief hazard in the use of cyanide jars is not so much the gas given off (which is toxic if inhaled in quantity) as the possibility of a broken jar and a cut hand. This hazard can be greatly reduced by covering the bottom part of the jar with adhesive, masking, or electric tape. *All* killing jars should be so taped, and all should be conspicuously labeled POISON.

Other materials that can be used in a killing jar are carbon tetrachloride and chloroform. Carbon tetrachloride is the easier to obtain but is more dangerous. A killing jar made with either of these materials (which are liquids) is made in the same way as an ethyl acetate jar, and must be recharged frequently.

The efficiency of a killing jar depends to a large extent on how it is used. Never leave it open any longer than necessary. The hazard of an open jar, even one made with cyanide, is not very great, but the escaping gas reduces the strength of the jar. Keep the inside of the jar dry. Moisture from the insects or from the plaster sometimes condenses on the sides, particularly if the jar is exposed to bright sunlight. Moisture can be reduced by keeping a few pieces of cleansing tissue in the jar. When butterflies or moths are put into a jar, many of their scales come off and remain there; other insects put into this jar will become covered with the scales and look dusty. It is advisable to have a special jar for butterflies and moths and put other insects into different jars. All killing jars should be wiped out occasionally.

The time required for a killing jar to kill an insect depends on both the jar and the insect, and may vary considerably. Insects can be left in an ethyl acetate jar for long periods without damage, but if left too long in a cyanide jar they may become discolored. Remove insects from a jar within an hour or two after they are killed.

Specimens removed from a killing jar in the field may be stored temporarily in small boxes or envelopes; the boxes should contain some pieces of cleansing tissue. Letter envelopes, or triangular ones (see illus. opp.), are better for insects with small bodies and large wings (butterflies, moths, dragonflies, etc.). Place the specimen in the envelope with the wings together above the body and write collecting data on the outside of the envelope.

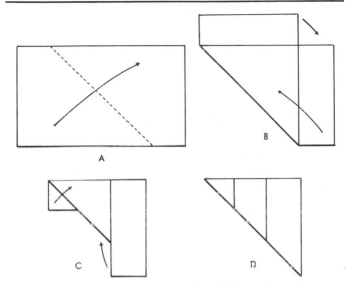

A triangular paper envelope, made by folding a piece of paper as shown in A, B, and C to form the completed envelope (D).

Beating. Many insects that occur on plants "play dead" and drop off the plant when it is jarred. You can collect such insects by placing a net or sheet under the plant and beating the plant with a stick. An insect net does not provide much collecting area. A better device is a sheet of white cloth about a yard square, held spread out by two slender pieces of wood along the diagonals of the cloth.

Sifters. Insects occurring in leaf litter or debris are most easily collected by some kind of sifting device. The simplest procedure is to shake a handful of the debris onto a sheet or cloth. Small animals present will be detected by their movement and can be picked up with forceps, an aspirator, or a wet brush. A more efficient method is to use a Berlese funnel, a large funnel with a circular piece of screen or hardware cloth in it. The material to be sifted is placed in the funnel on the screen, the funnel is held upright with a ringstand or other support, and a container (usually of alcohol) is placed below the funnel. As the material dries (this can be hastened by a light bulb over the material) the insects move downward, and eventually go into the container below the funnel. Generally it takes a few hours to get most of the insects out of the material by this method.

Aspirator. An aspirator is a device with which insects are sucked into a vial or other container. This container has intake and mouthpiece tubes entering it; the intake tube is usually a piece of ¼-in. glass tubing about 6 in. long, and the mouthpiece tube is ¼-in. glass tubing through the cork plus 1½ to 2 ft. of rubber tubing, with a piece of cloth over the inner end to prevent insects from being sucked into the mouth. Place the end of the intake tube close to the insect and suck through the mouthpiece. An aspirator is useful for collecting small, not too active insects, either from an insect net, a plant, or another situation. The insects so collected may be kept alive or transferred to a killing jar.

Traps. Insect traps can be constructed in various ways, and the attractant used will determine the types of insects likely to be caught. The attractant may be artificial light, decaying meat or fruit, or other things. The traps are constructed so that once the insects get in they cannot get out. Some traps (such as those with light as the attractant) may be constructed to direct the caught insects eventually into a container of alcohol or a killing jar; others (such as traps baited with decaying materials) may be constructed to direct the insects into a special chamber and not into the bait.

Insects attracted to lights or baits may often be collected by hand or with a net, without use of a trap. Insects attracted to a light, for example, can be collected directly into a killing jar or other container when they alight on some surface nearby (a light-colored wall or sheet). Thus you can collect only those specimens in which you are particularly interested.

Aquatic Collecting Equipment. Many aquatic insects can be collected by hand or with forceps from objects in the water; more can be collected with a net or other device. An aquatic net should be much sturdier than an aerial net, with a bag no deeper than the diameter of the rim. Kitchen strainers 4 to 6 in. in diameter make good dip nets.

If a dip net or strainer comes up full of mud and debris, the insects taken may be difficult to see unless they move. To locate them, dump the net contents into a white-bottomed pan of water. Insects are easily seen against this white background, and can be removed by hand, forceps, or eyedropper. Small free-swimming forms like mosquito larvae are best collected with a long-handled white enameled dipper, in which the insects similarly show up well. Remove them with an eyedropper.

Other Equipment and Methods. Many insects can be collected directly into a killing jar or other container, without the use of a net. This is the simplest way to collect insects that alight on a flat surface and do not fly too readily, and many insects on flowers. Various types of buildings often serve as insect traps; insects fly in and alight on the walls or ceiling and remain there, or they alight on windows — from which they can be collected directly into a jar.

A heavy knife is useful for cutting into logs or branches, opening galls, prying up bark, or digging into places that may harbor insects. Keep at hand a notebook and pencil for taking notes, and make triangular paper envelopes from the notebook pages.

Collecting some kinds of insects requires special equipment not mentioned above. An ingenious collector should be able to devise equipment and procedures for collecting such insects.

Mounting and Preserving Insects

Most insects are preserved dry, normally on pins, and once dry will keep indefinitely. Soft-bodied insects must be preserved in liquids, since they shrivel or become distorted if preserved dry. Minute insects that are hard-bodied may be mounted (dry) on "points," but many must be mounted on microscope slides for detailed study. Insects that are preserved dry should be pinned or mounted as soon as possible after they are collected; if allowed to dry first they become very brittle and may be broken in the process of mounting.

Relaxing Specimens. Dried specimens can be relaxed by placing them in a humid atmosphere for a few days; any airtight jar can be used as a relaxing chamber. Cover the bottom of the jar with wet sand (add phenol or ethyl acetate to prevent mold), put the insects into the jar in small open boxes or envelopes, and close the jar.

Pinning. Insects sufficiently hard-bodied to retain their shape when dry, and big enough to pin, are normally preserved by pinning. Common pins are too thick and too short, and they rust; insects should be pinned with *insect pins*, made especially for this purpose, which can be bought from a supply house. They are available in various sizes (thicknesses). The best sizes for general use are Nos. 1 (very slender), 2 (less slender), and 3 (thicker, for larger insects).

Most insects are pinned vertically through the thorax; a few are pinned sideways. Beetles and hoppers are pinned through the front part of the right wing, at a point where the pin on emerging from the underside of the body will not damage a leg. Bugs are pinned through the scutellum (p. 33) if it is large enough to take a pin or through the right wing, as for beetles. Grasshoppers and crickets are pinned through the rear edge of the pronotum, just to the right of the midline. A treehopper is pinned through the pronotum just to the right of the midline. Dragonflies and damselflies can be pinned vertically through the thorax with the wings horizontal, but it is better to pin them sideways, left side up, with the wings together above the body, the pin going through the thorax below the wing bases. If the wings are not together when the specimen is removed from the killing jar, place the specimen in an envelope (the wings together above the body) for a day or two until it has dried enough for the wings to stay in this position; then pin it.

The simplest way to pin an insect is to hold it between the thumb and forefinger of one hand and insert the pin with the other hand. All specimens and labels put on a pin should be at a uniform height; this is most easily accomplished with a pinning block.

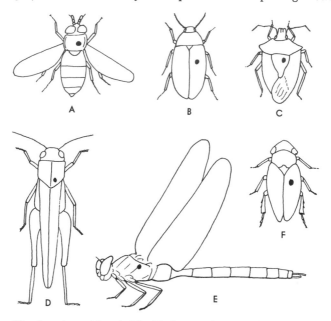

How insects are pinned. The black spots show the location of the pin in the case of flies (A), beetles (B), bugs (C), grasshoppers (D), dragonflies and damselflies (E), and leafhoppers, froghoppers, and planthoppers (F).

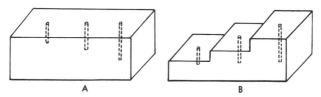

Pinning blocks. These may be made of a rectangular piece of wood (A) or one shaped like steps (B), with holes drilled to 1, ⅝, and ⅜ in. After placing a specimen on a pin, insert the pin in the 1-in. hole until it touches bottom. The ⅝-in. hole is used to position the locality-date label on the pin, and the ⅜-in. hole to position a second label, if there is one.

Mount the insect about an inch up on the pin. With large-bodied insects there should be enough of the pin above the insect to permit easy handling.

Sagging of the abdomen of a pinned insect (like a dragonfly) can be prevented in the following ways: (1) by sticking the pinned specimen onto a vertical surface, with the abdomen hanging down, and leaving it there until the abdomen dries; or (2) by placing a small piece of cardboard on the pin, just under and supporting the insect, and leaving it there until the insect dries; or (3) by supporting the sagging abdomen with crossed pins, the abdomen resting in the angle where the pins cross.

A sheet of cork, balsa wood, or other soft material is useful for the temporary storage of pinned specimens until they can be sorted and placed in boxes.

Mounting Small Insects on Pins. Insects hard-bodied enough to mount dry but too small to pin are usually mounted on "points." Points are small triangular pieces of cardboard, about 8 mm long and 3 or 4 mm. wide at the base; the pin is put through the base of the point and the insect is glued to the tip. Points can be cut with scissors or punched out with a special punch (obtainable from a supply house), or they can be purchased from a supply house.

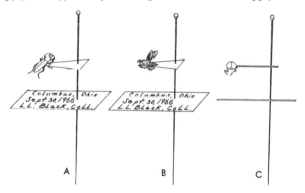

Mounting small insects on points. A, beetle, dorsal side up; B, fly, left side up; C, beetle mounted dorsal side up, attached by its side to the bent-down tip of the point.

Insects put on points should be glued so that the body parts to be examined in identifying the insect are not obscured. The best position for an insect is on its right side, with the head away from the pin (B, above). Flat insects that may be difficult to mount on their side are usually mounted dorsal side up at the extreme tip of the point.

Place an insect that is to be put on a point at the edge of a block,

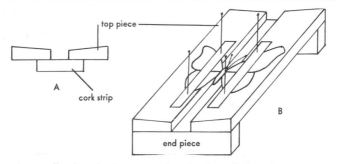

A spreading board. A, cross section showing relation of cork strip to groove in center of board; B, a board containing a spread specimen. Supply-house boards are generally 12 in. long and 4 to 6 in. wide; the top pieces are $\frac{3}{8}$ in. thick at the inner edge and $\frac{1}{2}$ in. thick at the outer edge. The center groove may be $\frac{1}{4}$ to $\frac{1}{2}$ in. or more wide, depending on the size of the insects to be spread on it; the width of this groove is adjustable in some supply-house boards. The top pieces should be of wood soft enough to take a pin easily.

grasp a pin containing a point by the sharp end, and touch the tip of the upper surface of the point to a drop of glue and then to the insect. Long and slender insects put on points should be put on 2 points that have their tips separated to form a V. Use glue or household cement, *not* mucilage.

Very small insects are sometimes mounted on "minuten" pins instead of points; they are short and very slender. Insert one end of the pin into the insect and the other into a small piece of cork on a regulation insect pin.

Spreading. The position of the legs and wings in a pinned insect is generally not important, as long as all parts can be seen and studied, but some insects (such as butterflies, moths, and perhaps some others) should have their wings spread before being put into the collection. An insect can be spread on a spreading board or spread upside down on a flat surface; the position of the wings depends on the type of insect. A spreading board (see illustration above), which can be bought from a supply house or can be home-made, is used for specimens going into a pinned collection. The specimens are usually spread dorsal side up. An insect to be mounted under glass, as in a Riker or glass mount (see pp. 19–20), is spread on any flat surface soft enough to take an insect pin and placed in an upside-down position. The spreading process consists of placing the wings in a standard position and fastening them there (with strips of paper), and leaving them to dry in that position.

The wings of a butterfly, moth, or other insect in which the front wings are more or less triangular are spread with the hind margin

of the front wings at right angles to the body, and the hind wings
far enough forward so that there is no large gap at the side between
the front and hind wings (see illustrations of such insects in this
book). In grasshoppers, damselflies, dragonflies, and other insects
whose front wings are elongate rather than triangular, the wings
are generally spread with the front edge of the hind wing at a right
angle to the body and with the front wings far enough forward to
just clear the hind wings. The front and hind wings of a butterfly,
moth, or mayfly overlap at the base, with the front edge of the hind
wing under the rear edge of the front wing, and should be so overlapped when spread. The front and hind wings of most other
insects are not overlapped when spread.

For the beginner and those interested primarily in mounting a
few large or showy insects for display, we recommend spreading
the insect in an upside-down position and mounting it under glass;
spreading is easier this way, and does not require a special spreading board. For the more advanced student, or anyone planning to
specialize in insects that are normally spread, we recommend the
use of a spreading board.

The steps in spreading a butterfly in an upside-down position
are illustrated on p. 16. Butterflies and moths must be handled
carefully (forceps preferred) to avoid rubbing scales off the wings.
To spread a butterfly, grasp it by the thorax, ventral side up, and
insert a pin through the thorax. Pin the insect on its back on a flat
surface; if the wings are together above its back, spread them
apart with forceps as the insect is lowered to the surface. Pin
strips of paper over the wings on each side (A). Remove the lower
pin on one side and, holding the strip fairly tight, raise the front
wing. Do this with a pin, placing the pin behind a heavy vein
near the front basal part of the wing; avoid pushing the pin through
the wing because this leaves a hole. If the body tends to swivel,
insert a pin alongside it at the base of the abdomen. When the
wing is raised to the proper position (rear edge at a right angle to
the body) insert a pin through the paper strip just in front of the
front edge of the wing (B); pin the lower end of the strip down,
anywhere behind the hind wing. Now do the same thing on the
other side (C). Next raise one hind wing so that the notch at the
side between the two wings is reduced, and pin the paper strip just
behind the rear edge of the hind wing (D). Repeat the process on
the other side (E). Next orient the antennae to a symmetrical
position and hold them in place by pins placed alongside them (E).
Now hold the body down with forceps at the pin through the body,
and carefully remove this pin (F). If the legs project upward very
far fasten them close to the body with a strip of paper across the
entire specimen, at right angles to the body. Data on when and
where the specimen was collected should be noted beside it, this
information accompanying the specimen when it is later put into
the collection.

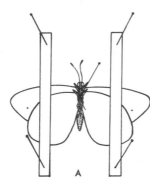

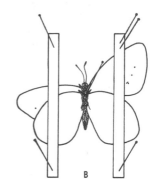

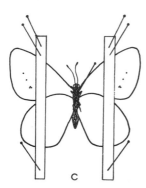

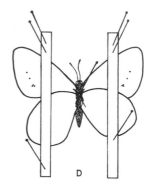

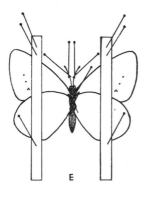

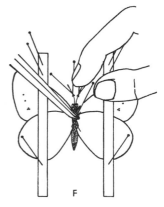

Steps in spreading a butterfly.

The time required for a spread specimen to dry will vary with its size and with temperature and humidity. A large butterfly or moth that might take several days to dry at normal room conditions can be dried in an hour or two with heat — in an oven or under an electric light. To determine if the specimen is dry, touch the abdomen gently with a pin; if the body is stiff the specimen is dry, but if the abdomen is still flexible the specimen is not yet dry.

A specimen spread on a spreading board is spread the same way, except that the insect is pinned through the thorax from above and placed at the standard height on the pin. The pinned specimen is then placed on the board, the pin going into (perhaps through) the cork strip at the bottom of the groove, until the base of the wings is even with the upper surface of the top piece of the board. Next the wings are spread and fastened down and the antennae oriented to a symmetrical position. Do not remove the pin in the insect's body after the spreading.

Mounting on Slides Small Insects, particularly soft-bodied forms that shrivel when preserved dry, and various insect parts (genitalia, wings, mouth parts, etc.) are often mounted on slides for detailed study. Insects and parts that are thick or dark-colored are usually cleared (made translucent) before mounting. Then they are mounted on microscope slides in some type of mounting medium (like Canada balsam) Such mounting involves treatment of the specimen with various reagents. Consult technical books for an explanation of the process.

Insect wings can be mounted on slides (preferably 2 x 2 in., which can be projected) without any mounting medium. If the wings are not folded they can be broken off a dried specimen, placed on a 2 x 2 slide and oriented, another slide put on top, and the slides taped together. Such a slide is permanent, and can be made in a few minutes. A folded wing must be relaxed and unfolded before it is mounted. This is done by putting the wing in alcohol or a special relaxing solution. The flattened wings are allowed to dry before the other slide is added. The wings of butterflies and moths can be mounted this way, but cannot be projected unless the scales are bleached or removed; directions for doing this are given on p. 221.

Preservation in Envelopes. Use envelopes for the permanent preservation of slender and relatively fragile insects such as dragonflies, damselflies, and crane flies. This method saves considerable space compared to mounting on pins, and the specimens are less apt to be broken (or if they are, the parts are not lost). The simplest type of envelope is a triangular one (see p. 9), labeled on the outside. If clear plastic envelopes are used, it is possible to study the specimens without removing them from the envelope. Labeling for specimens in plastic envelopes can be done on a white card (same size as the envelope) placed inside with the specimen.

Preservation in Fluids. Larvae, nymphs, and soft-bodied adults are usually preserved in fluids, since they shrivel when preserved dry. The best preservative is a 75–80 percent solution of ethyl alcohol. Alcohol is a killing agent for most insects but is unsatisfactory for killing larvae — it may distort or discolor them. Larvae should be killed by hot water or chemicals. A good killing solution for larvae is one containing 1 part of kerosene, 7–10 parts of 95 percent ethyl alcohol (less for very soft-bodied larvae), 2 parts of glacial acetic acid, and 1 part of dioxane. Isopropyl alcohol (rubbing alcohol) will serve if ethyl alcohol is not available.

Containers of specimens preserved in alcohol should be completely filled with alcohol and tightly stoppered to reduce or prevent evaporation. If a large number of specimens are killed in a small amount of alcohol, this alcohol should be replaced after a day or two. Rubber stoppers are better than cork ones. Periodically inspect containers of specimens in fluid and replace any evaporated fluid.

Arrangement and Care of the Collection

Labeling. All specimens should be labeled with at least the locality and date of their capture — specimens without such labels are of little value. The name of the collector and in some cases information on the habitat or food of the specimen are also helpful. For pinned insects put this information on one or two small pieces of paper on the pin below the insect; for specimens mounted on slides put it on a slide label, and on a piece of paper (in pencil or waterproof ink) placed inside the container for specimens preserved in fluid.

Labels on pinned specimens should be on fairly stiff white paper no larger than about $\frac{1}{4}$ x $\frac{3}{4}$ in. (or smaller), at a uniform height on the pin (achieved with a pinning block) and parallel with the insect (or point). Locality labels may be handwritten or printed. Printed labels can be obtained from a supply house or cut from a photograph of a sheet of typewritten labels. The locality given should indicate the capture site as closely as possible. The county may be sufficient in some cases, but if the county is large list the town or other site. Place these labels on the pins in the same way

Columbus Columbus Columbus Columbus	Lincoln Co., Lincoln Co., Lincoln Co.,
O. O. O. O.	Me. Me. Me.
Columbus Columbus Columbus Columbus	D.J. Borror D.J. Borror D.J. Borror
O. O. O. O.	Lincoln Co., Lincoln Co., Lincoln Co.,
Columbus Columbus Columbus Columbus	Me. Me. Me.
O. O. O. O.	D.J. Borror D.J. Borror D.J. Borror
Columbus Columbus Columbus Columbus	Lincoln Co., Lincoln Co., Lincoln Co.,
O. O. O. O.	Me. Me. Me.
	D.J. Borror D.J. Borror D.J. Borror

A   B

Sheets of locality labels, actual size, with locality alone (A) or locality and collector (B), each label with a space for writing in the date.

throughout the collection — whether they are read from the right or left is a matter of personal preference.

An insect collection should contain some identification labels. How these are best arranged will depend on the size of the collection and the extent to which it is identified. Most collections should be labeled at least to order and family, and the specimens so arranged that a single label can serve for all the specimens in a group. When specimens are labeled to species, an identification label is placed on each specimen or on the first in a group. This label is a plain piece of paper (about an inch square and at the base of the pin) containing the scientific name of the insect, the name of the person identifying it, and the date (month and year) the identification was made.

Boxes for Pinned Insects. Pinned insects are kept in boxes having a soft material in the bottom to permit easy pinning; they can be obtained from a supply house or be homemade. The most common supply-house type is a Schmitt box, a wooden box about 9 x 12 x 2½ in. with a tight fitting lid and the bottom lined with sheet cork or similar material. Such boxes cost from a few to several dollars. Similar boxes made of heavy cardboard are available from supply houses and cost one half to one third as much. Homemade boxes may be made of wood or heavy cardboard, and the bottom lined with sheet cork, balsa wood, Styrofoam, or corrugated cardboard. The material in the bottom should fit tightly, and if corrugated cardboard is used it should be soft enough to take an insect pin. Large collections may be housed in Schmitt boxes, or in cabinets containing trays or drawers constructed like Schmitt boxes.

Riker Mounts. A Riker mount is a glass-topped box containing cotton, with insects on the cotton just under the glass. Insects spread for mounting in a Riker mount are spread in the upside-down position described on p. 15. Riker mounts may be of various sizes, and also can be purchased from a supply house or be home-

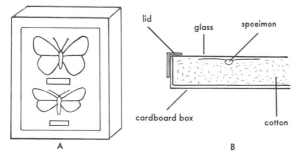

The Riker mount. A, a completed mount; B, sectional view showing a specimen in place under the glass.

made. Homemade boxes can be made from a cardboard box, a piece of glass cut to fit inside the lid, and binding tape. The box should be about an inch deep; cut deeper boxes down to this depth. Cut out a section of the lid, leaving about a $\frac{3}{8}$-in. margin around the edge. Covering the lid with tape will improve its appearance. Fasten the glass inside the lid with strips of tape placed so that they do not show from the outside. If you want to hang up the mount put 2 brass fasteners on the bottom and reinforce them inside with tape, then tie a string or wire between them on the outside. Before putting a thick-bodied insect into a Riker mount, tease the cotton to make a small depression for the body. Fasten the lid on the mount with pins or tape.

Specimens in Riker mounts are easily displayed, and the mounts can take considerable handling without damage to the specimens. A repellent can be placed under the cotton to protect the specimens from pests. Riker mounts have 2 disadvantages: only 1 side of the specimen can be seen and many moths will fade after prolonged exposure to light.

Glass Mounts. Glass mounts are similar to Riker mounts, but have glass on both top and bottom and contain no cotton; both sides of the specimen(s) are visible. They generally contain only one or a few specimens, and can be made in various ways. The following procedure is relatively simple:

1. Spread specimen in an upside-down position and fasten legs close to the body.

2. Cut 2 identical pieces of glass for top and bottom of the mount, allowing a margin of at least $\frac{1}{4}$ in. on all 4 sides of the specimen(s).

3. Cut enough supporting glass to provide room for body of the insect. These pieces will be as long as 1 dimension (usually the shorter) of top and bottom pieces, and a width that will leave a

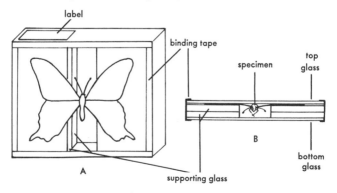

A glass mount. A, the finished mount; B, sectional view.

space in center of the mount 2 or 3 times as wide as body of the insect.

4. Clean the glass thoroughly, preferably with a glass cleaner; use a brush to remove lint from the glass.

5. Fasten the pieces of supporting glass to bottom piece with a small drop of cement (such as Duco household cement) on each outside corner; press each supporting piece into position, in line with the edges of bottom glass; remove any excess cement. Allow time for cement on one piece to set before the next piece is added.

6. When the supporting glass is in position, place specimen on the supporting pieces and center it. Put a small drop of cement on the 4 outer corners, then put on the top glass, being careful not to move the specimen. Press this glass down hard and put a small weight on it; leave the weight in place until cement sets (an hour or more).

7. Tape edges of the mount with slide-binding or electric tape; this covers any sharp edges of the glass, and gives the mount a finished look. The label is placed as shown.

Glass mounts are inexpensive, easy to make, and they provide a safe and attractive method for storage and display. If specimens put into these mounts are heat dried or otherwise made pest-free before mounting, they should remain pest-free. Two or more specimens can be put in a single mount by placing one above the other, or side by side (with 3 groups of supporting glass). The use of standard sizes will simplify glass cutting and mount storage.

Plastic Mounts. Plastic mounts may consist of 2 sheets of thick plastic with the insect mounted between them, or a plastic block in which the insect is embedded. If 2 sheets are used, each is bulged out where the insect's body will be, the 2 sheets are put together with the insect between them, and sealed around the edge with acetone or tape. Embedding an insect in a block of plastic is a rather involved process, but the final product is an attractive, durable, and permanent mount. Supply houses offer materials for this type of mounting and instructions for their use.

Protection of a Collection from Pests. Insect collections are subject to attack by dermestid beetles and other pests, which will ruin the collection if it is not protected. Insect boxes can be protected with a repellent such as naphthalene flakes or paradichlorobenzene (napthalene flakes last longer). The repellent in a box of pinned insects can be put in a small pillbox or wrapped in a piece of cloth firmly attached to a corner of the box. The repellent in a Riker mount is placed under the cotton. Collections should be examined periodically for signs of damage, and if pests are detected the collection should be fumigated or heat-treated. It is not possible to put repellent in a glass mount, so take care to make sure that specimens are pest-free when put in these mounts.

Handling Specimens. Insects are very brittle when dry. Careless handling of a pinned specimen can result in the loss of legs, antennae, or other parts. Broken-off parts often can be replaced by use of glue or cement. Experience (and accidental damaging of prize specimens) will impress on the collector the importance of care in handling mounted specimens.

Supply Houses

Much of the equipment necessary for making an insect collection can be homemade or purchased in local stores, but some items must be obtained from a supply house. A few of the leading supply houses are:

Bio Metal Associates, 316 Washington St., El Segundo, Calif. 90245. Eastern office: BioQuip East, 1115 Rolling Rd., Baltimore, Md. 21228. Western office: BioQuip West, P.O. Box 61, Santa Monica, Calif. 90406

Carolina Biological Supply Co., Burlington, N.C. 27216

General Biological Supply House, Inc., 8200 S. Hoyne Ave., Chicago, Ill. 60620

Powell Laboratories, Gladstone, Oregon 97027

Ward's Natural Science Establishment, Inc., P.O. Box 1712, Rochester, N.Y. 14603

Ward's of California, P.O. Box 1749, Monterey, Calif. 93942.

Work with Living Insects

THE STUDENT of insects who does nothing but collect, kill, and mount these animals and study the dead specimens will miss the most interesting part of insect study. Anyone who takes time to study *living* insects will find that they are fascinating and often amazing little animals. They can be studied in the field or in captivity. Many are very easy to keep in captivity, where they can be studied more easily and at closer range than in the field.

Some collectors try to catch an insect as soon as they see it. We suggest that you occasionally stop and watch an insect awhile before you try to collect it. In the case of fast flying insects such as dragonflies, you will probably not be able to catch one the instant you see it and may be compelled to watch. This may try your patience, but in observing it you may learn things about its habits that you would not learn if you caught it immediately.

If insects are kept in captivity a short period and released (or put into the collection), they require only simple containers and relatively little care. If collected in immature stages and kept until the adults appear, they require more care and sometimes special containers. If cultures are maintained throughout one or more generations, special containers and even more attention are necessary.

Field Observations

You can insect-watch much as you might bird-watch — by going into the field and keeping your eyes and ears open. A binocular is seldom needed, as for bird-watching, except occasionally for such insects as dragonflies. A hand lens, however, is very helpful.

A pond is an excellent place for watching insects. It probably has water striders and whirligig beetles moving about on its surface, various beetles, bugs, and other insects in shallow water where they can be watched from shore, and dragonflies, damselflies, and other insects flying or resting near or over the pond. Brief observation of the dragonflies, for example, will reveal that each species has a definite zone (height) of flight above the water surface, each flies and rests in a characteristic manner, and each has a particular method of laying eggs, flying in tandem, and chasing other individuals. An observer will discover that some of the larger dragonflies patrol definite territories; they will attempt to mate with any female of their species entering this territory, and will engage in characteristic dances or chases with other males of their species that enter the territory. An examination of the vegetation around the edge of a pond may reveal nymphs transforming into adults. This process — the adult emerging from the nymphal skin and expanding — takes from one-half hour to an hour, and at the right season it is not difficult to find emerging individuals.

Another good place to watch insects is at flowers. Many insects occur there, and although some of them are fairly small they can usually be observed at close range. Observation of the bees will reveal the basis of the expression "busy as a bee"; a butterfly may be seen uncoiling its proboscis and extending it down into the flower; and the sunlight striking the wings of a hovering syrphid fly will make it look like a small jewel. Some animals occurring on flowers are predators, and feed not on the flower but on the bees and flies that visit it. If you observe a flower containing a crab spider or an ambush bug for a while, you might see either animal catch its prey, and you will be surprised at how large an insect it can catch. The observant insect-watcher will see many instances of one animal eating another.

Ants are very interesting insects, especially at or near their nest, or on plants in association with aphids. The ground around a large mound nest may be almost alive with ants coming or going, perhaps carrying something. If a stone or board is lifted and an ant nest under it is exposed, the ants will busily transport their young to the shelter of their underground burrows. A few ants are often found around a large cluster of aphids on a plant; if you observe these for a time you may see the ants feeding on a watery fluid (honeydew) that issues from the end of the aphid's abdomen.

A plant selected at random and examined carefully will seldom fail to have some insects on it, and there are likely to be many kinds — each on a particular part of the plant and behaving in a characteristic fashion. The plant may also contain the eggs, larvae, or pupae of many types of insects.

If you visit a site at fairly regular intervals you may be able to follow the seasonal history and development of the insects there. The development of a wasp nest, galls, leaf miners, the activities of a group of webworms or tent caterpillars, or the development of the insects in a fallen log can be followed by such observations.

Keeping Insects in Captivity

Cages. There are many types of containers to use as insect cages. A few simple types are illustrated opposite (especially A–D). An insect can be kept for a time in an empty container (A or B), but it will generally live longer if something like natural conditions are provided.

Plants supplied as food must be kept fresh or regularly replaced.

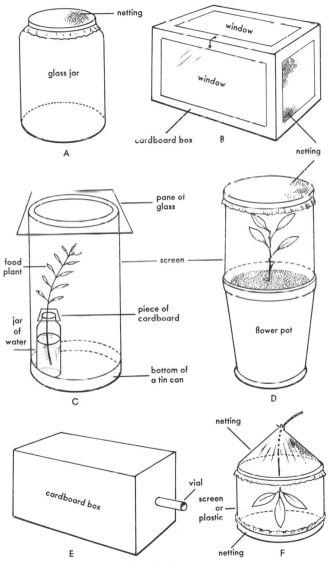

Cages for insects.

They can be kept fresh longer if placed in a jar of water, with a cover on the jar around the stem of the plant to prevent insects on the plant from falling into the water (C). Sometimes a plant can be grown in the cage, or a cylinder of screen can be placed over a potted plant (D). Since insects vary greatly in the type of food they eat, the kind provided for a caged insect will depend on the insect's food habits. Insects feeding on other living insects present the major food problems. Whatever the food, generally it must be replaced before it deteriorates.

Many insects, notably plant feeders, need not be provided with water, since they get enough in their food. Others, those feeding on drier foods, may require additional moisture, and a wet sponge or piece of cotton in the cage or a vial of water plugged with cotton and lying on its side can supply it. Take care to avoid an excess of moisture; this promotes the growth of mold, and droplets of water on the walls of the cage may trap insects.

It is often effective to approximate natural conditions in the cage by having sand, soil, or stones, and a plant or some object on which the insect can rest. If an insect being reared requires special conditions for pupation (soil or debris), these conditions should be provided.

Sometimes the best way to have caged insects under conditions as natural as possible is to cage them in the field in their normal habitat. Insects feeding on a plant can be caged in a bag or cylinder placed over the part of the plant on which they are feeding (F). Aquatic insects can be reared in cages partly submerged in their habitat; the screen of such cages should be fine enough to contain the insects but coarse enough to allow food material to get in.

Rearing. Rearing adults from immature stages enables one to follow the insect's life history, and provides the collector with excellent specimens for a collection. Many insects can be reared relatively easily.

A box like that shown in E can be used to rear adults from larvae living in debris or other materials. This material is put into the box and the box sealed; the emerging adults are usually attracted to light, and go into the vial.

Caterpillars are good insects to rear, since many are large and easily observed, and their transformations are very striking. The chief problem is providing suitable food, because most caterpillars feed on only a few kinds of plants. If the food plant is not known, you must either identify the caterpillar and determine its food from a reference book or try a number of plants in the hope of finding something the caterpillar will eat. The food plant can be kept fresh by placing it in water, as described above (C). If the caterpillar requires special condition for pupation (soil or debris), these conditions should be provided.

A caterpillar collected in the fall may overwinter before pupating, or it may overwinter as a pupa; sometimes it may not complete its development unless it is subjected to low temperatures. Cocoons collected in the fall and brought indoors may fail to develop, either because they dry out or require exposure to low temperature. Drying can be prevented by placing the cocoons in a container with a little soil and occasionally sprinkling the soil with water. Exposure to low temperature can be accomplished by placing the cocoons in a refrigerator for a few weeks, or by keeping them outdoors (for example, on the outside windowsill of a room).

Many aquatic insects, especially those living in stagnant water and feeding on microorganisms or debris, are easily reared indoors. They can be reared in some of the water from which they were collected, often without special equipment to aerate the water and without adding more food. If a stream-inhabiting insect is put into an aquarium, the water usually must be aerated and its temperature not allowed to go too high. Adult mosquitoes can be reared from larvae or pupae in containers as small as vials, cover the vials (with netting or a plug of cotton) to prevent the adults from escaping. Predaceous insects such as dragonfly or damselfly nymphs require other insects or small aquatic animals as food, and the aquarium must contain something extending out of the water — a stick or piece of screen — onto which the nymphs can climb when they are ready to transform into adults.

Meal-infesting insects generally are very easy to rear or maintain from generation to generation, since they normally live indoors and do not require extra moisture. They can be kept in containers of their food material; this material should be sifted at intervals, and the insects transferred to a fresh batch.

Spiders, and predaceous insects such as mantids or dragonfly nymphs, will prove interesting to keep in captivity if supplied with suitable insects as food. Many of these animals have unusual methods of capturing their prey. Caged spiders can be watched making their webs or egg sacs.

Adult crickets sing readily in captivity, and live a fairly long time; you can enjoy their songs and also see how the songs are produced. Chicken mash or ground-up dog food can be used as food, and a wet sponge or piece of cotton will provide adequate moisture.

With a little knowledge of an insect's food habits and habitat requirements, an ingenious student should be able to devise methods of rearing almost any type of insect. Anyone rearing adults from immature stages will sooner or later get parasites instead of the adults expected, and thus will learn something about the habits and hosts of parasitic insects.

The Projection of Living Insects

Living insects can be demonstrated to a group by means of projection, and the effect is much like a motion picture. Aquatic insects in a watch glass can be projected for small groups with a photographic enlarger or similar projector, or the insects can be put into a slide-sized water cell and projected for larger groups with a slide projector — the more practical method. The construction of such a water cell is illustrated below.

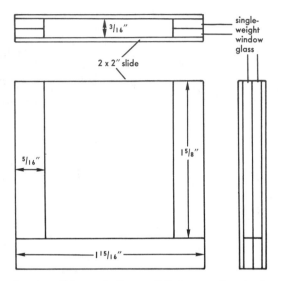

A water cell for projecting small living aquatic animals.

The Structure of Insects

A KNOWLEDGE of insect structure is essential to understanding descriptions and the characters distinguishing different groups. The following account is rather brief; more information is given in the accounts of some of the insect orders.

General Structure. The body of an insect is segmented and more or less elongated. The segments are grouped in 3 body regions, the head, thorax, and abdomen. The *head* bears the eyes, antennae, and mouth parts; the *thorax* bears the legs and wings; the *abdomen* usually bears no locomotor appendages but often has some appendages at its posterior end.

An insect's body wall is somewhat hardened (*sclerotized*). It serves as a shell to protect the internal organs and acts as a skeleton or, more properly, *exoskeleton* (muscles are attached to its inner surface). The surface of the body is divided by intersegmental and other lines into a number of platelike areas, or *sclerites*.

The Head. The head is the anterior capsulelike body region that bears the eyes, antennae, and mouth parts. It is usually quite hard. The surface is divided by sutures into a number of areas, each with a name; the same names are used in different insect groups — where the areas can be homologized — but special terms are used in some groups.

Insects generally have 2 kinds of eyes, simple and compound. For the majority there are 3 simple eyes (*ocelli*), located on the upper front part of the head. Some insects lack ocelli and others have only 2. The *compound eyes* are situated dorsolaterally on the head, each composed of many facets. In some insects they occupy most of the head and contain hundreds of facets.

The Antennae are usually located on the front of the head below the ocelli; they vary greatly in form and in the number of segments contained, and are often used to distinguish different insect groups. Various terms that describe the antennae are explained in the accounts of the groups in which antennal characters are used for identification.

Mouth Parts. The mouth parts of an insect generally are located on the ventral or anterior part of the head, and they vary a great deal in different insect groups. The mouth part structures typically present are a *labrum* (upper lip), a pair of jawlike *mandibles*, a pair of jawlike *maxillae*, a *labium* (lower lip), and a tongue-like structure called the *hypopharynx*.

Insect mouth parts are of 2 general types, chewing and sucking. Insects with chewing mouth parts have laterally moving mandi-

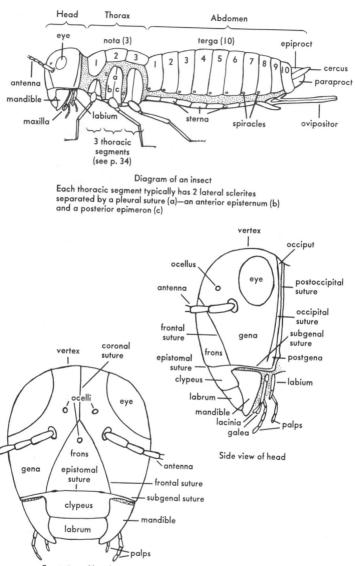

Diagram of an insect

Each thoracic segment typically has 2 lateral sclerites separated by a pleural suture (a)—an anterior episternum (b) and a posterior epimeron (c)

Side view of head

Front view of head

bles (see opp.) and generally chew their food; those with *sucking mouth parts* have the parts modified into a *beak* (proboscis) through which liquid food is sucked. The mandibles are either lacking in sucking mouth parts or are styletlike and form part of the beak; they do not move laterally. A few insects (such as bees) have laterally moving mandibles and a beaklike tongue, and suck liquid through the tongue; a few, like the larvae of dytiscid beetles (p. 154), have well-developed mandibles that move laterally but they suck their food in liquid form through channels in the mandibles.

Chewing mouth parts normally can be recognized by the laterally moving mandibles, visible on the lower front part of the head; most chewing insects do not have a beak. Scorpionflies have the head prolonged ventrally into a beaklike structure (see illus., p. 209), with mandibles that move sideways, and these insects chew. Snout beetles (see illus., p. 203) have the front of the head prolonged into a snout, but there are tiny laterally moving mandibles at the tip of the snout.

Mouth parts of sucking insects (see p. 32) vary in appearance and in how they operate. Many sucking insects (such as bug, froghopper, mosquito, and stable fly) pierce the tissue fed upon. What you see of the beak is usually (bug, mosquito, froghopper) not the part that does the piercing but a sheath enclosing the piercing structures. The piercing is done by a group of hairlike or swordlike *stylets*. When these stylets pierce something, the sheath folds up or back out of the way. If when about to be bitten by a mosquito you will let it alone and watch it in action, you will see that the beak bends in the middle as the 6 hairlike stylets inside it go into the skin. The proboscis of a butterfly or moth does no piercing; generally it is coiled like a watch spring on the ventral side of the head when not in use, and is uncoiled when the insect feeds. Some flies (like the House Fly) have a rather fleshy proboscis that is incapable of piercing.

The stylets of an insect with sucking mouth parts usually enclose the food and salivary channels (the proboscis of a butterfly or moth contains only a food channel). When you are bitten by a mosquito, for example, saliva is first injected through the salivary channel (this is what causes the irritation), then blood is sucked up through the food channel.

The maxillae and labium of most insects bear small feelerlike structures called *palps;* each maxilla bears 1, and the labium bears 2. Some insects lack 1 or both pairs of palps: the flies (Diptera) have only the maxillary palps, most butterflies and moths (Lepidoptera) have only the labial palps, and the bugs (Hemiptera) have no palps at all.

The Thorax (see opp. and p. 34). This, the middle section of the body (between head and abdomen), is divided into 3 segments: (1) *prothorax* (2), *mesothorax,* and (3) *metathorax.* Each segment

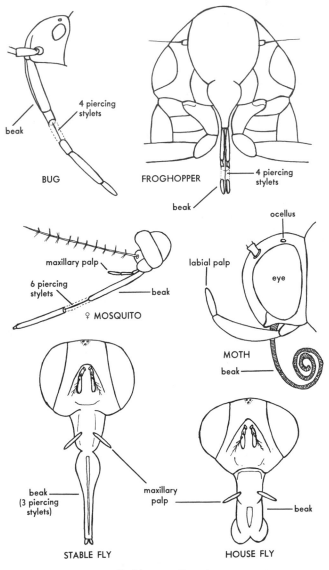

Sucking mouth parts.

typically bears a pair of legs lateroventrally, and the mesothorax and metathorax usually bear a pair of wings dorsolaterally. Some insects have only 1 pair of wings (generally borne by the mesothorax), and some are wingless.

Each thoracic segment bears 4 groups of sclerites (platelike areas), the *notum* (plural, nota) dorsally, a *pleuron* (plural, pleura) on each side, and the *sternum* (plural, sterna) ventrally. Any thoracic sclerite can be indicated as belonging to a particular segment by using the proper prefix. For example, the notum of the prothorax is the *pronotum*, the pleura of the mesothorax are the *mesopleura*, and so on. The pronotum in many insects is a large and conspicuous sclerite forming the dorsal surface of the body between the head and the base of the wings. The pleura of the mesothorax and metathorax are usually larger than the propleura.

The pronotum is a single sclerite, but may contain various grooves or ridges. Each of the other nota is usually divided into 3 sclerites the *scutum*, *scutellum*, and *postnotum*. The meso-*scutellum* in bugs and most beetles is a conspicuous and more or less triangular sclerite between the pronotum and the base of the wings, and is often simply called the scutellum. Each pleuron is usually separated into 2 sclerites by the *pleural suture*, a line of division which extends from the base of the leg to the base of the wing; the anterior sclerite is the *episternum*, and the posterior one is the *epimeron*. The pleura sometimes contain sutures setting off additional sclerites, and frequently there are 1 or more small sclerites in the membranous area between the pleura and the base of the wings. The sternum is often divided by sutures into 2 or 3 sclerites.

The Legs (see p. 34). An insect leg typically contains the following segments: *coxa* (basal segment), *trochanter* (generally small, just beyond the coxa), *femur*, *tibia*, and *tarsus*. The tarsus usually bears at its apex a pair of *claws* and 1 or more padlike structures. Since the legs vary considerably in different insects, leg characters are often used in identification. The legs may differ in the relative size and shape of the various segments, the number of subdivisions of the tarsus (usually called tarsal "segments," though the entire tarsus technically constitutes a single leg segment), and the trochanter, the character of the claws, pads, and other structures at the apex of the tarsus, and the spines or hairs on the legs.

The Wings. Insect wings also vary considerably, and much use is made of this variation in classification and identification. Many insect order names end in *ptera* (from the Greek, meaning "wing"). The wings are located dorsolaterally on the mesothorax and/or the metathorax. Most of the muscles that move the wings are attached to the walls of the thorax rather than to the base of the wings, and the wing movements are produced largely by changes in the shape of the thorax.

Insect wings vary in number, size, shape, texture, venation,

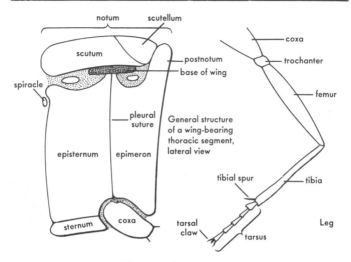

Thorax and leg structure.

and in the position held at rest. Most insects have 2 pairs of wings but many have only 1 pair and some are wingless. Although most insect wings are membranous (thin, like cellophane) some are thickened or leathery; many are covered with hair and some bear scales. Typically insects fold the wings over the abdomen at rest, but a few hold them vertically above the body and some hold them outstretched.

Certain terms are used in referring to different edges, regions, or angles of the wing: the anterior edge is the *costal margin,* and the base of the costal margin is the *humeral angle;* the posterior edge is the *anal margin,* and the posterior basal part of the wing is the *anal area;* an angle of the wing in the anal area is the *anal angle,* and an angle at the tip of the wing is the *apical angle,* or *apex.* There are 1 or 2 lobes in the anal area of the wing in some insects.

Wing Venation. Wing venation — the system of thickened lines in the wing — is frequently used in classification and identification. Although it is generally possible to homologize the veins of different insects and use a standard system of names for them, there are differences of opinion as to how these names should be applied for some insects. A special venational terminology is used in a few groups.

The most widely used terminology of wing venation is illustrated opposite. Longitudinal veins are indicated by capitalized abbreviations, and cross veins by small-letter abbreviations. Branches of the longitudinal veins are indicated by subscript numerals.

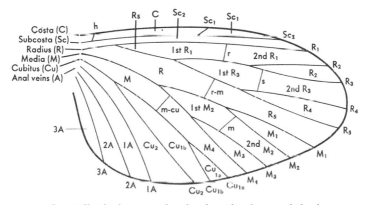

Generalized wing venation (explanation in text, below).

The vein forming the costal wing margin is the *costa* (C). The 1st vein behind the costa is the *subcosta* (Sc), which typically has 2 branches, Sc_1 and Sc_2. The next vein is the *radius* (R), which gives off basally a posterior branch, the *radial sector* (Rs); the radial sector usually forks twice. The anterior branch of R is R_1; the anterior branch of Rs is R_{2+3}, and its terminal branches are R_2 and R_3; the posterior branch of Rs is R_{4+5}, and its terminal branches are R_4 and R_5. The next vein is the *media* (M), which generally forks twice; the anterior branch beyond the first fork is M_{1+2}, and its terminal branches are M_1 and M_2; the posterior branch is M_{3+4}, and its terminal branches are M_3 and M_4. The next vein is the *cubitus* (Cu), which has 2 branches, Cu_1 and Cu_2; Cu_1 often has 2 terminal branches, Cu_{1a} and Cu_{1b}. The remaining veins are *anal veins* (A), and are numbered from anterior to posterior (1A, 2A, 3A). The usual *cross veins* are: the *humeral* (h) between C and Sc near the base of the wing; the *radial* (r) between R_1 and the anterior branch of Rs; the *sectorial* (s) between 2 branches of Rs; the *radiomedial* (r-m) between the radius and media; the *medial* (m) between 2 branches of M; and the *medio-cubital* (m-cu) between M and Cu.

Cells in the wing (spaces between veins) are named from the vein on the anterior side of the cell. Cells at the base of the wing are usually named without a subscript numeral. A cell behind a 2-parted or fused vein is named from the posterior component of that vein. Two or more cells with the same name are individually designated by numbering from the base of the wing (1st M_2 cell, 2nd M_2 cell). An *open cell* extends to the wing margin; a *closed cell* does not reach the margin.

A given insect may have more veins than shown in the figure above as a result of additional cross veins or additional branches of the longitudinal veins; it may have fewer veins because some fail to branch, fuse together, or are lost. Special terms are used for particular features of the venation in some orders and are discussed in this book under the accounts of those orders.

The Abdomen (see p. 30). The insect abdomen typically consists of 11 segments, but the last segment is usually represented by appendages only; so the maximum number of complete segments in most insects is 10. Many insects have fewer abdominal segments because of fusing or telescoping of some segments.

Each abdominal segment generally contains 2 sclerites, a dorsal tergum and a ventral sternum. The *terga* usually extend down the sides of the segments and overlap the sterna.

Most insects lack appendages on the abdomen except at the posterior end. The terminal appendages may be lacking or drawn into the body and hidden. When terminal appendages are present (top figure, p. 30) they usually consist of a pair of dorsolateral *cerci* (singular, cercus), a median dorsal *epiproct*, a pair of lateroventral *paraprocts*, and the *genitalia*. Cerci when present may be feelerlike or clasperlike; the epiproct, if present, may be short, or elongate and threadlike. The anal opening is at the posterior end of the abdomen, just below the epiproct if the latter is present.

The genitalia are structures associated with the genital openings — those of the male transfer sperm to the female and those of the female lay the eggs. Male genitalia are extremely variable and often quite complex; they provide valuable taxonomic characters in many groups. Some or all of the male genitalia may be withdrawn into the body and not be apparent without dissection. An *ovipositor* (egg-laying organ) is present in many insects. It is formed by structures on the 8th and 9th abdominal segments and extends beyond the body in some insects and in others is withdrawn into the body when not in use.

The sexes in many groups can be distinguished by the genital structures at the end of the abdomen. In groups having internal genitalia the sexes may differ in other ways (size or color), or they may be indistinguishable without dissection.

Internal Anatomy. Although space does not permit a detailed account here, we believe that some features of the internal anatomy of insects are of sufficient interest to warrant at least brief mention.

The *breathing system* of insects is very different from that of man. Insects have a system of tubes (*tracheae*) that open externally at the *spiracles* and branch internally to supply all parts of the body. There are usually 2 pairs of spiracles on the thorax and several pairs on the abdomen; abdominal spiracles are generally located on the lateral edges of the terga. Oxygen goes from the outside directly to the tissues by way of the tracheal tubes, and is not transported in the hemoglobin of the blood as in vertebrates.

Insects' blood is ordinarily not red, and it does not reach all parts of the body in blood vessels. The *heart* is a tube located in the abdomen above the digestive tract. Blood is pumped anteriorly from the heart through a dorsally situated vessel called the *aorta*, and in the neck region empties into the body cavity. It flows through the body cavity, and reenters the heart through lateral openings called *ostia*.

The *excretory system* of insects consists of a number of tubes (*malpighian tubules*) that empty into the alimentary tract. Wastes from the blood enter these tubes and pass into the alimentary track and to the outside by way of the anus.

The *nervous system* of insects consists of a ganglion called the brain, located dorsally in the head, a pair of connectives passing around the alimentary canal, and a ventral nerve cord; nerves extend from the brain and nerve cord to various parts of the body. There are many sense organs, located mainly in the body wall. Some of these respond to tactile stimuli, some to chemical stimuli, some to sound, some to light, and some to other stimuli. Chemical receptors (organs of taste and smell) are located principally on the mouth parts, antennae, and feet. Special auditory organs, when present, generally consist of drumlike structures (tympana) or special hairs sensitive to sound waves. Many insects can detect sounds pitched far above the hearing range of man.

The Growth and Development of Insects

Eggs. Insect eggs differ in shape and color, and some are ornamented with ridges, spines, or other processes. Most insects lay their eggs in a situation where the young on hatching will have conditions suitable for development. Many lay their eggs in characteristic masses, and a few cover their eggs with a protective material of some sort. The eggs of some insects develop internally, and the young are born alive.

Insect eggs ordinarily develop only if they have been fertilized but some undergo *parthenogenesis*, that is, they develop without fertilization. Fertilization sometimes determines sex. In Hymenoptera, for example, an unfertilized egg usually develops into a male and a fertilized egg into a female. Unfertilized eggs of most parthenogenetic insects develop into females, and in some species no males are known. A few Hymenoptera undergo *polyembryony* (a single egg develops into more than 1 young). This sometimes occurs in man, producing identical twins or triplets. In polyembryonic insects, from 2 to more than 1000 young may develop from 1 egg.

Growth. The growth of an insect is accompanied by a series of molts, in which the exoskeleton (outer shell) is shed and renewed. Insects change in form as they grow, and the amount and character of this differ from group to group. This change is called *metamorphosis*.

The exoskeleton of insects is generally rather hard, and the extent to which it can stretch is limited. An insect cannot grow continuously but must shed the exoskeleton at intervals and replace it with a larger one. This shedding process is called molting, or *ecdysis*. Molting involves a shedding of the outer surface of the body, the linings of the tracheae, and the lining of the anterior and posterior parts of the alimentary canal. It begins with a splitting of the old exoskeleton, usually on the dorsal side of the head or thorax. In some cases (caterpillars, for example) the shed exoskeleton shrinks into a small irregular mass, but in others it retains the shape of the insect.

The stages between molts are called *instars*. The number of molts is generally 4 to 8, but may be as many as 20 in some insects. Molting usually stops when the adult stage is reached. Only a very few insects (like bristletails) continue to molt after becoming adult.

Metamorphosis. Successive instars differ not only in size but in other features as well. This change during growth (metamorphosis) is relatively slight in some insects, very marked in others. There are 2 principal types of metamorphosis — simple and complete.

Simple Metamorphosis. In this type the wings (if present in the adult) develop externally during the early instars, compound eyes are present in the early instars if they are present in the adult, and there is no prolonged resting stage before the last molt. The immature instars of insects with this type of metamorphosis are called *nymphs*.

Nymphs usually resemble the adults except in size, body proportions, and the development of the wings; they generally live in the same habitat as the adult, and feed on the same foods. If the adults are wingless, the chief difference between nymphs and adults is in size. If the adult has wings, the wings are relatively small through the last nymphal instar, and expand to their adult size after the last molt.

Nymphs of mayflies, stoneflies, dragonflies, and damselflies differ from adults slightly more than in other insects with simple metamorphosis. They live in water and have gills, and when full-grown come to the surface of the water or crawl out of the water for their final molt.

Complete Metamorphosis. The eggs of insects with complete metamorphosis hatch into a wormlike stage called a *larva*. The larvae of insects vary in appearance: some have legs and others are legless and some lack a well-developed head. Larvae do not have compound eyes (but may have ocelli), and if the adult is winged the wings begin their development in the larval stage but develop internally. The larval stage lasts from a few to several instars, increasing in size and sometimes changing in color or other characters. After the molt of the last larval instar the insect changes to what is called a *pupa*. Pupae are usually inactive. They do not feed, and are sometimes enclosed in a protective covering, which may be a *cocoon* formed by the last larval instar before it molted or may be a *puparium* (formed of larval exoskeleton).

Insect larvae vary considerably in form. Several terms (eruciform, scarabaeiform, campodeiform, vermiform, elateriform, etc.) are used to describe them. *Eruciform larvae* are caterpillarlike, with a well-developed head, thoracic legs, and abdominal *prolegs* (see illus., p. 219); they occur in the Lepidoptera, Mecoptera, and some Hymenoptera (Symphyta). *Scarabaeiform larvae* are grublike, with thoracic legs but without abdominal prolegs, and are usually pale-colored and rather sluggish; they occur in certain Coleoptera (such as Scarabaeidae). *Campodeiform larvae*, which resemble diplurans in the family Campodeidae (see illus., p. 63), are elongate and somewhat flattened, with the antennae, cerci, and thoracic legs well developed, and are generally fairly active; they are found in the Neuroptera and many Coleoptera. *Vermiform larvae* are wormlike or maggotlike, without legs, and with or without a well-developed head; they occur in the Diptera, Siphonaptera, and most Hymenoptera (Apocrita), and in a few insects in other orders. *Elateriform larvae* are elongate and cylindrical, hard-bodied, and short-legged; they occur in certain Coleoptera (like Elateridae).

Most insect pupae look somewhat like mummified adults, with the appendages free and visible. Such pupae are called *exarate*, and occur in most insects with complete metamorphosis except the Diptera and most Lepidoptera. Some pupae have the appendages

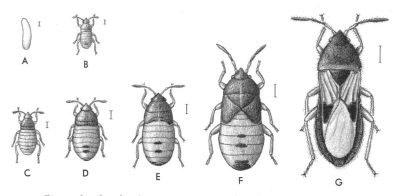

Stages in the development of a bug (simple metamorphosis). A, egg; B–F, nymphal instars; G, adult.

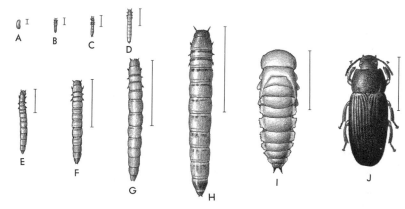

Stages in the development of a beetle (complete metamorphosis). A, egg; B–H, larval instars; I, pupa; J, adult.

glued to the body, and look much less like the adults. Such pupae are called *obtect*, and occur in the Lepidoptera and some Diptera (Nematocera). The pupa of many Diptera (Brachycera and Cyclorrhapha) is enclosed in a puparium, and is termed *coarctate;* puparia are oval and brownish, and look rather like the fecal pellet of a small rodent.

The different larval instars of some insects with complete metamorphosis are not of the same type: the 1st instar is campodeiform and the remaining instars scarabaeiform or vermiform. This type of development is called *hypermetamorphosis*, and occurs in some parasitic insects. The active 1st instar larva seeks out and enters a host and, once there, molts to a less active type of larva.

The larvae and adults of insects with complete metamorphosis are usually so different that one unfamiliar with their life history would not believe them to be the same insect. They often live in dissimilar habitats and feed on dissimilar foods. The transformation of an insect with complete metamorphosis is a remarkable phenomenon — something well worth following in the field or with caged individuals.

Changes in the Adult. Immediately after its molt to the adult stage an insect is soft-bodied and pale-colored. If the adult has wings, these are usually small immediately after the molt — about the same size as they were in the pupal (or last nymphal) instar — and must expand to their full adult size. This expansion may occur in a few minutes, or take a half hour or more. The coming-out of the adult at the final molt is called *emergence.* Darkening to the adult coloration generally takes place in a short time (an hour or less), but in some insects it may be a week or more before the adult has developed its full coloration. Some adults (such as Hymenoptera) that emerge from cocoons may undergo these changes before emerging from the cocoon. Others (a butterfly) have the coloration fully developed at the time of the final molt and the wings expand to their full size *after* emergence. Once the adult has expanded to its full size it does not grow any more, and (with rare exceptions) does not undergo any more molts.

Life History. Most insects in our area have a single generation a year. The adults are present for a limited time during some part of the year, and the winter is passed in a dormant state. Insects overwinter in different stages — some as eggs, some as nymphs, some as larvae, some as pupae, and some as adults. A period of dormancy at low temperature is often an essential feature of the life cycle. Many insects (particularly those occurring in the northern part of the country) will not complete their development unless exposed to low temperature.

Some insects regularly have 2 generations a year and others may have several — continuing to reproduce as long as weather conditions are favorable. A few require more than a year to complete their development. Many of the larger insects in northern areas take 2 or 3 years; the record holders are some of the periodical cicadas, which take 17 years.

Adults of most insects live only a short time, ordinarily from a few days to a few weeks. An overwintering adult lives several months, and the queens of some social insects can live several years. Many insects that are short-lived as adults do not feed in the adult stage.

Classifying and Naming Insects

Classification. There are about a million kinds of described animals in the world, and their study calls for some plan of dividing them into groups. Zoologists classify animals chiefly on the basis of structure: those with certain structures in common are placed in one group and those with other structures are put in other groups. These groups are divided and subdivided. The result is a system of categories, each with certain structural features in common, and a name.

The animal kingdom is divided into a number of major groups called *phyla* (singular, phylum); each phylum is divided into *classes*, each class into *orders*, each order into *families*, each family into *genera* (singular, genus), and each genus into *species*. In many larger groups there are additional categories, such as subclasses, suborders, superfamilies, subfamilies, and tribes. The species is the basic category; it is a kind of animal. That is, it consists of individuals fundamentally similar in structure which interbreed to produce offspring but do not ordinarily interbreed with other groups. Sometimes species are divided into *subspecies*. Subspecies are generally geographic races that differ from one another only slightly and are capable of interbreeding.

The arrangement of animals into these categories is arbitrary: it is the opinion of the specialist that determines the limits of a category. Although specialists do not always agree on the limits of some categories, differences of opinion mostly are minor. This system is an indispensable tool in the study of animals. Anyone studying animals must be familiar with it.

Nomenclature. Animals have 2 types of names — scientific and common. Scientific names are the names used by scientists; they are used throughout the world, and every animal or group has one. Common names are vernacular names. They are less precise than scientific names, and many animals lack them. Some common names are used for more than 1 species or group, and a given animal or group may have several common names.

Scientific Nomenclature. The scientific naming of animals follows certain rules, only a few of which can be mentioned here. Scientific names are Latinized. They may be derived from various languages or from the names of people and places, but most are from Latin or Greek and refer to characteristics of the animal or group named. Names of groups above genus are single words in the nominative plural; names of genera are single words in the nominative singular; names of species are 2 words — the name of the genus plus a specific name; and names of subspecies are 3 words — the name of the species (2 words) plus a subspecific name. Specific and subspecific names may be adjectives or participles (in which case they must agree in gender with the genus name) or nouns in the nominative or genitive.

Names of genera, species, and subspecies are written in italics and are usually followed by the name of a person (the *author*). If the author name follows a species name it indicates the person who proposed the specific name; if it follows a subspecies name it indicates the person who proposed the subspecific name; if it is in parentheses it means that the author's species or subspecies was originally placed in a genus other than the present one. For example:

Musca domestica Linn. — the House Fly. Linnaeus (abbreviated "Linn." in this *Field Guide*) first described the House Fly and gave it the specific name *domestica* and placed it in the genus *Musca*.

Automeris io (Fabricius) — the Io Moth. Fabricius first described this moth and gave it the specific name *io*, and placed it in a genus other than *Automeris*.

Diabrotica undecimpunctata howardi Barber — the Spotted Cucumber Beetle. Barber proposed the name *howardi* and placed it in *Diabrotica*. There is no way of knowing from this name whether Barber originally described *howardi* as a subspecies of *undecimpunctata*, or as a subspecies of another species of *Diabrotica*, or as a species of *Diabrotica*.

Whenever a new group (from subspecies to superfamily) is described, the describer is supposed to designate a *type*. This type provides a reference if a question arises as to what the group includes, and type genera provide the basis of the name of certain higher categories. The type of categories from tribe to superfamily is a genus, the type of a genus is a species, and the type of a species or subspecies is a specimen.

Names of some categories have standard endings: -oidea for superfamily, -idae for family, -inae for subfamily, and -ini for tribe. The names are formed by adding the ending to the root of the type genus name. If 1 of these groups is divided into 2 or more subgroups, the subgroup containing the group's type genus will have the same name as the group except for the ending. An illustration: *Colletes* is the type genus of the family Colletidae (plasterer and yellow-faced bees); this family is divided into 2 subfamilies, the Colletinae (plasterer bees; with *Colletes*) and the Hylaeinae (yellow-faced bees; named for *Hylaeus*, the genus designated as the type of this subfamily).

This same principle applies when a species is divided into 2 or more subspecies — the subspecific name of one of the subspecies (the one containing the type of the species) will be the same as the specific name of the species. For example, the dragonfly *Tetragoneuria cynosura* (Say) is divided into 2 subspecies, *Tetragoneuria cynosura cynosura* (Say) and *Tetragoneuria cynosura simulans* Muttkowski. *T. c. cynosura* contains the type of *T. cynosura* and another specimen was designated as the type of *T. c. simulans*.

A given animal (or group) may be described by different people and thus may have more than 1 name. The first name proposed in such cases (provided the author followed certain rules) is the correct one and the other names become *synonyms*. It is not always easy to determine which of 2 or more names for an animal is the correct one: different names may be used by authorities who do not agree on which name has priority.

It sometimes happens that a person describing a new genus will use for it a name previously used for another genus. When this is discovered the later genus must be renamed, since the rules state that no 2 genera of animals may have the same name. Similarly, no 2 species or subspecies in the same genus may have the same specific or subspecific name. These rules relating to *homonyms* (cases of the same name being used for different groups) apply only within a particular category group, and they except the case mentioned above of 1 subspecies having the same specific as subspecific name. The category groups are: phylum through order, superfamily through subgenus, and species through subspecies. A name used in 1 of these 3 category groups can be used in another without violating the rules. For example, a specific or subspecific name may be the same as the name of a genus, or the same name may be used for both a genus and an order.

As our knowledge of animals increases it often becomes necessary to change scientific names. A group may be subdivided or combined with another group; a name widely used may become a synonym because of the discovery of an older name; or a group may be renamed because of the discovery of an earlier use of its name for another group. Problems encountered in scientific nomenclature are sometimes solved differently by taxonomists. For these reasons, the classification and nomenclature used by authorities may not always be the same. We follow in this book the opinions of most present-day entomologists. Other names and groupings most often encountered are included in the Index.

Common Names. Common names of insects mainly apply to groups rather than to individual species. The few species having common names are generally of some economic importance or are particularly striking in appearance. The name "beetle," for instance, refers to all the Coleoptera, of which there are some 290,000 world species (about 28,600 in the U.S.); the name "leaf beetle" refers to all the Chrysomelidae (about 25,000 world species, nearly 1400 in the U.S.); the name "tortoise beetle" refers to all the Cassidinae, a subfamily of the Chrysomelidae (about 3000 world species and about 24 in the U.S.); the name "Argus Tortoise

Beetle" refers to the single species *Chelymorpha cassidea* (Fabricius), which is of economic importance as a pest of sweet potato and other plants.

Most 1-word common names used for insects (beetle, bug, fly, termite, caddisfly, and others) refer to entire orders. Some (damselfly, grasshopper, lacewing, and others) refer to suborders or groups of families. Only a few (ants, Formicidae; cockroaches, Blattidae; and others) refer to families. Most common names of families — where the family has a common name — consist of 2 or more words, the last being the name of the larger group and the other(s) descriptive (robber flies, Asilidae; leaf beetles, Chrysomelidae; metallic wood-boring beetles, Buprestidae; and others).

The majority of the common names used in this book are in rather wide use; some are used by relatively few people, and a small number are used here for the first time. Many entomologists prefer scientific to common names because they are more precise and are widely used (at least among entomologists), and are sometimes easier to remember. An adjectival form of a group's scientific name is often used as a common name — a member of the order Orthoptera could be called an "orthopteran," or a member of the family Libellulidae could be called a "libellulid." If a family has no common name (and sometimes even when it does), it is standard practice to use the adjectival form of the family name as a common name. In a few cases (mantids, Mantidae; syrphid flies, Syrphidae; etc.) such names are widely used as common names.

"Fly" and "bug" are a part of the common names of many different insects. The "fly" of the name is written as a separate word if the insect is in the order Diptera (horse fly, robber fly, black fly), and together with the descriptive word if the insect is in another order (butterfly, dragonfly, scorpionfly). The "bug" of the name is written as a separate word if the insect is in the order Hemiptera (stink bug, bed bug, leaf bug), and together with the descriptive word if the insect is in another order (mealybug, lightningbug, ladybug).

Systematic
Chapters

Arthropods: Phylum Arthropoda
(Insects and Their Relatives)

Identification: Body segmented, segments usually grouped in 2 or 3 fairly distinct body regions. Paired, segmented appendages usually present. Body wall more or less hardened; forms an exoskeleton (external skeleton) periodically shed and renewed.

Similar phyla: Annelida (earthworms, leeches, various marine worms) have body segmented but with little or no differentiation of body regions. Appendages unsegmented or lacking. Body wall does not form an exoskeleton. Legless insect larvae differ from annelids in their internal anatomy (generally have tracheae and malpighian tubules, which annelids lack), and the body usually contains fewer segments (13 or fewer in insect larvae, ordinarily more in annelids).

Immature stages: Most arthropods other than insects undergo little or no metamorphosis and the young resemble the adults. A few (some crustaceans) have a larval stage markedly different from the adult; some (millipedes, some centipedes, some arachnids) have fewer legs in immature stages than in adult stages.

Habits: Very diversified. Practically every animal habitat contains some arthropods, and different arthropods vary greatly in their feeding habits.

Importance: The importance of insects has been outlined briefly in the Preface (p. vii). Other arthropods are important in similar ways. Many crustaceans (lobsters, crabs, shrimps) are used as food by man, and their collection and distribution constitute a sizable industry. Insects are rarely used as food by man (at least in this country), though certain insect products (honey) are.

Classification: Present-day arthropods are arranged in 2 subphyla, the Mandibulata and the Chelicerata, which differ principally in the number and character of the appendages. Each subphylum is further divided into classes.

No. of species: World, 840,000; N. America, 104,000.

Mandibulate Arthropods:
Subphylum Mandibulata

One or 2 pairs of antennae. 1st pair of appendages behind antennae are mandibles. Number of legs and body regions variable.

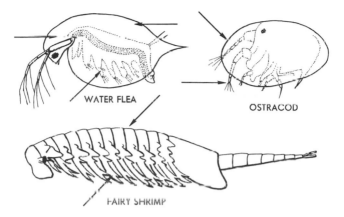

WATER FLEA

OSTRACOD

FAIRY SHRIMP

CRUSTACEANS Class Crustacea

Identification: Appendages vary in number, but 2 pairs of antennae (1 pair may be quite small), a pair of appendages on each segment of cephalothorax (head and thorax) and sometimes on each abdominal segment; usually at least 5 pairs of loglike appendages. 2 body regions generally distinct (cephalothorax and abdomen), the cephalothorax often with a hardened covering (carapace). Mostly aquatic, gill-breathing.

Two major groups of crustaceans are usually recognized, Entomostraca and Malacostraca. Entomostraca include a large and varied assemblage of small aquatic forms, generally less than 5 mm., which lack appendages on the abdominal segments. Malacostraca are mostly larger forms, with appendages on the abdominal segments. Only the more common types in each group can be mentioned here.

Entomostraca. Members of this group occur in both salt- and freshwater; important chiefly as food of larger animals. Three types of entomostracans have a *bivalved carapace:* water fleas, clam shrimps, and ostracods. **Water fleas** (Cladocera) — sometimes extremely abundant in freshwater pools — have *4 to 6 pairs of flattened thoracic legs;* head is *outside the carapace.* **Clam shrimps** (Conchostraca) and **ostracods** (Ostracoda) have entire body *enclosed in the carapace,* and resemble tiny clams; clam shrimps have 10–32 pairs of legs and are sometimes fairly large (to 10 mm.), whereas ostracods have only *3 pairs of legs* and are always quite small. **Fairy shrimps** (Anostraca) are elongate, usually pale-colored, *without a carapace,* and have *11 pairs of flattened legs;* often abundant in temporary pools. **Tadpole shrimps** (Notostraca) have an oval carapace covering anterior part of body, 35–71 pairs of thoracic appendages, and 2 long threadlike appendages at posterior end of body; they are fairly large for entomostracans (½–2 in.); occur only in the West. **Copepods** (Copepoda) generally have a more or less elongate body that lacks a carapace, and 4–6 pairs of legs; females of many species carry their eggs in *2 laterally located egg sacs* near posterior end of body; copepods include freshwater and marine forms; most are free-living, but some (called fish lice) are parasites of fish. **Barnacles** (Cirripedia) are marine forms that as adults live attached to objects in the water (rocks, pilings, seaweeds, other marine animals, or boat bottoms); body usually enclosed in a shell of some sort.

Malacostraca. Most members of this group are marine, but a few occur in freshwater and a few are terrestrial. The **decapods** (Decapoda) include some of the largest and best-known malacostracans — lobsters, crayfish, crabs, and shrimps; entire cephalothorax is *covered by a carapace;* 5 pairs of cephalothoracic appendages are leglike; 1st pair of legs usually bears a *large claw.* Crayfish are the only decapods likely to be encountered in freshwater; most of the others are marine. **Amphipods** (Amphipoda) *lack a carapace* and have the body laterally flattened; 7 pairs of cephalothoracic appendages are *leglike.* This group includes freshwater and marine forms; some of the marine forms (beach fleas) are common along the seashore, where they live under stones or seaweed; freshwater amphipods are called scuds or sideswimmers. **Isopods** (Isopoda) *lack a carapace* and are dorsoventrally flattened, and the 7 pairs of cephalothoracic appendages are *leglike.* Most isopods are marine, but the group includes the sowbugs (pillbugs), which are common under stones or bark.

MILLIPEDES Class Diplopoda
Identification: Wormlike, cylindrical or only slightly flattened. Many-legged, with 2 pairs of short legs *on most body segments. 1 pair of short, usually 7-segmented* antennae. Terrestrial.

Millipedes are common animals usually found in soil and debris or under stones and bark. They are slow-moving, and most of them feed on plants or decaying materials; they do not bite. Our species vary from a few mm. to about 4 in. Most are blackish, sometimes with light markings. Many give off an ill-smelling fluid from openings along sides of the body.

CENTIPEDES Class Chilopoda
Identification: Elongate, wormlike. Similar to millipedes but body more flattened. Legs (15 or more pairs) arranged *1 pair per body segment.* Antennae with 14 *or more segments.* 1st pair of appendages behind head clawlike and functioning as *poison jaws.*

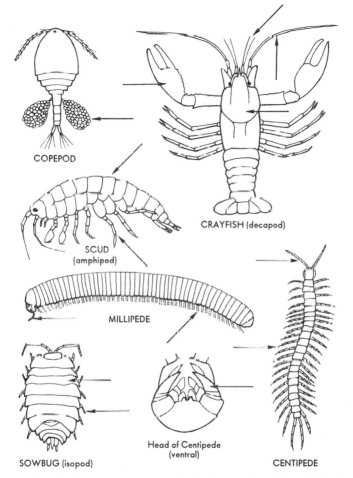

COPEPOD

CRAYFISH (decapod)

SCUD
(amphipod)

MILLIPEDE

SOWBUG (isopod)

Head of Centipede
(ventral)

CENTIPEDE

Centipedes are common animals found in soil and debris, under bark, in rotting wood, and in similar protected places. They are active and fast-running and are predaceous on insects and other small arthropods. They have poison jaws with which they paralyze their prey; the smaller centipedes are relatively harmless to man but some of the larger ones can inflict a painful bite. Our species are generally pale yellowish to dark brown, and vary from a few mm. to about 6 in.

PAUROPODS   Class Pauropoda   **Not illus.**
   Identification: Similar to centipedes but minute (1.0–1.5 mm.).
   Only 9 pairs of legs. Antennae branched.
       Pauropods are whitish and occur in soil and debris. They
   are not common.

SYMPHYLIDS   Class Symphyla   **Not illus.**
   Identification: Similar to centipedes but with 10–12 pairs of
   legs. 1–8 mm. Antennae not branched.
       Symphylids occur under stones, in rotting wood, and in the
   soil. They are not common.

INSECTS   Class Insecta
   Identification: Three pairs of legs. 3 body regions (head, thorax,
   abdomen). Often 1 or 2 pairs of wings. 1 pair of antennae.
   For additional characters, see p. 56.
       Since the bulk of this book is concerned with insects, nothing
   more need be said about them here.

Chelicerate Arthropods:
Subphylum Chelicerata

Usually 6 pairs of appendages: 1st pair (chelicerae) jawlike or
fanglike; 2nd pair (pedipalps) somewhat feelerlike (sometimes
clawlike, rarely leglike); remaining pairs leglike. Antennae absent.
Usually 2 body regions, cephalothorax (bearing the appendages)
and abdomen. Legs often with an extra segment (patella) between
femur and tibia.

SEA SPIDERS   Class Pycnogonida   **Not illus.**
   Identification: Long-legged, spiderlike marine animals, with a
   small cephalothorax (head and thorax) and a very small abdomen. Usually 5 pairs of legs.
       Sea spiders have a leg spread of 1 to several cm., and generally
   occur beneath low tidemark. They have a sucking proboscis and
   feed on other small animals.

HORSESHOE CRABS   Class Xiphosura
   Identification: Body with a *broadly oval shell* and a *long slender
   tail.* Abdomen with leaflike gills on ventral side. Large animals,
   up to 1½ ft.
       Horseshoe crabs are marine animals with a very distinctive
   appearance. They are fairly common along the seashore.

WATER BEARS   Class Tardigrada   **Not illus.**
   Identification: Minute animals, 1 mm. or less, with 4 pairs of
   unsegmented legs, each leg with several claws.

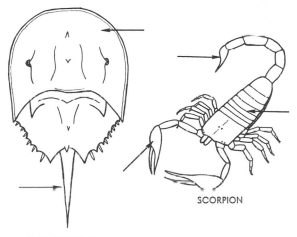

SCORPION

HORSESHOE CRAB

Water bears occur in fresh- and saltwater, mud, sand, and various other damp places. They are not often encountered.

TONGUEWORMS Class Linguatulida Not illus.

Identification: Immature stages with 4–6 pairs of legs. Adults wormlike and legless, living as parasites in the mouth or respiratory track of various vertebrates.

This is an aberrant group, and may or may not be correctly placed in the Chelicerata. Tongueworms are not common, and are unlikely to be encountered by users of this book.

ARACHNIDS Class Arachnida

Identification: Adults nearly always with 4 pairs of segmented legs (rarely the pedipalps are leglike). Usually not wormlike.

This is the largest and most often encountered chelicerate class. Its members are very widely distributed. Several types within the class merit separate consideration.

Scorpions, Order Scorpionida. Relatively large arachnids (to 4 or 5 in.). Pedipalps *large and clawlike.* Abdomen *distinctly segmented* and ending in a *sting* usually curved upward. Scorpions occur in the South and West and are largely nocturnal. They feed chiefly on insects and spiders. The sting of a scorpion can be quite painful, and in some cases (2 species occurring in Arizona) may be fatal.

Whip-scorpions, Orders Microthelyphonida, Pedipalpida, Schizopeltida, and Amblypygi. Scorpionlike, but abdomen *oval,* *segmented*, and lacking a sting. Pedipalps *not clawlike*. 1st pair of legs *longer than others*. Some species have *a long whiplike tail*. Principally tropical; our species occur only in the southern states. They vary from a few mm. to 4 or 5 in. and are predaceous. A few of the larger species give off a vinegarlike odor when disturbed.

Wind-scorpions, Order Solpugida (not illus.). Spiderlike, but with body only slightly constricted behind cephalothorax, chelicerae very large, abdomen segmented. Solpugids occur in the desert areas of the Southwest; they are nocturnal, spending the day in burrows or under stones, cow chips, and other objects and foraging at night. They are fast-running animals about an inch long or less and are predaceous.

Pseudoscorpions, Order Chelonethida. Small (generally 5 mm. or less), flattened, oval-bodied arachnids with *large clawlike pedipalps*. Normally occur under bark or in debris. They are moderately common.

Daddy-long-legs or Harvestmen, Order Phalangida. Body oval and compact. Legs *extremely long and slender*. Abdomen *segmented*. These common and well-known animals are found in wooded areas or in fairly dense vegetation. Most feed on dead insects or on plants. The few long-legged spiders and mites that might be confused with daddy-long-legs have the abdomen unsegmented, and spiders have a strong constriction between the cephalothorax and abdomen.

Mites and Ticks, Order Acarina. Abdomen *unsegmented* and *broadly joined to cephalothorax*. Body more or less oval, usually minute (some ticks may be several mm.). This is the largest order in the class, and its members occur almost everywhere, often in considerable numbers. Many are free-living and many are parasites of other animals. The free-living **mites** are probably most abundant in the soil and in debris, where their populations may number several million per acre. Some parasitic forms are important pests of man and domestic animals (chiggers or harvest mites, scab and mange mites, ticks): **chiggers** are annoying pests of man, and a few act as disease vectors; **scab and mange mites** are pests of both man and animals; some of the **ticks** are also important disease vectors. **Spider mites** are serious pests of various cultivated plants, especially orchard trees and greenhouse plants. **Water mites,** many of which are reddish or orange, are common inhabitants of ponds. A few mites are gall makers, usually forming small pouchlike galls on leaves.

Spiders, Order Araneida. Abdomen *strongly constricted at base*, nearly always *unsegmented*, with a group of fingerlike spinnerets at posterior end. Spiders are common and well-known animals occurring in many habitats. They feed on insects and other small animals, paralyzing their prey with venom from glands opening on the chelicerae. Spiders are venomous but rarely bite man; only a few are dangerously venomous. Silk spun from the spinnerets is used in the construction of webs, snares, shelters, and egg sacs. Many spiders do not build webs; they forage or lie in wait for their prey. Webs of different spiders vary in form: many (called orb webs) are more or less circular, with spirally wound strands of sticky silk and radiating strands that are not sticky; some are sheetlike or funnel-like; others are irregular. Sexes of spiders can usually be distinguished by the pedipalps — slender in the female, somewhat clubbed in the male. The male is often smaller than the female and not as frequently seen. In some species the female kills and eats the male after mating.

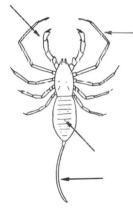

WHIP-SCORPION

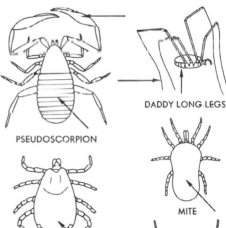

PSEUDOSCORPION

DADDY LONG-LEGS

TICK

MITE

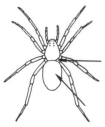

GROUND SPIDER

CRAB SPIDER

ORB-WEAVER SPIDER

Insects: Class Insecta

Identification: Three pair of legs. 3 body regions (head, thorax, and abdomen). Often 1 or 2 pairs of wings. 1 pair of antennae (antennae rarely absent). Mouth parts typically consist of a labrum, a pair of mandibles, a pair of maxillae, a hypopharynx, and a labium. Genital ducts open near posterior end of body. Winged insects differ from all other invertebrates in the possession of wings; wingless insects differ from most other arthropods in having 3 pairs of legs and a pair of antennae. Caterpillars appearing to have more than 3 pairs of legs have the first 3 pairs behind the head normally leglike in structure and the remaining pairs (prolegs) stout and fleshy and quite different in structure.

Similar groups: (1) Arachnida: larval mites with only 3 pairs of legs lack antennae, the body is not differentiated into 3 regions, and the abdominal region is not segmented. (2) Diplopoda: newly hatched millipedes, with 3 pairs of legs, have the head structure characteristic of millipedes (see illus., p. 51) and a single liplike structure (gnathochilarium) behind the mandibles instead of maxillae and a labium. (3) Annelida: some resemble legless insect larvae but have more than 13 body segments and lack a tracheal system.

Classification: The class Insecta is divided into 2 subclasses, the Apterygota and Pterygota. The **Apterygota** include the orders Protura, Thysanura, and Collembola, and the **Pterygota** include the remaining orders. The Apterygota are wingless, and most Pterygota have wings. The wingless Pterygota are thought to have evolved from winged ancestors because they have certain features of thoracic structure (for example, the thoracic pleura divided by a pleural suture into episternum and epimeron, and the meso- and metanotum divided by sutures) correlated with the development of wings; Apterygota have a simpler thoracic structure (these sclerites not divided by sutures). The Apterygota usually have stylelike appendages on the pregenital segments of the abdomen; such appendages are lacking in the Pterygota. The orders of insects are separated principally by the characters of the wings, mouth parts, legs, and the metamorphosis.

No. of species: World, 703,500; N. America, 88,600.

Some of the outstanding features of the 26 orders of insects are outlined in the following table.

CHARACTERS OF THE INSECT ORDERS

Order	Examples	Wings	Mouth parts	Usual no of tarsal segments	Meta-morphosis	Usual size	Where usually found
Protura	proturans	none	sucking	1	simple	minute	in debris
Thysanura	bristletails	none	chewing	1–5	simple	medium	in debris, buildings
Collembola	springtails	none	chewing	1	simple	minute	in debris
Ephemeroptera	mayflies	4 (rarely 2); HW small	vestigial	3–5	simple; nymphs aquatic	medium	near water
Odonata	dragonflies and damselflies	4; many-veined; HW as large as FW	chewing	3	simple; nymphs aquatic	large	near water
Orthoptera	grasshoppers, crickets, cockroaches, mantids, walkingsticks	0–4; FW narrow and thickened; HW folded fanwise	chewing	3–5	simple	large	on ground or vegetation
Isoptera	termites	0–4; FW and HW similar in size	chewing	4	simple	small	in ground or wood
Plecoptera	stoneflies	4; FW narrow; HW with a large anal lobe	chewing	3	simple; nymphs aquatic	medium	near water
Dermaptera	earwigs	0–4; FW short and thickened; HW folded	chewing	3	simple	medium to small	in debris
Embioptera	webspinners	0–4; HW a little smaller than FW	chewing	3	simple	small	in debris

57

CHARACTERS OF THE INSECT ORDERS (contd.)

Order	Examples	Wings	Mouth parts	Usual no. of tarsal segments	Meta-morphosis	Usual size	Where usually found
Psocoptera	booklice and barklice	0–4; HW smaller than FW	chewing	2–3	simple	small to minute	in debris, buildings
Zoraptera	zorapterans	0–4; HW smaller	chewing	2	simple	minute	in debris
Mallophaga	chewing lice	none	chewing	1–2	simple	minute	on birds or mammals
Anoplura	sucking lice	none	sucking	1	simple	minute	on mammals
Thysanoptera	thrips	0–4; wings long, narrow, fringed	rasping-sucking	1–2	intermediate between simple and complex	minute	in debris or on vegetation
Hemiptera	bugs	0–4; FW thick-ened at base	sucking	2–3	simple	medium	on debris, on vegeta-tion, on or in water
Homoptera	cicadas, hoppers, whiteflies, aphids, scale insects	0–4; FW membran-ous or thickened, uniform in texture	sucking	1–3	simple	minute to large	on vegetation
Neuroptera	dobsonflies, fishflies, alderflies, lacewings, antlions	4; many-veined; HW as large as FW	chewing	5	complete	medium to large	on vegetation, often near water

58

CHARACTERS OF THE INSECT ORDERS (contd.)

Order	Examples	Wings	Mouth parts	Usual no of tarsal segments	Meta-morphosis	Usual size	Where usually found
Coleoptera	beetles	0–4; FW thickened and veinless; HW membranous and folded	chewing	3–5	complete	minute to large	all habitats
Strepsiptera	twisted-winged parasites	0–2 (FW clublike); HW fanlike; only male with wings	vestigial	2–5	complete, with hypermetamor-phosis	minute	in other insects, chiefly bees and hoppers (Homoptera)
Mecoptera	scorpionflies	0–4; FW and HW about same size	chewing; beaklike	5	complete	medium	on vegetation
Trichoptera	caddisflies	4; hairy	chewing; reduced	5	complete; larvae aquatic	small to medium	near water
Lepidoptera	butterflies and moths	0–4; scaly	sucking, with coiled proboscis	5	complete	minute to large	on vegetation
Diptera	flies	0–2 (HW reduced to halteres)	sucking	5	complete	minute to large	on vegetation
Siphonaptera	fleas	none	sucking	5	complete	minute	on birds or mammals
Hymenoptera	sawflies, ichneumons, chalcids, ants, wasps, bees	0–4; HW smaller than FW	chewing, chewing-sucking	5	complete	minute to large	on ground or vegetation

Proturans: Order Protura

Identification: Minute whitish insects, 0.6–1.5 mm. Eyes, wings, cerci, and *antennae lacking.* Front legs carried in an elevated position like antennae. Abdomen of adult 12-segmented, with a pair of short *styli* (fingerlike processes) on each of the 3 basal segments. Metamorphosis simple.

Similar orders: Other minute wingless insects have less than 12 abdominal segments, and either have antennae or lack styli on the 3 basal abdominal segments.

Immature stages: Similar to adult stage, but with fewer abdominal segments (a segment added each molt).

Habits: Proturans occur in moist soil, moss, leaf mold, under bark, and in rotting wood. They are rather rare, and are very infrequently collected.

Importance: This order is considered to be a very primitive one. Its members are not of economic importance.

Classification: Three families, separated principally on the basis of the presence or absence of a tracheal system and the character of the abdominal appendages.

No. of species: World, 118; N. America, 18.

Key to Families

1.   Tracheae present, with 2 pairs of spiracles on thorax; abdominal appendages with a terminal vesicle
                                                              Eosentomidae
1'. Tracheae and spiracles absent; abdominal appendages with or without a terminal vesicle   2
2(1'). At least 2 pairs of abdominal appendages with a terminal vesicle   **Protentomidae**
2'. Only the 1st pair of abdominal appendages with a terminal vesicle   **Acerentomidae**

Bristletails: Order Thysanura

Identification: Elongate (rarely oval) wingless insects with 2 or 3 tail-like appendages at end of abdomen. Some abdominal segments with a pair of *styli* (fingerlike processes). Antennae long and many-segmented. Metamorphosis simple.

Similar orders: (1) Protura: no antennae or tails; minute size (Thysanura usually more than 1.5 mm.). (2) Collembola (p. 62): no tails, but usually a forked appendage at end of abdomen; no styli on abdomen, but generally a single short tubular structure (collophore) on 1st segment; antennae short. (3) Orders of Pterygota (larvae and wingless adults), p. 56: no styli on abdominal segments.

Immature stages: Similar to adult stage.
Habits: Most species occur in leaf litter, under bark and stones, or in debris; some species may be found in buildings.
Importance: A few sometimes are pests in houses.
Classification: Two suborders, Ectognatha and Entognatha, differing in the number of terminal abdominal appendages and the segmentation of the tarsi. Many authors give these groups order rank, the Thysanura including only the Ectognatha and the Entognatha being placed in the order Diplura.
No. of species: World, 700; N. America, 50.

Common Bristletails: Suborder Ectognatha

Three tail-like appendages at end of abdomen (cerci and a median caudal filament). Compound eyes usually present. Body generally covered with scales. Tarsi 3- to 5-segmented. Mostly active, fast-running or jumping insects.

SILVERFISH Family Lepismatidae
Identification: Compound eyes *small and widely separated.* Ocelli absent. Tarsi *3 or 4 segmented.* Coxae without styli.

The Silverfish, *Lepisma saccharina* Linn., and Firebrat, *Thermobia domestica* (Packard), the most commonly encountered members of this group, often are pests in houses and other buildings, where they feed on all sorts of starchy substances. They are 10–12 mm. The Silverfish is silvery, usually occurs in cool damp situations. The Firebrat is brownish, inhabits warm situations around furnaces and steam pipes.

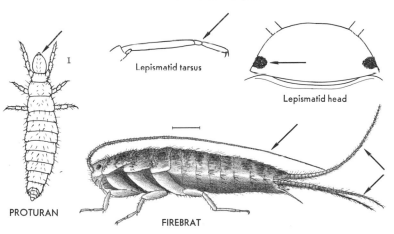

Lepismatid tarsus

Lepismatid head

PROTURAN

FIREBRAT

NICOLETIIDS Family Nicoletiidae   **Not illus.**
   Identification: Similar to Lepismatidae but compound eyes
   lacking and body sometimes not covered with scales.
      Nicoletiids may be elongate or oval: elongate forms lack scales,
   occur in caves and mammal burrows; oval forms have scales,
   live in ant and termite nests. All are quite rare but have been
   found in Florida and Texas.

PRIMITIVE BRISTLETAILS   **Not illus.**
Family Lepidotrichidae
   Identification: Similar to Lepismatidae but body lacks scales,
   ocelli are present, and tarsi are 5-segmented.
      A single rare species occurs in n. California, under bark and
   in rotting wood of fallen Douglas fir. Yellowish gray, 12 mm.,
   with antennae and tails quite long.

JUMPING BRISTLETAILS Family Machilidae
   Identification: Similar to Lepismatidae but compound eyes
   large and touching and there are styli on middle and hind coxae.
   Tarsi 3-segmented.
      Machilids are active jumpers usually found in leaf litter,
   under bark and stones, or among rocks along the seashore.
   Most are brownish.

Diplurans: Suborder Entognatha

Only *2 appendages* (cerci) at end of abdomen. Body without
scales. Compound eyes lacking. Tarsi 1-segmented. Slender
whitish insects about 6 mm. or less. Usually found in soil and leaf
litter, under bark, or under stones and logs.

DIPLURANS
Families Campodeidae, Anajapygidae, and Japygidae
   Identification: Campodeidae: cerci and antennae *about equal
   length*; 4–6 mm. Anajapygidae: similar, but cerci shorter than
   antennae and fewer-segmented; less than 4 mm. Japygidae:
   cerci *1-segmented and forcepslike.*
      The most commonly encountered diplurans are the campo-
   deids. The other 2 families are small and rare (anajapygids are
   represented by only 1 species, occurring in California).

Springtails: Order Collembola

Identification: Minute wingless insects, most less than 6 mm.
Body elongate or oval. Abdomen with 6 or fewer segments and
without cerci. Usually *a forked structure (furcula)* on 4th or 5th

abdominal segment, and *a small tubular structure* (*collophore*) on 1st abdominal segment. Antennae *short*, 4- to 6-segmented. Metamorphosis simple.

Similar orders: (1) Protura (p. 60): no antennae or furcula. (2) Thysanura (p. 60): cerci present; some abdominal segments with styli. (3) Small wingless Pterygota (p. 56): no furcula.

Immature stages: Similar to adult stage.

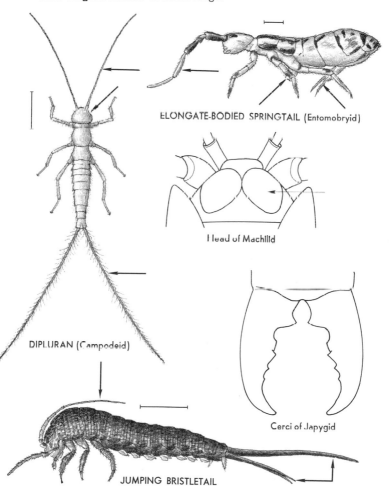

ELONGATE-BODIED SPRINGTAIL (Entomobryid)

Head of Machilid

DIPLURAN (Campodeid)

Cerci of Japygid

JUMPING BRISTLETAIL

Habits: Springtails occur in soil and leaf litter, under bark and in decaying wood, in fungi, and on the surface of water (freshwater ponds and along the seashore); a few occur on vegetation. Species with a furcula are jumpers; the furcula is normally folded forward under the abdomen, and the insect jumps by suddenly extending the furcula ventrally and posteriorly.

Importance: Important chiefly as scavengers, rarely as pests. Often very abundant, their populations sometimes numbering several million individuals per acre.

Classification: Two suborders, Arthropleona and Symphypleona, differing in body shape and segmentation.

No. of species: World, 2000; N. America, 314.

Suborder Arthropleona

Body elongate. Abdomen distinctly 6-segmented.

ELONGATE-BODIED SPRINGTAILS
Families Poduridae and Entomobryidae

   Identification: Poduridae: minute, usually gray or black, wit *short appendages;* integument (body covering) granular or tuberculate, without scales; pronotum *well developed* and visible from above; antennae *4-segmented.* Entomobryidae: similar, but integument usually smooth, pronotum reduced and usually not visible from above, and antennae *4- to 6-segmented.*

   The family Entomobryidae is the largest in the order, and includes most of the springtails found in leaf litter and soil, in fungi, under bark, and in similar situations. The Marsh Springtail, *Isotomurus palustris* (Müller), is a common entomobryid occurring in moist woodlands. The Seashore Springtail, *Anurida maritima* (Guérin), is a slate-colored podurid occurring along the seashore, often in large numbers; it is frequently found in dense clusters on the surface of small pools between the tidemarks. *Podura aquatica* Linn. is common on the surface of freshwater ponds and streams. The Snow Flea, *Achorutes nivicolus* (Fitch), is a dark-colored podurid that sometimes occurs in large numbers on the surface of snow.

Suborder Symphypleona

Body oval or somewhat globular. Basal abdominal segments more or less fused.

GLOBULAR SPRINGTAILS   Family Sminthuridae
   Identification: By the characters of the suborder.
      Sminthurids are minute insects, generally yellowish or mottled, with black eyes. They are often abundant on vegetation.

Mayflies: Order Ephemeroptera

Identification: Small to medium-sized, elongate, very soft-bodied, usually found near water. FW *large, triangular, many-veined*. HW small and rounded (rarely absent). Wings held together above body at rest. Abdomen with 2 or 3 hairlike tails. Antennae *small, bristlelike,* inconspicuous. Tarsi 3- to 5-segmented. Mouth parts vestigial. Metamorphosis simple.

Similar orders: (1) Odonata (p. 68): HW as large as FW or larger; terminal abdominal appendages relatively short; harder-bodied. (2) Hymenoptera (some ichneumons), p. 312: harder-bodied; antennae long; wings with fewer veins; tarsi 5-segmented. (3) Plecoptera (p. 92): HW with an anal lobe; wings held flat over abdomen at rest; antennae long and conspicuous.

Immature stages: Leaflike gills along sides of abdomen, and 3

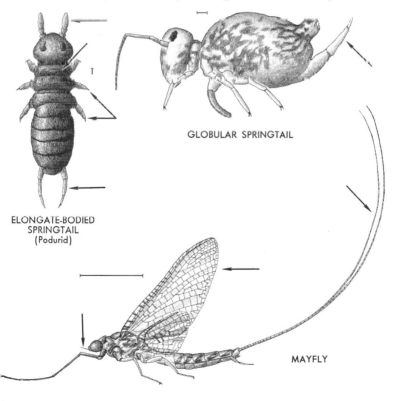

GLOBULAR SPRINGTAIL

ELONGATE-BODIED
SPRINGTAIL
(Podurid)

MAYFLY

hairlike tails. Common inhabitants of ponds and streams. Food consists of small aquatic organisms and organic debris.

Habits: The last nymphal instar molts to the winged form at the water surface or on an object just out of water. This winged stage, usually rathe⁻ dull in appearance and somewhat pubescent, is not yet adult and is called a *subimago;* it molts once more to become adult (mayflies are unique among insects in undergoing a molt after the wings become functional). Adults seldom live more than a day or two, and do not feed. Males of many species engage in swarming flights, the members of the swarm flying up and down in unison. Eggs are attached to stones and other objects in the water, or are washed off the abdomen onto the water's surface. Adults often emerge in large numbers from lakes and ponds, and sometimes actually pile up along the shore.

Importance: Adults and nymphs are an important food of many freshwater fish. Artificial flies used by fishermen are often modeled after mayflies.

Classification: Three families, separated chiefly by wing venation and tarsal segmentation.

No. of species: World, 2100; N. America, 585.

BURROWING MAYFLIES
Family Ephemeridae
   Identification: Most have clear wings; a few have spotted wings. Base of M_2 in FW *extends toward Cu_1, then bends abruptly distad*. R_{4+5} in HW *not forked*. Hind tarsi 4-segmented.
   Nymphs occur in ponds, lakes, and large rivers, and are usually burrowing in habit. This group includes our largest mayflies. Some often emerge from lakes and rivers in enormous numbers.

STREAM MAYFLIES   Family Heptageniidae
   Identification: Base of M_2 in FW *nearly straight*. Cubital intercalaries in FW in 2 parallel pairs. R_{4+5} in HW *forked*. Hind tarsi 5-segmented.
   Nymphs are flattened and streamlined, and occur on the underside of stones in streams. Adults are medium-sized to small, and usually clear-winged.

SMALL MAYFLIES   Family Baetidae
   Identification: M_2 in FW *as in Heptageniidae*. Cubital intercalaries in FW variable, but not in 2 parallel pairs. R_{4+5} in HW variable. Venation in HW often reduced (HW sometimes lacking). Hind tarsi 3- or 4-segmented.
   Nymphs are more cylindrical than those of Heptageniidae, and occur in various aquatic habitats. Adults vary in size and appearance but are usually less than 15 mm.

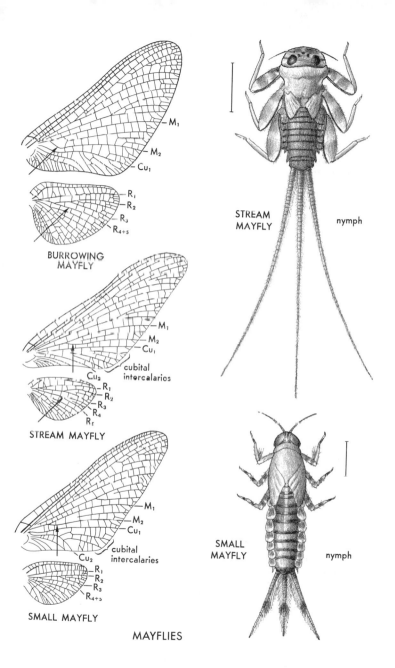

BURROWING MAYFLY

M_1
M_2
Cu_1

R_1
R_2
R_3
R_{4+5}

STREAM MAYFLY

M_1
M_2
Cu_1
cubital intercalaries
Cu_2

R_1
R_2
R_3
R_4
R_5

SMALL MAYFLY

M_1
M_2
Cu_1
cubital intercalaries
Cu_2

R_1
R_2
R_3
R_{4+5}

STREAM MAYFLY   nymph

SMALL MAYFLY   nymph

MAYFLIES

Dragonflies and Damselflies:
Order Odonata

Identification: Two pairs of elongate, membranous, many-veined wings. Wings at rest usually held outstretched (Anisoptera) or together above body (Zygoptera). FW and HW similar in size and shape (Zygoptera), or HW *broader at base than FW* (Anisoptera). Abdomen *long and slender*. Compound eyes large, often occupying most of head. Antennae *very short, bristlelike*, and inconspicuous. Prothorax small, the other 2 thoracic segments making up most of thorax. Tarsi *3-segmented*. Copulatory organs of ♂ located on ventral side of 2nd abdominal segment. Cerci present, 1-segmented, in male functioning as clasping organs during mating. Mouth parts chewing. Metamorphosis simple.

Similar orders: (1) Neuroptera (p. 140): antennae long; tarsi 5-segmented; wing venation different. (2) Hymenoptera (p. 312): antennae long; tarsi 5-segmented; HW smaller than FW; wings with fewer veins. (3) Diptera (p. 260): 1 pair of wings. (4) Ephemeroptera (p. 65): 2–3 long tails; HW smaller than FW; very soft-bodied.

Immature stages: Nymphs are aquatic, and occur in ponds and streams. They feed on other insects, which are captured with a peculiarly modified labium. When not in use, the labium is folded under the head, and when used is thrust forward very quickly to catch prey in a pair of clawlike structures at its apex. Labium when extended is sometimes as long as ⅓ body length. Gills of the nymph are located in rectum (Anisoptera) or are in form of 3 leaflike tails (Zygoptera).

Habits: Adults usually found near water (in which nymphs live), but many are strong fliers and can range many miles. Often fly in tandem, the male holding female by back of head or the prothorax with the appendages at end of his abdomen. Eggs generally laid in aquatic vegetation or are washed off end of the abdomen when female flies low over water. Adults relatively large insects (about 1–3½ in.), and many are brightly colored. Most are good fliers, and spend a large part of their time on the wing. They feed on other insects they catch on the wing.

Importance: All stages are predaceous, feeding on mosquitoes, midges, and other small insects, and help keep them under control. Adults attempt to bite when handled, but only the larger dragonflies can inflict a painful pinch; they do not sting.

Classification: Two suborders, Anisoptera and Zygoptera, which differ in wing shape, position of wings at rest, appendages at end of abdomen, and *characters of nymphs*. Principal characters separating families are those of wing venation.

No. of species: World, 4950; N. America, 400.

Dragonflies: Suborder Anisoptera

Relatively stout-bodied, about 1–3½ in. HW *broader at base than FW*, the wings at rest held outstretched. ♂ with 3 terminal abdominal appendages, 2 above and 1 below. ♀ with only 2 (dorsal) terminal appendages. ♀ of some groups with an ovipositor, located on ventral side of terminal abdominal segments, giving end of abdomen a somewhat swollen appearance. Nymphs robust, with gills in rectum; breathing is accomplished by drawing water into rectum through anus, and then expelling it; this expulsion of water serves as a means of locomotion, the insect thus moving by "jet" propulsion.

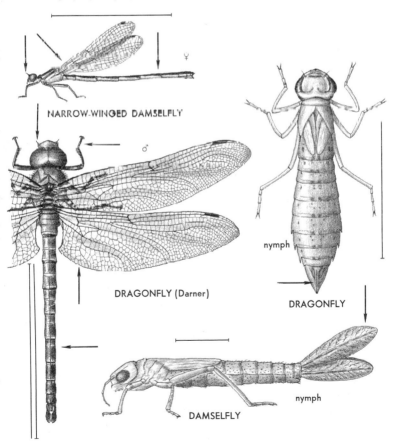

NARROW-WINGED DAMSELFLY

♀

♂

DRAGONFLY (Darner)

nymph

DRAGONFLY

nymph

DAMSELFLY

Graybacks, Clubtails, Darners, and Biddies:
Superfamily Aeshnoidea

Triangles in FW and HW similar in shape and location. Most antenodal cross veins between C and Sc not in line with those between Sc and R. *A brace vein at proximal end of stigma* (except Cordulegastridae). Wings nearly always clear, without spots or bands. Ovipositor present or absent.

GRAYBACKS Family Petaluridae
Identification: Large grayish-brown or blackish dragonflies, about 3 in. Compound eyes do not meet on dorsal side of head. Median lobe of labium *notched*. Stigma at least 8 mm. Ovipositor well developed.

Two species of graybacks occur in the U.S., 1 in the East and 1 in the West; both are rare and local. Eastern species, *Tachopteryx thoreyi* (Hagen), is grayish brown and found along small streams in wooded valleys. Western species, *Tanypteryx hageni* (Selys), is blackish, and occurs at high elevations in mountains. Graybacks often alight on tree trunks, where their color blends with that of the bark.

CLUBTAILS Family Gomphidae   **See also Pl. 1**
Identification: Compound eyes *do not meet* on dorsal side of head. Medium lobe of labium not notched. Stigma less than 8 mm. Terminal segments of abdomen sometimes dilated. ♀ lacks ovipositor.

Members of this large group occur along streams or shores of large lakes. Most species are 2–3 in., and dark brown with yellowish or greenish markings. Flight is usually steady, without periods of hovering; some occasionally fly with a very undulating flight. Adults often alight on a flat surface.

DARNERS Family Aeshnidae   **See also Pl. 1**
Identification: Compound eyes *in contact* for a considerable distance on dorsal side of head. Ovipositor *well developed*.

This group includes our largest dragonflies. Most species are $2\frac{1}{4}$–$3\frac{1}{4}$ in., and a few may reach $3\frac{1}{2}$ in. or more. All are strong fliers, difficult to catch. Most species are dark brown, often with bluish or greenish markings. The Green Darner (Pl. 1), *Anax junius* (Drury), a common species found around ponds, has a light green thorax, bluish abdomen and a targetlike mark on upperpart of the face. *Aeshna* is a large and widely distributed genus whose members are mostly dark-colored, with bluish markings. Darners generally occur around ponds and swamps.

BIDDIES Family Cordulegastridae
Identification: Large brownish to blackish dragonflies with yellowish markings. *No brace vein.* Ovipositor of ♀ conspicuous.
Biddies resemble river skimmers (*Macromia*, family Macromiidae), but have a slightly different wing venation (compare illustrations, pp. 71 and 73). The compound eyes are slightly separated on dorsal side of head or meet at a single point only. A small group, and its members are not common. They occur chiefly along small woodland streams.

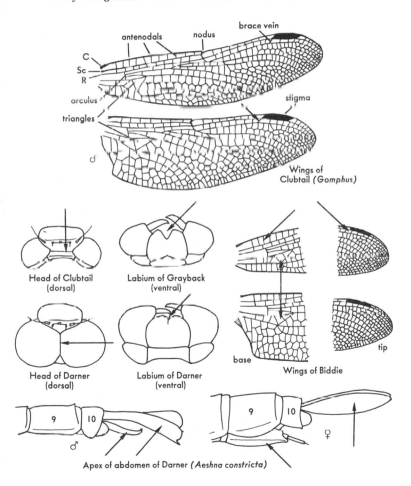

Wings of
Clubtail (*Gomphus*)

Head of Clubtail
(dorsal)

Labium of Grayback
(ventral)

Head of Darner
(dorsal)

Labium of Darner
(ventral)

base

tip

Wings of Biddie

Apex of abdomen of Darner (*Aeshna constricta*)

Skimmers: Superfamily Libelluloidea

Triangles in FW and HW *different in shape*, and triangle in FW
farther beyond arculus than triangle in HW. Most antenodal cross
veins between C and Sc *in line with those between Sc and R. No
brace vein.* Ovipositor lacking.

BELTED and RIVER SKIMMERS Family Macromiidae
  Identification: Anal loop of HW rounded, without a bisector.
  ♂ with inner margin of HW *slightly notched*, and a small lobe on
  each side of 2nd abdominal segment. Wings never with spots
  or bands.
  Belted skimmers (*Didymops*) are brownish, with light markings on the thorax, and 2¼–2¾ in. They occur along boggy
  pond shores but are not common. River skimmers (*Macromia*)
  are 2½–3¼ in., blackish, with yellow markings on thorax and
  abdomen, and the eyes in life are bright green; adults are fast
  fliers and are found along large streams and lakes.

GREEN-EYED SKIMMERS Family Corduliidae
  Identification: Anal loop in HW elongate, with a bisector,
  usually *not particularly foot-shaped.* ♂ with inner margin of
  HW *slightly notched*, and a small lobe on each side of 2nd abdominal segment. Hind margin of compound eyes *slightly lobed.*
  Corduliids are usually blackish or metallic, and seldom have
  conspicuous light markings. The eyes in life are bright green.
  Most species have the wings entirely clear or with only a small
  dark spot at the base, and are 1½–2¾ in. Members of the
  genus *Epicordulia* have a dark spot at base, middle, and tip of
  each wing, and are 2½–3¼ in. Corduliids generally occur about
  swamps and ponds; only a few are found along streams.

COMMON SKIMMERS Family Libellulidae   **See also Pl. 1**
  Identification: Anal loop in HW elongate, with a bisector, and
  usually *foot-shaped.* Inner margin of HW *rounded in both sexes.*
  ♂ without lateral lobes on 2nd abdominal segment. Hind
  margin of compound eyes *straight* or only very slightly lobed.
  Wing color variable.
  This is a large group and many species are very common.
  They occur chiefly about ponds and swamps; most dragonflies
  one sees in such places belong to this family. The majority are
  1–2½ in. (a few may be somewhat larger), and the wingspread
  is noticeably greater than the body length. Many are brightly
  colored, often with spots or bands on the wings; some have the
  sexes differently colored. Their flight is fast, sometimes interrupted by periods of hovering.

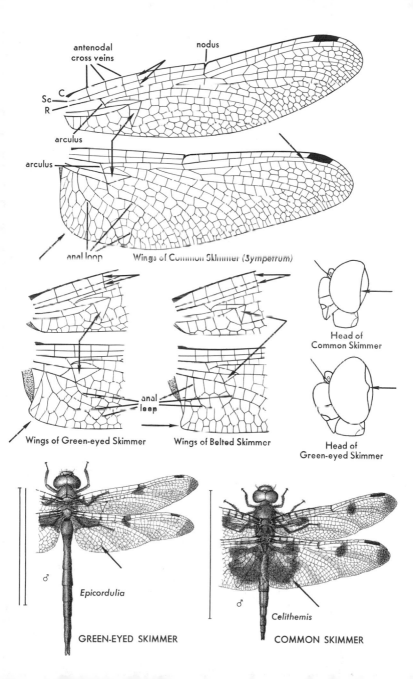

antenodal
cross veins

nodus

Sc
C
R

arculus

arculus

anal loop

Wings of Common Skimmer (*Sympetrum*)

anal loop

Wings of Green-eyed Skimmer

Wings of Belted Skimmer

Head of
Common Skimmer

Head of
Green-eyed Skimmer

Epicordulia

♂

GREEN-EYED SKIMMER

Celithemis

♂

COMMON SKIMMER

Damselflies: Suborder Zygoptera

FW and HW similar in size and shape, at rest held together above body or somewhat divergent. Abdomen *very slender*. ♂ with 4 terminal appendages. ♀ with a well-developed ovipositor. Nymphs with *3 leaflike gills* at end of abdomen (illus., p. 69); swim by body undulations.

BROAD-WINGED DAMSELFLIES   See also Pl. 1
Family Calopterygidae
> **Identification:** Wings *gradually narrowed at base*, with *10 or more antenodal cross veins;* usually blackish or with blackish markings *(Calopteryx)* or clear with a reddish spot at base *(Hetaerina)*; wings at rest held together above body.
>
> These are large damselflies that occur along streams. The male Black-winged Damselfly (Pl. 1), *Calopteryx maculata* (Beauvais), a common eastern species, has blackish wings and a metallic greenish-black body; female has dark gray wings with a white stigma and body is not metallic. The American Ruby-spot, *Hetaerina americana* (Fabricius), another common eastern species, is reddish, with a bright red spot at base of wings.

SPREAD-WINGED DAMSELFLIES   Family Lestidae
> **Identification:** Wings *stalked at base; 2 antenodul cross veins,* M_3 rises *closer to arculus* than to nodus. Wings clear, usually held diverging above body at rest.
>
> Spread-winged damselflies are $1\frac{1}{4}$–2 in., and are common around swamps and ponds. Most belong to the genus *Lestes*. They can be recognized in the field by *the way they hold their wings at rest*.

NARROW-WINGED DAMSELFLIES   See also p. 69 and Pl. 1
Family Coenagrionidae
> **Identification:** Similar to Lestidae, but M_3 rises *behind nodus*, and wings at rest *held together above body*.
>
> This group includes most of our damselflies (about 75 species). The majority occur around ponds and swamps (where often abundant), but a few are found along streams. Most species are 1–$1\frac{1}{4}$ in. and clear-winged. *Argia fumipennis* (Burmeister), which occurs in the southern states, has smoky-brown wings. Many are very brightly colored. Most bluets *(Enallagma*, the largest genus in the family) are light blue with black markings, species of *Amphiagrion* are blackish with a red abdomen, and males of the Violet Dancer (Pl. 1), *Argia violacea* (Hagen), are largely violet. Color pattern usually differs in the sexes, males being more brightly colored than females.

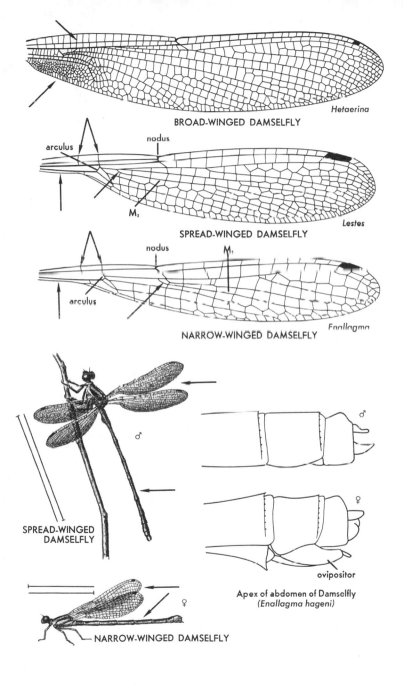

BROAD-WINGED DAMSELFLY

Hetaerina

arculus

nodus

M_3

SPREAD-WINGED DAMSELFLY

Lestes

arculus

nodus

M_1

NARROW-WINGED DAMSELFLY

Enallagma

♂

SPREAD-WINGED
DAMSELFLY

♂

♀

ovipositor

Apex of abdomen of Damselfly
(*Enallagma hageni*)

♀

NARROW-WINGED DAMSELFLY

Grasshoppers, Katydids, Crickets, Mantids, Walkingsticks, and Cockroaches: Order Orthoptera

Identification: Usually 2 pairs of wings: FW long, narrow, many-veined, and somewhat thickened (called *tegmina*); HW membranous, broad, with many veins, and folded fanwise at rest under FW; 1 or both pairs of wings sometimes small or absent. Antennae many-segmented, often long and hairlike. Cerci present, sometimes short and clasperlike, sometimes long and feelerlike. ♀ usually has ovipositor, which may be long and slender or short. Tarsi generally 3- to 5-segmented. Mouth parts chewing. Metamorphosis simple.

Similar orders: (1) Coleoptera (p. 146): FW thickened but veinless; cerci lacking; antennae rarely with more than 11 segments; HW longer than FW, with relatively few veins and not folded fanwise at rest. (2) Hemiptera (p. 112): mouth parts sucking; FW usually with base thickened and tip membranous; HW with few veins; antennae with 5 or fewer segments. (3) Dermaptera (p. 98): FW thickened but short; cerci forcepslike. (4) Homoptera (p. 128) (hoppers): rarely over 12 mm.; mouth parts sucking; antennae short and bristlelike. (5) Isoptera (p. 88): FW and HW of winged forms similar in size and shape, HW not folded at rest; cerci short and inconspicuous; antennae short, threadlike, or beadlike; tarsi 4-segmented.

Immature stages: Similar to adult but wings short or absent.

Habits: Many Orthoptera "sing" by rubbing one body part against another. Long-horned grasshoppers (p. 80) and crickets (p. 82) rub a sharp edge (*scraper*) of one front wing over a filelike ridge (*file*) on underside of other front wing. Slant-faced grasshoppers rub hind legs against the tegmina. Band-winged grasshoppers snap hind wings in flight. Males generally do the singing; females of a few species produce soft noises. Song most often heard ("calling" song) functions mainly in getting the sexes together. Each species has a distinctive song and some Orthoptera can produce more than one type of sound.

Importance: Most orthopterans are plant feeders and some are very destructive to cultivated plants; a few species sometimes increase to enormous numbers and migrate long distances, completely destroying large areas of crops on the way. A few are predaceous and a few rather omnivorous. Some orthopterans (like cockroaches) may be pests in buildings.

Classification: Six suborders, separated chiefly by characters of legs, antennae, body form, and ovipositor.

No. of species: World, 22,500; N. America, 1015.

Suborder Caelifera

Grasshopperlike jumping insects, with hind femora more or less *enlarged*. Tarsi with 3 or fewer segments. Antennae *relatively short*. Tympana (eardrums) usually present on sides of 1st abdominal segment. Ovipositor short.

PYGMY MOLE CRICKETS Family Tridactylidae
Identification: Length *10 mm.* or less. Front tibiae *enlarged* and

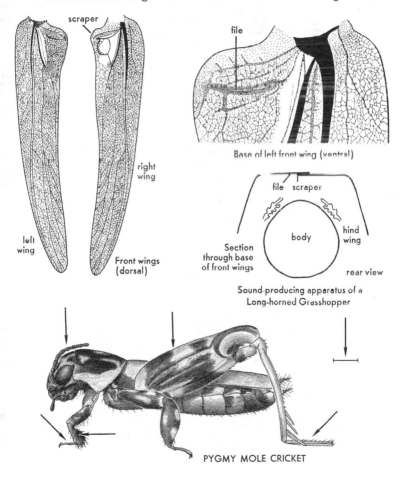

scraper

file

Base of left front wing (ventral)

right wing

file   scraper

body

hind wing

left wing

Section through base of front wings

Front wings (dorsal)

rear view

Sound-producing apparatus of a
Long-horned Grasshopper

PYGMY MOLE CRICKET

fitted for digging. Front and middle tarsi *2-segmented*, hind tarsi *1-segmented* or lacking. Antennae *11-segmented*. Body not pubescent.

These insects occur chiefly in moist sandy situations along shores of ponds and streams. They burrow in the ground but are sometimes found on the surface. They are not common.

PYGMY GRASSHOPPERS   Family Tetrigidae

Identification: Pronotum *extends back over abdomen* and is pointed posteriorly. Hind tarsi *3-segmented*, other tarsi *2-segmented*. Tegmina (FW) very short. 18 mm. or less.

Pygmy grasshoppers overwinter as adults and are most often encountered in spring and early summer. They are moderately common but are not of much economic importance.

SHORT-HORNED GRASSHOPPERS   See also Pl. 2
Family Acrididae

Identification: Pronotum *not prolonged back over abdomen*. Wings usually well developed. Tarsi *3-segmented*.

This group contains our most common grasshoppers. Many are important pests of cultivated plants. Most of them oviposit in the ground and overwinter in the egg stage.

Spur-throated Grasshoppers, Subfamily Cyrtacanthacridinae (see also Pl. 2). Spine or *tubercle* on prosternum. Pronotum flat dorsally and *broadly rounded posteriorly*. Face usually vertical. HW generally clear. This group contains many common species, including most of the pests. A few species sometimes increase to epidemic proportions and migrate long distances; these migrating swarms contain millions of grasshoppers and cause enormous damage.

Slant-faced Grasshoppers, Subfamily Acridinae. Similar to Cyrtacanthacridinae but with face *slanting backward* and without prosternal spine. Often occur in wet meadows or near marshes. Less abundant than other Acrididae.

Band-winged Grasshoppers, Subfamily Oedipodinae (see also Pl. 2). HW usually *colored*. A median *longitudinal keel* on pronotum. Posterior margin of pronotum *triangularly extended backward*. Face vertical or nearly so. Bandwings are often very common in sparse vegetation and along roadsides. They are conspicuous in flight because of color of the hind wings and the crackling noises they sometimes make but very inconspicuous when they alight, because the hind wings are concealed and the front wings are usually colored like the background.

EUMASTACIDS and MONKEY GRASSHOPPERS   Not illus.
Families Eumastacidae and Tanaoceridae

Identification: Similar to Acrididae but wingless. Medium-sized to small, usually brownish. Tympana generally absent.

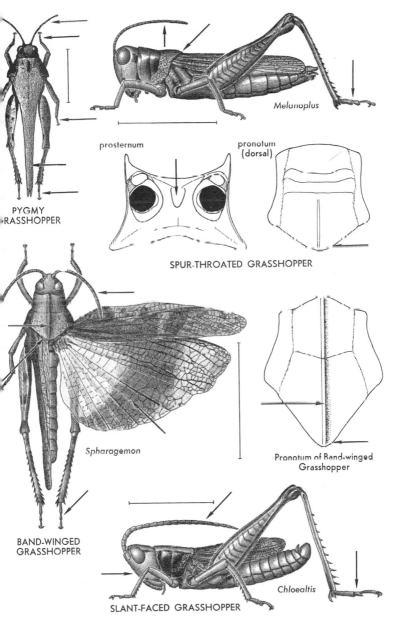

PYGMY
GRASSHOPPER

Melanoplus

prosternum

pronotum
(dorsal)

SPUR-THROATED GRASSHOPPER

Spharagemon

Pronotum of Band-winged
Grasshopper

BAND-WINGED
GRASSHOPPER

Chloealtis

SLANT-FACED GRASSHOPPER

79

Antennae shorter than front femora (Eumastacidae) or considerably longer (Tanaoceridae).

These 2 families contain several species of relatively uncommon grasshoppers that are found in the chaparral country of the Southwest, where they usually occur on the uppermost branches of bushes.

Suborder Ensifera

Similar to Caelifera but antennae *long, slender*, usually as long as body or longer. Tympana (eardrums), when present, located *at base of front tibiae*. Ovipositor *long and slender*, sometimes as long as body or longer. Tarsi 3- or 4-segmented.

LONG-HORNED GRASSHOPPERS   See also Pl. 2
Family Tettigoniidae
   Identification: Large insects, usually greenish (sometimes brownish). Tarsi *4-segmented*. Ovipositor sword-shaped. Wings present (sometimes small), with less than 8 principal longitudinal veins. ♂ with sound-producing structures on FW (p. 77).
   The best-known members of this group are greenish and the males are noted songsters. Most of the front wing surface slopes over the sides of the body, only a small portion being horizontal and dorsal; the base of the left tegmen is usually uppermost. Most species are plant feeders, and lay their eggs on or in plant tissues; winter is generally passed in the egg stage. Our species are grouped into 8 subfamilies, the most important of which are the following.
   True Katydids, Subfamily Pseudophyllinae. FW *broadly oval, somewhat convex.* Prosternum with pair of short spines. Pronotum *about as long as wide* and with 2 transverse grooves. Arboreal and not often seen, but their *katy-did, katy-didn't* song is well known; they sing only at night. Principally eastern in distribution.
   Bush and Round-headed Katydids, Subfamily Phaneropterinae. Usually green and over 1 in. FW *flat,* long and narrow or *elongate-oval.* HW usually longer than FW. Dorsal surface of 1st tarsal segment smoothly rounded (laterally grooved in the following subfamilies). Vertex somewhat rounded. Prosternal spines lacking. These katydids occur on weeds, bushes, and trees. Bush katydids (*Scudderia*, Pl. 2) have long, narrow, and nearly parallel-sided tegmina (FW). Angular-winged katydids (*Microcentrum*) have tegmina *widened* and *somewhat angulate in the middle,* and hind femora extend only a little beyond middle of tegmina. Round-headed katydids (*Amblycorypha*) have oval tegmina and hind femora extend almost to tips of tegmina. The songs of these katydids usually consist of a series of high-pitched lisps or ticks.
   Cone-headed Grasshoppers, Subfamily Copiphorinae. Green-

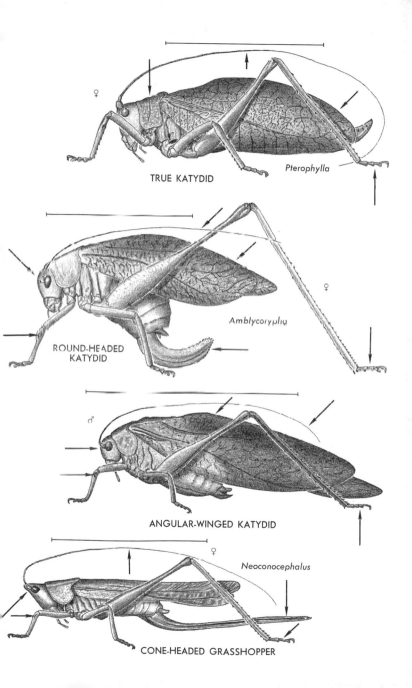

TRUE KATYDID

Pterophylla

ROUND-HEADED KATYDID

Amblycorypha

ANGULAR-WINGED KATYDID

Neoconocephalus

CONE-HEADED GRASSHOPPER

ish, usually over 1 in. Head *conical,* vertex *extending well beyond basal antennal segment.* Tegmina long and narrow. Ovipositor *nearly or quite as long as body.* Usually found in high grass or weeds. Songs high-pitched and buzzy.

Meadow Grasshoppers, Subfamily Conocephalinae (see also Pl. 2). Slender, greenish, seldom over 1 in. Vertex *does not extend beyond basal antennal segment.* Prosternum usually with a pair of small spines. Found principally in wet meadows or in grassy areas near ponds and streams. Songs usually consist of relatively long buzzes separated by zips (*bzzzzzz-zip-zip-zip-zip-bzzzzzz*).

Shield-backed Grasshoppers, Subfamily Decticinae. Brownish to black, usually 1 in. or longer, generally short-winged, with pronotum *extending back to abdomen.* Most eastern species belong to the genus *Atlanticus,* and occur in dry upland woods. Some western species often are serious pests of field crops; the most important of these is the Mormon Cricket, *Anabrus simplex* Haldeman.

CAMEL CRICKETS and OTHERS Family Gryllacrididae
Identification: Similar to Tettigoniidae but usually wingless, gray or brown. Wings if present with 8 or more principal longitudinal veins, and FW of ♂ lacking sound-producing structures. Auditory organs generally lacking.

Cave or Camel Crickets, Subfamily Rhaphidophorinae. Brownish, somewhat humpbacked appearance. Antennae *contiguous at base or nearly so.* Hind femora *long.* Occur in caves, cellars, under logs and stones, and in similar dark moist places. Most of our species belong to the genus *Ceuthophilus.*

Leaf-rolling Grasshoppers, Subfamily Gryllacridinae (not illus.). Tarsi lobed, somewhat flattened dorsoventrally. Hind femora extend beyond apex of abdomen. Ovipositor upturned. Our only species, *Camptonotus carolinensis* (Gerstaecker), which occurs in the East, is brownish and 13–15 mm. Nocturnal, feeding on aphids and spending the day in a leaf it has rolled up and tied with silk.

Jerusalem or Sand Crickets, Subfamily Stenopelmatinae. Large, robust, somewhat brownish. Tarsi *not lobed,* and *more or less flattened laterally.* Hind femora *do not extend beyond apex of abdomen.* These insects are western, occurring chiefly along the Pacific Coast. Nocturnal; spend the day under stones or in loose soil.

CRICKETS Family Gryllidae   See also Pl. 2
Identification: Somewhat flattened insects. Tarsi *3-segmented.* Ovipositor usually long and cylindrical. Cerci long and feeler-like.

This group contains many common insects, and the males are well-known songsters. Most of the tegminal (FW) surface is

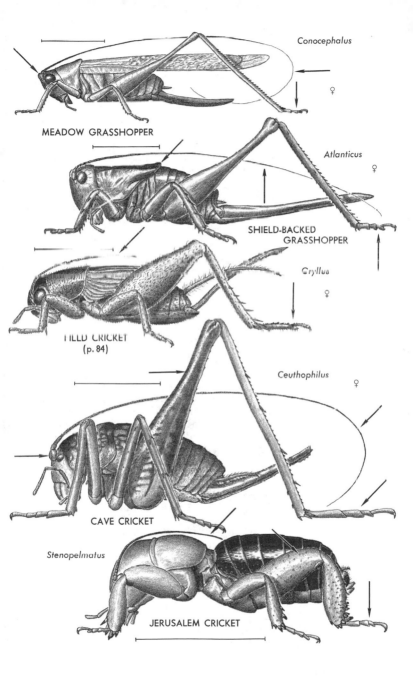

Conocephalus

♀

MEADOW GRASSHOPPER

Atlanticus

♀

SHIELD-BACKED
GRASSHOPPER

Gryllus

♀

FIELD CRICKET
(p. 84)

Ceuthophilus

♀

CAVE CRICKET

Stenopelmatus

JERUSALEM CRICKET

dorsal in position, with a narrow lateral portion bent down abruptly; the right tegmen is usually uppermost at rest. The front wings of females are thickened and leathery; those of males contain large membranous areas and are often wider. Songs generally are rapid trills or chirps — more musical and less lisping than songs of Tettigoniidae. Only the more common subfamilies can be mentioned here.

Mole Crickets, Subfamily Gryllotalpinae. Brownish, very pubescent. Usually 1 in. or longer. Front legs *broad and spade-like.* Antennae *relatively short.* Ovipositor not visible externally. Tegmina usually short, covering only about half of abdomen. Mole crickets burrow in the ground, ordinarily in moist places, and are not often encountered.

Bush Crickets, Subfamilies Eneopterinae and Trigonidiinae (see also Pl. 2). Most bush crickets are less than 9 mm. and brownish. 2nd tarsal segment *heart-shaped and flattened dorsoventrally* (small and flattened laterally in the remaining subfamilies). Eneopterinae have *small teeth on hind tibiae* between the spines and the ovipositor is cylindrical and nearly straight. Trigonidiinae lack teeth between the tibial spines and the ovipositor is somewhat sword-shaped (Trigonidiinae are sometimes called sword-bearing crickets). Found in bushes; they resemble the more common ground crickets (Nemobiinae) but do not live on the ground. Relatively uncommon in the North and more common in the South.

Tree Crickets, Subfamily Oecanthinae. Differ from the following subfamilies in lacking ocelli. Most species are pale green, have *small teeth between the spines on hind tibiae.* Many species are very common and all are excellent singers. Some occur on trees and shrubs, and others on high grass and weeds. The tree- and bush-inhabitants generally sing only at night, whereas the weed-inhabitants usually sing both day and night. Song of most species is a prolonged trill, but a few chirp. The chirping of the Snowy Tree Cricket, *Oecanthus fultoni* Walker, a bush inhabitant, is a common night sound in much of the country, and is at a very regular rate. All insects sing more slowly as the temperature drops (in the case of crickets the pitch of the song also falls with decreasing temperature), and the chirp rate of the Snowy Tree Cricket provides a means of estimating the temperature: the number of chirps in 13 seconds plus 40 gives a good estimate of the temperature in degrees Fahrenheit.

Ground Crickets, Subfamily Nemobiinae. Length 12 mm. or less; brownish. Spines on hind tibiae *long* and movable. Ground crickets are common insects, and occur on the ground in pastures, lawns, and in wooded areas. The songs are soft and high-pitched, and usually consist of pulsating trills or buzzes.

Field and House Crickets, Subfamily Gryllinae. Spines on hind tibiae *short, stout,* fixed. 12 mm. or longer. Field crickets (*Gryllus,* p. 83) are common and widely distributed; occur in fields, pastures, lawns, along roadsides, and in woods. The different species of *Gryllus*, which were formerly thought to represent a single species, are very similar in appearance but differ in seasonal life history, habitat, and song. The House Cricket, *Acheta domesticus* (Linn.), is a species introduced from

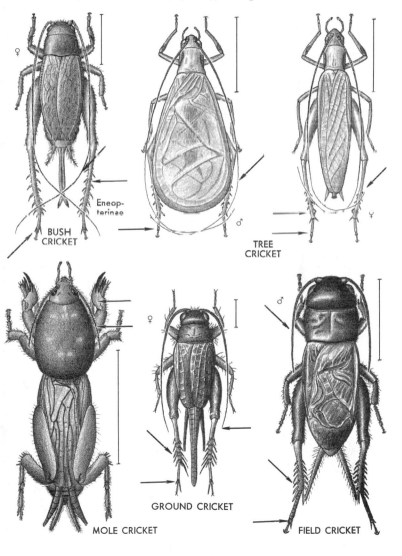

BUSH CRICKET

Eneopterinae

TREE CRICKET

MOLE CRICKET

GROUND CRICKET

FIELD CRICKET

Europe that often enters houses; it differs from field crickets in having the head light-colored with dark crossbands. Gryllinae sing both day and night, and most species chirp.

Suborders Mantodea, Phasmatodea, Blattodea, and Grylloblattodea: Nonjumping Orthoptera

MANTIDS Family Mantidae

Identification: Large insects, usually over 1 in., with a distinctive appearance. Prothorax and front coxae *greatly lengthened*. Front femora and tibiae armed with spines and *fitted for grasping prey*. Middle and hind legs slender, their coxae shorter. Tarsi *5-segmented*. Antennae *short*.

Mantids are predaceous, and usually lie in wait for their prey with the front legs upraised. Eggs are laid in papier-mâché-like cases attached to weeds or twigs; the eggs overwinter in the case. Mantids mostly are tropical. Our native species occur mainly in the South. The only ones common in the North are 2 introduced species: the European Mantid, *Mantis religiosa* Linn. (pale green, about 2 in.), and the Chinese Mantid, *Tenodera aridifolia sinensis* Saussure (3–4 in.). The latter is rather widely distributed and the former is largely restricted to the eastern states. In the West, *Litaneutria minor* (Scudder) occurs all the way to British Columbia.

WALKINGSTICKS Family Phasmatidae

Identification: Body and legs *very long and slender*. Wingless (1 species in s. Florida has very short wings). Tarsi usually *5-segmented* (3-segmented in *Timema* — small, stout-bodied, earwiglike forms occurring in the far West). Cerci 1-segmented. Antennae usually *long and slender*.

These insects strongly resemble twigs, and are usually found on trees or shrubs. They are plant feeders. Largely tropical and more common in the South; only a few species occur in the North. Most of them are at least 2 in.; 1 southern species reaches 6 in. or more.

COCKROACHES Family Blattidae   **See also Pl. 2**

Identification: Body flattened and oval, the head *concealed from above by pronotum*. Wings usually present. Antennae *long and slender*. Tarsi 5-*segmented*. Cerci many-segmented.

The best-known cockroaches are those that invade houses, where they may be serious pests. They feed on a variety of foods, and have an unpleasant odor, hide in cracks during the day, and feed at night. All are active, fast-running insects; many species seldom if ever fly. Cockroaches are much more abundant in the South than in the North; the few northern species that occur out of doors (wood cockroaches) are usually found in woods under dead logs and stones.

GRYLLOBLATTIDS Family Grylloblattidae **Not illus.**
Identification: Elongate, cylindrical, wingless, 15–30 mm.
Antennae about half as long as body or less. Tarsi 5-segmented.
Cerci long, 8- or 9-segmented. Ovipositor long, sword-shaped.

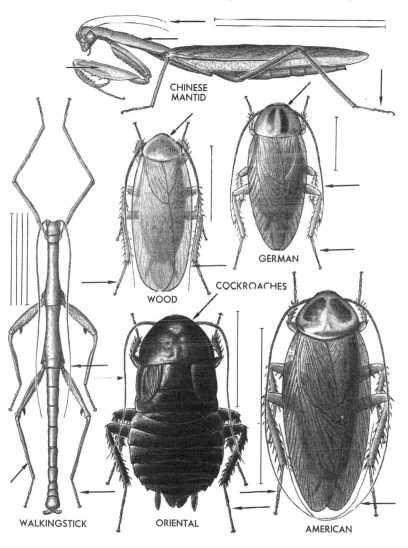

CHINESE MANTID

COCKROACHES

WOOD

GERMAN

WALKINGSTICK

ORIENTAL

AMERICAN

Compound eyes small or absent. Ocelli absent. Head relatively large, mouth parts situated anteriorly.

Grylloblattids occur in the mountains of nw. U.S. (California, Montana, Idaho, Washington, and Oregon) and w. Canada, usually at high elevations. They are generally found under logs or stones or at the edge of glaciers. Probably feed as scavengers.

Termites: Order Isoptera

Identification: Small, soft-bodied, usually pale-colored. Social, with caste differentiation. Antennae generally *short, thread- or beadlike.* Tarsi 4-segmented. Cerci usually short. Winged or wingless; winged forms with *2 pairs of wings similar in size and shape,* relatively long and narrow (as long as body or longer), at rest held flat over abdomen; wings eventually broken off along a basal fracture. Mouth parts chewing (vestigial in nasuti, p. 91). 2 ocelli or none. Metamorphosis simple.
Similar orders: (1) Hymenoptera (ants, p. 344): abdomen constricted at base (not so in termites); hard-bodied, dark-colored; antennae elbowed; HW smaller than FW. (2) Embioptera (p. 100): tarsi 3-segmented; basal segment of front tarsi greatly enlarged. (3) Zoraptera (p. 101): tarsi 2-segmented; antennae 9-segmented. (4) Orthoptera (p. 76): cerci usually longer; antennae hairlike. (5) Psocoptera (p. 102): tarsi 2- or 3-segmented; FW larger than HW; wings at rest held rooflike over body; cerci absent.
Immature stages: Similar to adult but wings short or absent.
Habits: Termite colonies occur in ground or in wood. Food consists principally of wood or other vegetable materials. The 4 castes usually present are the reproductive caste, supplementary reproductives, workers, and soldiers; some species have a 5th (nasutus) caste. Reproductives have fully developed wings, compound eyes, and are generally dark-colored; they are produced in large numbers at certain seasons and leave the colony in a swarm. Mating occurs at this time, and individual pairs establish new colonies; reproductives shed their wings after mating. Female reproductives (queens) do all or most of the egg laying and ordinarily are long-lived. Supplementary reproductives have shorter wings, smaller eyes, and are generally lighter in color; they are able to reproduce. Most workers are sterile adults; they are pale-colored and lack compound eyes, and their mandibles are small. They do the main work of the colony — collecting food, feeding the queen, soldiers, and young, constructing galleries, and the like. Some termites lack a worker caste and nymphs of other castes act as workers. Soldiers are usually sterile adults with *large heads and mandibles;* they attack intruders in the colony. A few termites lack a soldier caste. Some termites have a nasutus caste; nasuti have the head drawn out anteriorly into a slender snout, and they function in the defense of the colony.

Importance: Many species cause considerable damage to buildings, furniture, utility poles, fence posts, and other materials; infested timbers are hollowed out and may eventually collapse. Termites are beneficial in their important role of converting dead trees and other plant materials into substances useful to plants.

Classification: Four families in the U.S. Winged forms are separated mainly on basis of head and wing characters. Soldiers are separated on basis of head shape, leg and antennal characters, and characters of the mandibles.

Identification of Termites: The *fontanelle* is a depressed pale spot on front of head between the eyes. *Radius* is a longitudinal vein behind the costal margin of the wing. Wingless termites are best identified by association with winged forms and by habits. Termites found in the nest often can be identified by character and location of the nest, and by area in which it is found.

No. of species: World, 2100; N. America, 41.

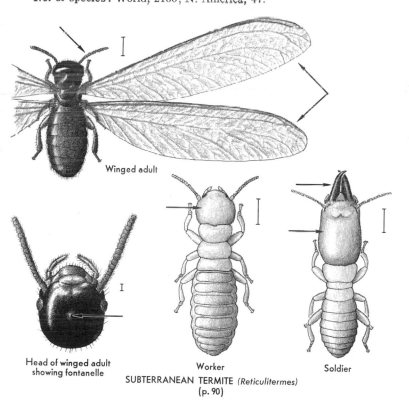

Winged adult

Head of winged adult
showing fontanelle

Worker

Soldier

SUBTERRANEAN TERMITE *(Reticulitermes)*
(p. 90)

SUBTERRANEAN and DAMP-WOOD TERMITES   p. 89
Family Rhinotermitidae

Identification: Widely distributed with colonies in the ground (subterranean termites), or occurring in Florida in damp wood (damp-wood termites). Winged adults with *a fontanelle.* Scale of FW (basal portion left after wing breaks off) longer than pronotum. R usually without anterior branches.

Subterranean termites nest in the ground and burrow into wood, usually fallen trees and stumps. They enter the wood of buildings or other structures where it comes in contact with the ground, or they may get into buildings by way of tunnels through cracks in the foundation. The Eastern Subterranean Termite, *Reticulitermes flavipes* (Kollar), a widely distributed and destructive species, is the only termite occurring in the Northeast; winged forms appear in spring. The damp-wood termites in this family (*Prorhinotermes*) occur in moist dead wood or tree roots and are found only in Florida.

SOLDIERLESS, DESERT, and NASUTIFORM TERMITES
Family Termitidae

Identification: Occur in Southwest. Colonies in wood with a ground contact, or in ground. Winged adults similar to Rhinotermitidae, but scale of FW shorter than pronotum.

Soldierless termites (lacking a soldier caste) burrow under logs or cow chips. Desert termites nest in the ground, and sometimes damage the wood of buildings and poles. Nasutiform termites attack wood in contact with the ground; they have a nasutus caste.

DRY-WOOD, DAMP-WOOD, and POWDER-POST TERMITES
Family Kalotermitidae

Identification: Occur in s. and w. U.S., the colonies in dry wood above ground (dry-wood and powder-post termites), or in moist dead wood or tree trunks (damp-wood termites). Winged adults without a fontanelle but *with ocelli,* and R usually with *anterior apical branches.*

Dry-wood termites (*Incisitermes* and others) attack buildings, furniture, utility poles, and piled lumber; they are important in the South, from S. Carolina to Texas. Powder-post termites (*Cryptotermes, Calcaritermes*), which occur in the South, attack dry wood and reduce it to a powder. Damp-wood termites (*Neotermes, Paraneotermes*) occur in Florida and the western states.

ROTTEN-WOOD TERMITES   Not illus.
Family Hodotermitidae

Identification: Similar to Kalotermitidae but winged adults without ocelli. Occur in western and southwestern states. Colonies in dead wood.

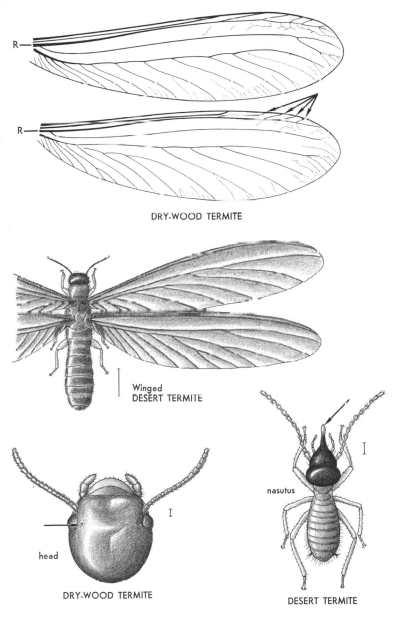

R

DRY-WOOD TERMITE

Winged
DESERT TERMITE

nasutus

head

DRY-WOOD TERMITE

DESERT TERMITE

This group includes only 3 species of *Zootermopsis*, which often damage buildings, utility poles, and lumber. They can live in wood without a ground contact but there must be some moisture in the wood.

Stoneflies: Order Plecoptera

Identification: Elongate, flattened, soft-bodied, usually found near streams. 4 membranous wings (a few males have wings reduced or lacking): FW long and narrow, HW shorter and nearly always with *a large anal lobe;* wings often with many veins, FW with a series of *cross veins between M and Cu_1 and between Cu_1 and Cu_2;* wings at rest held flat over abdomen, with anal lobe of HW folded fanwise. Antennae *long*, threadlike. Tarsi *3-segmented.* Cerci present, *often long* and many-segmented. Usually 3 (rarely 2) ocelli. Mouth parts chewing, sometimes reduced. Metamorphosis simple, nymphs aquatic.

Similar orders: (1, 2) Trichoptera and Neuroptera (pp. 210, 140): tarsi 5-segmented; HW without an anal lobe. (3) Orthoptera (cockroaches, p. 86): tarsi 5-segmented: generally not found near streams. (4) Ephemeroptera (p. 65): HW small or absent, wings held together above body at rest; antennae short and bristlelike.

Immature stages: Nymphs are elongate and flattened, with *long antennae and cerci*, and with *filamentous gills* on thorax about bases of legs (some nymphs lack gills); mayfly nymphs have 3 tails, and the gills are leaflike and located along the sides of the abdomen. Some nymphs are plant feeders, others are predaceous. When a nymph is full-grown it climbs out of the water onto a stone or other object, and there molts to the adult. Nymphal development takes about a year in most species, 2 years in some.

Habits: Most stoneflies are poor fliers, and seldom found far from water. Many are nocturnal, spending the day resting on stones, trees, under bridges and on bridge abutments, and other places near water; many are attracted to lights. Most adults appear in summer, but a few emerge and mate during fall and winter. Eggs are usually deposited in masses on surface of water. Many adults have poorly developed mandibles and do not feed; others, with well-developed mandibles, feed on blue-green algae and various plant materials. In general, stoneflies appearing in fall and winter are diurnal and feed as adults; those appearing in summer are chiefly nocturnal and do not feed as adults.

Importance: Of no economic importance except as food for fish and other animals.

Classification: Two suborders, Systellognatha and Holognatha, separated chiefly by mouth-part structure. Families are separated by characters of the wing venation, cerci, tarsi, gill remnants on the ventral side of the thorax, genitalia, and size. The characters likely to cause trouble for the beginner are those of the *gill remnants*, since these structures are often shriveled in dry pinned specimens and are difficult to evaluate; they are more easily studied in specimens preserved in alcohol. Length is measured from front of head to wing tips (wings in a resting position).

No. of species: World, 1550; N. America, 400.

Suborder Systellognatha

Cerci much longer than greatest width of pronotum. First tarsal segment much shorter than 3rd. Usually no forked vein rising from basal anal cell in FW. Adults appear in late spring and summer, usually nocturnal and nonfeeding.

COMMON STONEFLIES   Family Perlidae
  Identification: Color variable but usually yellowish to brownish and not green. Mostly 15–40 mm. *Remnants of branched nymphal gills* on ventral side of thorax, usually immediately behind bases of legs.
    This family is the largest in the order, and its members are the stoneflies most often collected. Most nymphs are predaceous. One of the largest and commonest genera is *Acroneuria;* some species in this genus are quite large (to 40 mm) and resemble pteronarcids but lack the rows of cross veins in the anal area of the front wings.

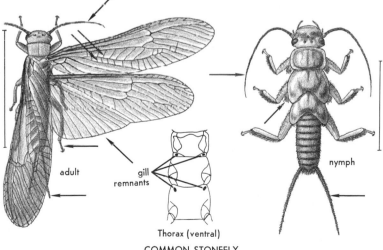

adult

gill
remnants

Thorax (ventral)

nymph

COMMON STONEFLY

GREEN-WINGED STONEFLIES Family Isoperlidae
Identification: Body greenish or yellowish, wings usually greenish. 6–15 mm. No remnants of nymphal gills on thorax. Front corners of pronotum acute or *narrowly rounded*. Anal lobe of HW *well developed*, with 4 or more veins reaching wing margin. ♂ without a lobe on 8th sternum, and 10th tergum usually *not notched*. ♀ subgenital plate generally small.

These are fairly common stoneflies, and adults are often seen running about on foliage near streams; many are pollen feeders. Nymphs vary in habits. This family contains a single N. American genus, *Isoperla*.

PERLODID STONEFLIES Family Perlodidae
Identification: Similar to Isoperlidae but 10–25 mm. and brownish or blackish; wings not greenish. Either no remnants of nymphal gills on ventral side of thorax or a remnant of a finger-like gill on each side of base of labium. ♂ usually with a lobe on posterior margin of 7th abdominal sternum, and 10th tergum with *a deep median notch*. ♀ subgenital plate generally large.

Nymphs occur in medium-sized to large streams, and adults appear in the spring or early summer. Most species are northern or western, and they are not common.

GREEN STONEFLIES Family Chloroperlidae
Identification: Length 6–24 (mostly 6–15) mm. Usually yellowish or greenish. No remnants of nymphal gills on ventral side of thorax. Front corners of pronotum *broadly rounded*. Either *a forked vein rising from basal anal cell in FW or anal lobe in HW reduced* (with 3 or fewer veins) or absent. ♂ without a lobe on either 7th or 8th abdominal sternum.

Nymphs usually occur in small streams, and adults appear in spring.

Suborder Holognatha

Cerci variable in length but usually short, no longer than greatest width of pronotum. Often a forked vein comes off basal anal cell in FW. Mandibles generally well developed. Most adults diurnal and plant feeding.

GIANT STONEFLIES Family Pteronarcidae
Identification: Length 1½–2½ in. Usually brown or gray. Anal area of FW with *2 or more rows of cross veins*.

Nymphs are plant feeders and occur in medium-sized to large rivers. Adults appear in spring and early summer. Adults are largely nocturnal and do not feed; often are attracted to lights.

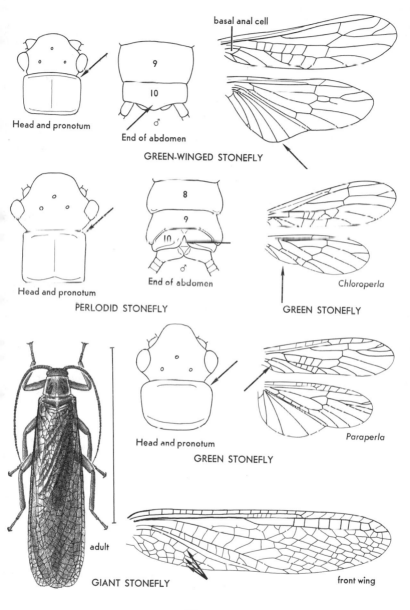

Head and pronotum

9
10
♂

End of abdomen

basal anal cell

GREEN-WINGED STONEFLY

Head and pronotum

PERLODID STONEFLY

8
9
10
♂

End of abdomen

Chloroperla

GREEN STONEFLY

Head and pronotum

GREEN STONEFLY

Paraperla

adult

GIANT STONEFLY

front wing

95

ROACHLIKE STONEFLIES Family Peltoperlidae

Identification: Brownish. 12–18 mm. FW with *10 or more costal cross veins*, and *without a forked vein rising from basal anal cell*. Cerci short. *2 ocelli.*

The common name refers to the roachlike appearance of the nymphs. The group is small, and chiefly northern and western. All the N. American species are in the genus *Peltoperla*, and only 2 of these occur in the East.

WINTER STONEFLIES Family Taeniopterygidae

Identification: Dark brown to blackish. *Usually 15 mm.* or less. Cerci short, 1- to 6-segmented. FW *with a forked vein rising from basal anal cell.* 2nd tarsal segment *about as long* as other tarsal segments.

Nymphs occur in large streams and rivers, and adults appear from January to April. Both nymphs and adults are plant feeders; adults sometimes feed on flowers.

ROLLED-WINGED STONEFLIES Family Leuctridae

Identification: Body appears very slender because wings at rest are bent down over sides of body. Generally 10 mm. or less. Cerci short, 1-segmented. FW *with a forked vein rising from basal anal cell*, and *without a cross vein just behind costa* in apical portion of wing; 2nd tarsal segment *much shorter* than other segments.

Leuctrids are brownish or blackish, and are most common in hilly or mountainous areas. Nymphs usually occur in small streams, and adults appear from December to June.

SPRING STONEFLIES Family Nemouridae

Identification: Similar to Leuctridae but sometimes larger (to 15 mm.). FW flat at rest and *with a cross vein just behind costa* in apical portion of wing.

Nymphs are plant feeders and occur chiefly in small streams with sandy bottoms. Adults are brownish or blackish and appear from April to June. There are about 2 dozen N. American species, all in the genus *Nemoura*.

SMALL WINTER STONEFLIES Family Capniidae

Identification: Length usually *10 mm. or less*. Blackish. Cerci longer than greatest width of pronotum. 1st tarsal segment *about as long as 3rd*. Venation reduced. FW with few cross veins.

Nymphs usually occur in small streams, adults appear from November to April. Nymphs are plant feeders, and adults feed on blue-green algae. In some members of this group (species of *Allocapnia*) the anal lobe of hind wing is *nearly as long as rest of the wing*.

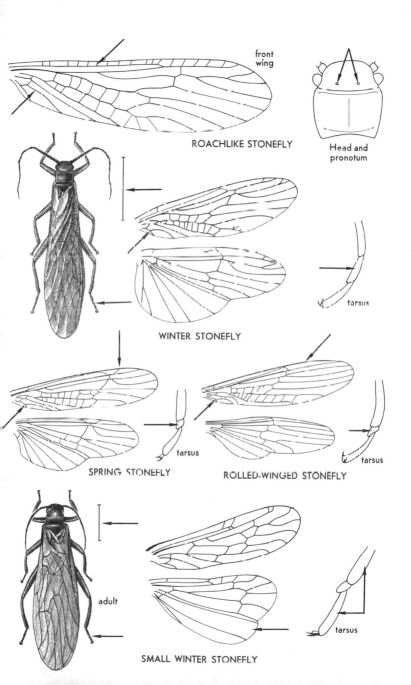

front wing

ROACHLIKE STONEFLY

Head and pronotum

WINTER STONEFLY

tarsus

SPRING STONEFLY

tarsus

ROLLED-WINGED STONEFLY

tarsus

adult

SMALL WINTER STONEFLY

tarsus

Earwigs: Order Dermaptera

Identification: Small to medium-sized, elongate, flattened. 4 wings (rarely wingless): FW (elytra) thickened, leathery, *short, meeting in a straight line down back;* HW membranous, at rest folded beneath FW. Cerci well developed and *forcepslike.* Antennae *threadlike, about half as long as body or less.* Tarsi 3-segmented. Mouth parts chewing. Metamorphosis simple.

Similar orders: (1) Coleoptera (p. 146) with short elytra: no forcepslike cerci. (2) Orthoptera (p. 76): cockroaches have short feelerlike cerci and 5-segmented tarsi; walkingsticks in the western genus *Timema* (which have 3-segmented tarsi) have antennae more than half as long as body.

Immature stages: Similar to adult but wings small or absent.

Habits: Earwigs are nocturnal, spending the day in debris, under bark, and in other protected situations; they are chiefly plant feeders or scavengers. Some species when disturbed eject a foul-smelling liquid from glands near base of the abdomen; this serves as a means of protection. Eggs are laid in burrows in the ground or in debris, and the female usually guards the eggs until they hatch. Cerci usually differ in shape in the sexes, being straight, stout, and closely approximated in the female, and more slender, curved, and pincerlike in the male.

Importance: The name "earwig" comes from an old superstition that these insects get into people's ears; this belief is without foundation. A few earwigs occasionally are pests in buildings and some may damage cultivated plants. The pincerlike cerci are used in defense, and can sometimes inflict a painful pinch.

Classification: Four families in N. America, separated chiefly by form of tarsi and antennae.

No. of species: World, 1100; N. America, 18.

COMMON EARWIGS   Family Forficulidae

Identification: Brownish. 2nd tarsal segment *lobed beneath,* expanded laterally, and *prolonged distally beneath 3rd segment.* Antennae *12- to 15-segmented.* Widely distributed.

The most common species in this family is the European Earwig, *Forficula auricularia* Linn., 10–15 mm. It sometimes damages cultivated plants.

BLACK EARWIGS   Family Chelisochidae   **Not illus.**

Identification: Similar to Forficulidae but black, and 2nd tarsal segment not expanded laterally.

Our only representative of this family is *Chelisoches morio* (Fabricius), a large earwig occurring in California.

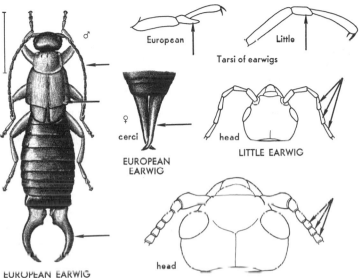

Tarsi of earwigs

♂ European   Little

cerci
EUROPEAN EARWIG

head   LITTLE EARWIG

head   LONG-HORNED EARWIG

EUROPEAN EARWIG
(Forficulidae)

LONG-HORNED EARWIGS Family Labiduridae

Identification: Second tarsal segment cylindrical and not prolonged distally beneath 3rd segment. Antennae 16- to 30-
segmented; *segments 4–6 together rarely longer than 1st segment.*
Wings present or absent.

A common wingless species in this group is the Seaside Earwig,
Anisolabis maritima (Géné), which occurs along Atlantic and
Pacific Coasts; it is 18–20 mm. and has 24-segmented antennae.
A common winged labidurid is the Striped Earwig, *Labidura
bidens* (Olivier), 13–20 mm., found in the eastern states. The
wingless labidurids (*Anisolabis* and *Euborellia*), which have the
male cerci asymmetrical (the right one more curved), are sometimes placed in a separate family, the Psalididae.

LITTLE EARWIGS Family Labiidae

Identification: Tarsi *as in Labiduridae.* Antennae 11- to 15-
segmented; *segments 4–6 together longer than 1st segment.* Wings
usually present. Generally 4–7 mm.

The most common little earwig is *Labia minor* (Linn.), which
is 4–5 mm. and light brown; it has been introduced into this
country from Europe. The Handsome Earwig, *Prolabia pulchella* (Serville), 6.0–6.5 mm. and dark brown, is fairly common
in the southern states.

Webspinners: Order Embioptera

Identification: Small (mostly 4–7 mm.), slender-bodied, usually yellowish or brownish. Tarsi *3-segmented*, basal segment of front tarsi *greatly enlarged*. Legs short, hind femora *thickened*. Antennae *short, threadlike, 16- to 32-segmented*. Ocelli absent. Wings present or absent in ♂ and always absent in ♀; 4 wings when present are membranous, HW slightly smaller than FW, and venation weak, *each vein in the middle of a brown band;* wings at rest held flat over body. Cerci present, *1- or 2-segmented*, usually asymmetrical in ♂ and always so in ♀. Mouth parts chewing. Metamorphosis simple.
Similar orders: (1, 2, 3) Isoptera, Zoraptera, and Psocoptera (pp. 88, 101, 102): basal segment of front tarsi not enlarged; Isoptera have 4-segmented tarsi, and their galleries are not silk-lined; Zoraptera have 2-segmented tarsi and 9-segmented antennae; Psocoptera have 2- or 3-segmented tarsi, and lack cerci.
Immature stages: Similar to adult but wings small or absent.
Habits: Webspinners live in colonies in silk-lined galleries in soil or debris and among mosses or lichens. Silk is spun from glands in basal segment of front tarsi; both adults and nymphs have silk glands. These insects are active and run rapidly, usually backward; sometimes play dead when disturbed. They feed chiefly on dead plant materials. The eggs are laid in the galleries and are often covered with chewed food particles; eggs are attended by the females. Both winged and wingless males occur in some species. Webspinners are not common, and are restricted to the southern states.
Importance: Not of economic importance.
Classification: Three families in N. America, which differ in wing venation and in the character of cerci and mandibles.
No. of species: World, 149; N. America, 9.

Key to Families

1. R_{4+5} in wings of ♂ forked; left cercus 2-segmented; Gulf Coast, Florida to s. Texas   **Teratembiidae**

1′. R_{4+5} in wings of ♂ not forked; if wings are absent, left cercus is 1-segmented   2

2. Mandibles without apical teeth; 10th tergum of ♂ completely divided by a median membranous area that reaches 9th tergum; left cercus of ♂ usually with peglike spines on inner side of basal segment; lower Mississippi Valley and Southwest   **Anisembiidae**

2′. Mandibles with distinct apical teeth; 10th tergum of ♂ incompletely divided by a median membranous area that does not reach 9th tergum; left cercus of ♂ smooth on inner side; southern states   **Oligotomidae**

Zorapterans: Order Zoraptera

Identification: *Minute* insects, 3 mm. or less. Tarsi *2-segmented*. Antennae *threadlike or beadlike, 9-segmented*. Wings present or absent; if present, 4, membranous, FW a little larger than HW, and with few veins; wings eventually shed, leaving short stubs attached to thorax. Wingless forms without eyes, winged forms with compound eyes and 3 ocelli. Cerci *short, 1-segmented*, and terminating in a long bristle. Mouth parts chewing. Metamorphosis simple.

Similar orders: Isoptera, Psocoptera, Embioptera (pp. 88, 102, 100), and others: differ in form of tarsi, antennae, and cerci.

Immature stages: Similar to adult but wings small or absent.

Habits: Zorapterans are usually found under slabs of wood buried in piles of old sawdust, under bark, or in rotting logs; often occur in colonies. They feed chiefly on mites and other small arthropods. Zorapterans occur in the southeastern states from Maryland, Illinois, Arkansas, and Oklahoma south to Florida and Texas. They are not common.

Importance: Not of economic importance.

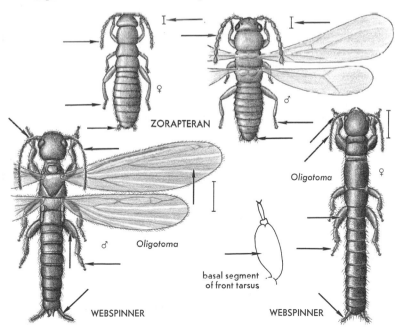

♀

♂

ZORAPTERAN

Oligotoma

♀

Oligotoma

♂

Oligotoma

basal segment
of front tarsus

WEBSPINNER

WEBSPINNER

Classification: One family, the Zorotypidae.
No. of species: World, 22; N. America, 2.

Booklice and Barklice: Order Psocoptera

Identification: Small, soft-bodied, usually less than 5 mm. Wings present or absent, if present 4 in number, membranous, FW larger than HW, and held rooflike over body at rest; wing venation reduced. Face somewhat bulging. Ocelli present or absent. Antennae long and slender. Tarsi 2- or 3-segmented. Cerci absent. Mouth parts chewing. Metamorphosis simple.

Similar orders: (1, 2) Mallophaga and Anoplura (pp. 106, 108): ectoparasites of birds and mammals; tarsi 1- or 2-segmented; antennae short. (3) Isoptera (p. 88): tarsi 4-segmented; antennae usually short; FW and HW of winged forms similar in size. (4) Zoraptera (p. 101): antennae 9-segmented; cerci present, 1-segmented. (5) Embioptera (p. 100): tarsi 3-segmented, basal segment of front tarsi greatly enlarged. (6) Small Neuroptera (p. 140): tarsi 5-segmented.

Immature stages: Similar to adult but wings small or absent.

Habits: Feed chiefly on dry organic matter, molds, or fungi. They occur in debris, under bark or stones, and on bark of trees; a few are found in buildings. Some species are gregarious and live in silken webs on trunks or branches of trees. Winged forms are commonly called barklice, and wingless forms booklice; most are active, fast-running insects.

Importance: Some species occasionally are pests in buildings, where they may damage books by feeding on starchy materials in bindings.

Classification: Three suborders and 11 families; suborders are separated by the characters of the antennae and labial palps, and families chiefly by characters of the legs, wings, and mouth parts. Some mouth-part characters cannot be seen in dried material and can be satisfactorily studied only in material preserved in alcohol or mounted on microscope slides.

No. of species: World, 1100; N. America, 143.

Suborder Trogiomorpha

Antennae *long and hairlike*, with 20 or more segments. Labial palps *2-segmented*. Tarsi *3-segmented*.

TROGIID BOOKLICE Family Trogiidae
    Identification: Wings usually lacking or rudimentary, if well developed then FW broadly rounded apically. Body and wings without scales. Hind femora *slender*.
    Trogiids most likely to be encountered are wingless forms

occurring in houses, barns, or granaries. They differ from liposcelid booklice in having slender hind femora.

SCALY BARKLICE   Family Lepidopsocidae   **Not illus.**
   Identification: Minute forms having the body and wings covered with scales.
   Only 3 rare species in this group occur in the U.S. They are found out of doors.

PSYLLIPSOCIDS   Family Psyllipsocidae   **Not illus.**
   Identification: Pale-colored. Head in profile short and vertical. Wings and body not covered with scales.
   Psyllipsocids occur in dark damp places such as cellars and caves; 1 species often occurs in cellars around wine or vinegar barrels. A few have 1 or both pairs of wings short.

Suborder Troctomorpha

Antennae with 17 or fewer segments. Labial palps *2-segmented*. Tarsi *3-segmented*.

PACHYTROCTIDS   Family Pachytroctidae   **Not illus.**
   Identification: Body short and arched. Legs long and slender, hind femora not thickened. Compound eyes in winged individuals composed of many facets.
   The 2 U.S. species in this group are quite rare.

LIPOSCELID BOOKLICE   Family Liposcelidae
   Identification: Body elongate and flattened. Legs short, hind femora *thickened*. Usually wingless, but if winged then compound eyes are composed of 2–8 facets.
   The most common member of this group is *Liposcelis divinatorius* (Müller), the booklouse most often found in buildings

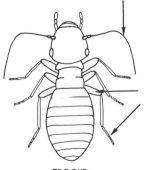

labial palps

Labium of Trogiid

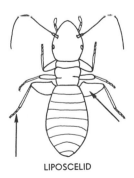

LIPOSCELID

TROGIID   **BOOKLICE**

It is a minute wingless insect recognized generally by its enlarged hind femora.

Suborder Eupsocida

Antennae with less than 20 segments. Labial palps 1-*segmented*. Tarsi 2- or 3-segmented.

POLYPSOCIDS   Family Polypsocidae
Identification: Tarsi *2-segmented*. Labial palps triangular, tapering distally. Laciniae tapering to a point apically, *without teeth*.
The members of this small group occur out of doors, and are not common.

EPIPSOCIDS   Family Epipsocidae
Identification: Tarsi 2-segmented. Labial palps oval. Laciniae *widened distally*, with *several small apical teeth*. Labrum with *2 diagonal internal ridges* that often unite anteriorly.
The 2 U.S. species in this group are relatively rare.

COMMON BARKLICE   Family Psocidae
Identification: Tarsi *2-segmented*. Labial palps oval. Laciniae *straight*, with *a few relatively large apical teeth*. Labrum with only *2 small tubercles internally*. Cu_{1a} in FW *fused with M*.
Most barklice with 2-segmented tarsi belong either to this family or to the Pseudocaeciliidae. Cu_{1a} in the front wing is nearly always separate from M (rarely completely lacking) in the Pseudocaeciliidae. The most common barklice belong to the Psocidae.

PSEUDOCAECILIID BARKLICE   Family Pseudocaeciliidae
Identification: Similar to Psocidae but Cu_{1a} in FW *nearly always separate from M* or lacking.
The majority of these barklice occur out of doors; only a few may occur indoors. One southern species makes rather unsightly webs on tree trunks and branches.

MESOPSOCID and MYOPSOCID BARKLICE   Not illus.
Families Mesopsocidae and Myopsocidae
Identification: Differ from other families in this suborder in having 3-segmented tarsi, and from families in the other 2 suborders in having the labial palps 1-segmented. Cu_{1a} in FW fused with M in the Myopsocidae, and usually free from M in the Mesopsocidae.
The 6 species of Mesopsocidae and 4 of Myopsocidae in the U.S. occur principally out of doors. They are not common.

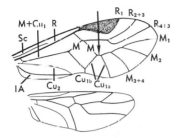

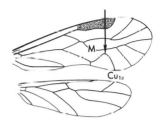

COMMON BARKLOUSE

PSEUDOCAECILIID BARKLOUSE

Labium of Polypsocid

Labrum of Epipsocid

Labrum of Common Barklouse

Lacinia of Polypsocid

Lacinia of Epipsocid

Lacinia of Common Barklouse

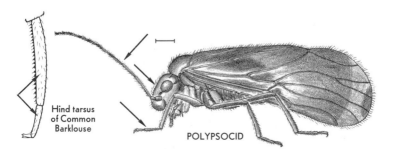

Hind tarsus of Common Barklouse

POLYPSOCID

Chewing Lice: Order Mallophaga

Identification: Small (mostly less than 5 mm.), wingless, flattened ectoparasites of birds and mammals. Mouth parts with *chewing mandibles*. Antennae short, 3- to 5-segmented, sometimes concealed in grooves on head. Compound eyes small. Head as wide as or wider than thorax. Legs short, the tarsi 1- or 2-segmented, with 1 or 2 claws. Cerci lacking. Metamorphosis simple.

Similar orders: Anoplura (p. 108): parasites of mammals only, including man (Mallophaga do not attack man); head generally narrower than thorax; 1 large tarsal claw; mouth parts sucking.

Immature stages: Similar to adult stage.

Habits: Chewing lice spend their entire life on their host; transmission from one host to another usually occurs when hosts come in contact, since lice are unable to survive long off a host. Each species attacks one or a few related species of hosts, and lives on a particular part of host's body. Eggs are laid on the host, usually attached to hair or feathers.

Importance: Lice are quite irritating to their hosts; heavily infested animals are often emaciated. No chewing lice attack man but several are important pests of domestic animals, especially poultry; they feed on feathers or hair of the host.

Classification: Two suborders and 6 families. Suborders differ in structure of the antennae and maxillary palps. Families are separated by various head characters and number of tarsal claws. These characters are most easily studied in specimens mounted on microscope slides. The host of a louse provides an important clue to its identity.

No. of species: World, 2675; N. America, 318.

Suborder Amblycera

Antennae clubbed, usually 4-segmented, *concealed in grooves* on head. Maxillary palps *4-segmented*.

GUINEA PIG LICE Family Gyropidae   **Not illus.**
  Identification: Tarsi with 1 claw or none. Parasites of guinea pigs.
    Most members of this group occur in Cent. and S. America. Only 2 species occur in the U.S.

BIRD LICE Family Menoponidae
  Identification: Tarsi with *2 claws*. Head *broadly triangular*, and *expanded behind eyes*. Antennae *in grooves on sides of head*. Parasites of birds.

This is a large group whose members attack various types of birds. Two species are important pests of poultry.

BIRD LICE   Family Laemobothriidae
Identification: Similar to Menoponidae but head less triangular, with *a swelling on each side at base of antennae;* antennal grooves *open ventrally*. Parasites of birds.
These lice attack water birds and birds of prey.

BIRD LICE   Family Ricinidae
Identification: Similar to Laemobothriidae but *without swellings on head at bases of antennae*. Parasites of birds.
The members of this small group attack chiefly sparrows and hummingbirds.

Suborder Ischnocera

Antennae threadlike, *not concealed in grooves*, and 3- to 5-segmented. Maxillary palps lacking.

BIRD LICE   Family Philopteridae   **Not illus.**
Identification: Tarsi with 2 claws. Antennae 5-segmented. Parasites of birds.
This is the largest family in the order, and its members attack a wide variety of birds. Two species are important pests of poultry.

MAMMAL CHEWING LICE   Family Trichodectidae
Identification: Tarsi with *1 claw*. Antennae usually 3-segmented. Parasites of mammals.

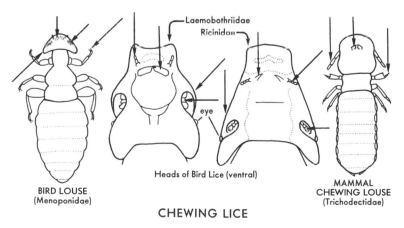

Laemobothriidae
Ricinidae

eye

Heads of Bird Lice (ventral)

BIRD LOUSE
(Menoponidae)

MAMMAL
CHEWING LOUSE
(Trichodectidae)

CHEWING LICE

These lice attack a number of different mammals, but not man. Some are important pests of domestic animals.

Sucking Lice: Order Anoplura

Identification: Small (usually less than 4 mm.), flattened, wingless ectoparasites of mammals. Mouth parts sucking, withdrawn into head when not in use; when used, everted through a short beaklike structure at front of head. Antennae short, threadlike or tapering distally, 3- to 5-segmented. Head small, nearly always narrower than thorax. Without ocelli and with compound eyes small or absent. Tarsi 1-segmented, with *1 large claw.* Cerci lacking. Metamorphosis simple.

Similar orders: Mallophaga (p. 106): mouth parts chewing; head usually as wide as or wider than thorax; 1 or 2 small tarsal claws; parasites of birds and mammals.

Immature stages: Similar to adult stage.

Habits: Sucking lice feed on the blood of their host, and their bites are often very irritating. Each species usually attacks one or a few related species of hosts, and generally lives on a particular part of the host's body. Eggs are usually attached to hairs of the host; eggs of the Body Louse are laid on clothing. Sucking lice spend their entire life on their host, and do not survive long away from the host.

Importance: These lice are irritating pests of man and animals, and 1 of the species that attacks man is an important disease vector.

Classification: Three families, separated chiefly by body shape and the development of compound eyes; these characters are best seen in specimens mounted on microscope slides.

No. of species: World, 250; N. America, 62.

SPINY SUCKING LICE Family Echinophthiriidae **Not illus.**
  Identification: Body thickly covered with short stout spines or with spines and scales. Parasites of marine mammals.
  The members of this small group attack seals, sea lions, and walruses.

MAMMAL SUCKING LICE Family Haematopinidae
  Identification: *Eyes lacking.* Parasites of various mammals (but not man or other primates).
  This is a large group, and includes most of the sucking lice that attack mammals other than man. Some species are pests of domestic animals.

HUMAN LICE Family Pediculidae
  Identification: *Eyes or eye tubercles present.* Parasites of man.
  This group contains 2 species, *Phthirus pubis* (Linn.), the Crab

or Pubic Louse, and *Pediculus humanus* Linn., the Head and Body Lice. These are the only lice that attack man.

The Crab Louse is 1.5–2.0 mm., broadly oval, with the head *much narrower* than thorax, and *with lateral lobes on the abdominal segments*. It usually occurs in the pubic region, but in very hairy individuals may occur almost anywhere on the body. Eggs are attached to body hairs.

P. humanus is 2.5–3.5 mm., more elongate, with head *only slightly narrower* than thorax, and *without lateral lobes on the abdominal segments*. Head and Body Lice are similar in appearance but differ somewhat in habits. The Head Louse (*P. h. capitis* De Geer) occurs on the head, and attaches its eggs to hairs. The Body Louse (*P. h. humanus* Linn.) occurs on the body, and its eggs are laid on clothing.

People who bathe and change clothes frequently seldom encounter these lice, lousiness generally occurs in people who live in crowded conditions and go for long periods without bathing or changing clothes. Head lice may be transmitted from one person to another on combs, hair brushes, or hats. Body lice may be transmitted by clothing or bedding, or they may migrate at night from one batch of clothing to another.

Lice are very annoying pests, and the Body Louse acts as a disease vector. The most important disease it transmits is epidemic typhus, which often occurs in epidemic form and has a high mortality rate. A louse becomes infective after biting a typhus patient; infection of another individual results from scratching the feces of this louse, or the louse itself, into the skin. The Body Louse also transmits relapsing fever. Infection in this case results from a crushed louse being scratched into the skin.

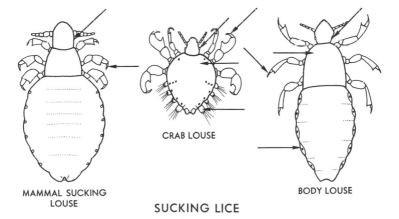

MAMMAL SUCKING LOUSE

CRAB LOUSE

BODY LOUSE

SUCKING LICE

Thrips: Order Thysanoptera

Identification: Slender, minute (mostly 0.5–2.0 mm.), pale to blackish. Antennae *short, 6- to 9-segmented*. Wings when present 4 in number, *long and narrow*, with few or no veins and *fringed with long hairs*. Legs short. Tarsi 1- or 2-segmented and swollen at tip. Mouth parts sucking, asymmetrical, *in form of a conical beak* at base of head on ventral side. Metamorphosis intermediate between simple and complex.

Similar orders: Thrips are not likely to be confused with insects in other orders. They can be recognized by the characteristic form of the mouth parts and wings.

Immature stages: First 2 instars similar to adult but wingless, and called larvae; remaining 2 or 3 preadult instars usually with short wing pads, inactive and nonfeeding; last preadult instar (pupa) sometimes enclosed in a cocoon.

Habits: Most thrips are plant feeders, and many are abundant on vegetation or in flowers. A few are predaceous on other small arthropods, and many feed on fungus spores.

Importance: Many plant-feeding thrips damage cultivated plants by their feeding. A few act as vectors of plant diseases.

Classification: Two suborders, Terebrantia and Tubulifera, differing in shape of abdomen and development of ovipositor.

No. of species: World, 4500; N. America, 600.

Suborder Terebrantia

Terminal abdominal segment conical in ♀, *rather broadly rounded* in ♂. ♀ nearly always with an ovipositor. FW, when present, with 2 longitudinal veins. Antennae 6- to 9-segmented.

BANDED THRIPS Family Aeolothripidae   **Not illus.**
  Identification: Bicolored or dark. Wings usually dark, banded or longitudinally striped; FW relatively broad, rounded apically. Ovipositor upcurved. Antennae 9-segmented. Body not flattened.
    Most banded thrips, especially larvae, feed on small insects and mites. They are sometimes common on flowers.

MEROTHRIPIDS Family Merothripidae   **Not illus.**
  Identification: Ovipositor curved downward, often reduced. Pronotum with a dorsal longitudinal suture on each side. FW if present narrow, pointed. Front and hind femora thickened.
    This is a small group, and the only common species is one occurring in the East in fungi and debris or under bark.

COMMON THRIPS Family Thripidae
Identification: Ovipositor usually well developed, curved downward. FW pointed apically. Body somewhat flattened. Antennae *6- to 9-segmented*, sensory areas (sensoria) on segments 3 and 4 conelike or fingerlike and projecting.

Thripids make up the bulk of the Terebrantia, and nearly all species are plant feeders; some are serious pests of cultivated plants. Most are 1.5 mm. or less.

HETEROTHRIPIDS Family Heterothripidae **Not illus.**
Identification: Similar to Thripidae, but antennae 9-segmented, and 3rd segment with an apical band of small blisterlike sensoria.

Seldom encountered in our area. One fairly common species feeds in the flowers of jack-in-the-pulpit.

Suborder Tubulifera

Terminal abdominal segment *tubular* in both sexes. ♀ without an ovipositor. FW when present veinless or with a short median vein. Antennae with *4-8 (usually 8) segments*.

TUBE-TAILED THRIPS Family Phloeothripidae
Identification: By the characters of the suborder.

This is a large group, and its members vary in habits. Some are plant feeders and are often abundant on vegetation, some are predaceous, and many feed on fungus spores.

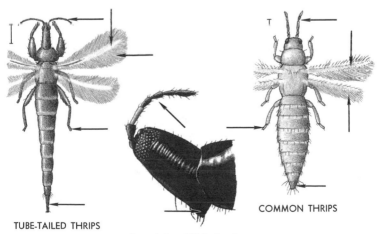

TUBE-TAILED THRIPS

COMMON THRIPS

Lateral view of Thrips head

Bugs: Order Hemiptera

Identification: FW usually *thickened at base, membranous at tip.*
HW *membranous,* shorter than FW. Wings at rest held flat over
body, tips of FW usually overlapping (some bugs are wingless, and
a few have FW uniformly thickened). Mouth parts sucking. Beak
generally *rises from anterior part of head,* and consists of 4 hairlike
stylets in a segmented sheath. Palps lacking. Antennae of 5 or
fewer principal segments, long and conspicuous or short and concealed. 2 ocelli or none. Tarsi with 3 or fewer segments. Metamorphosis simple.
Similar orders: (1) Homoptera (p. 128): FW uniform in texture,
membranous or thickened; beak rises from hind part of head;
antennae short and bristlelike or long and containing more than
5 segments. (2) Coleoptera (p. 146): FW uniformly thickened,
nearly always meeting in a straight line down back; mouth parts
chewing; antennae usually with 8 or more segments; tarsi often
4- or 5-segmented. (3) Orthoptera (p. 76): mouth parts chewing;
FW uniformly thickened; HW with many veins; antennae many-
segmented.
Immature stages: Similar to adult but wings small or absent.
Habits: This is a large order and its members vary in habits. Most
bugs are terrestrial, but many are aquatic and a few are external
parasites of vertebrates. Many bugs are plant feeders, and many
are predaceous on other insects.
Importance: A number of bugs are pests of cultivated plants; a
few are bloodsucking and are irritating pests, and some of these
are disease vectors. Some predaceous bugs are of value in keeping
pest species under control.
Classification: Two suborders, Cryptocerata and Gymnocerata,
differing in the character of the antennae. The principal characters
used in separating families of Hemiptera are those of the beak,
antennae, front wings, and legs.
No. of species: World, 23,000; N. America, 4500.

Short-horned Bugs: Suborder Cryptocerata

Antennae *short and usually concealed* in grooves on ventral side of
head. Ocelli generally absent. Aquatic or shore-inhabiting.

WATER BOATMEN Family Corixidae
  Identification: Elongate-oval, aquatic, usually 12 mm. or less.
  Dorsal surface of body flattened, with *narrow dark crosslines.*
  Front legs short, tarsi *1-segmented and scoop-shaped.* Hind legs
  elongate and functioning as oars.
    Water boatmen are common insects in ponds. A few occur in
  streams, and a few are found in the brackish water of pools along

the seashore above the high tidemark. They swim rapidly and somewhat erratically, and spend much time clinging to submerged vegetation. Most species feed on algae and other minute aquatic organisms. They do not bite man.

BACKSWIMMERS Family Notonectidae

Identification: Resemble Corixidae but dorsal surface of body convex and *often light-colored*, without dark crosslines; front tarsi *not scoop-shaped*. 15 mm. or less.

Backswimmers are so named because they swim upside down. They are common insects in ponds. They swim in a less erratic fashion than water boatmen and spend much time resting at the surface, the body at an angle and the head down. They feed on

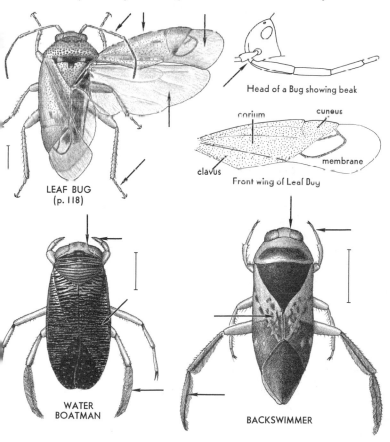

LEAF BUG
(p. 118)

Head of a Bug showing beak

corium   cuneus

clavus   membrane

Front wing of Leaf Bug

WATER
BOATMAN

BACKSWIMMER

other aquatic insects, and sometimes on small vertebrates. If handled carelessly they will bite, and the effect is rather like that of a bee sting.

WATERSCORPIONS Family Nepidae
Identification: *Slender and elongate* with long slender legs (*Ranatra*) or elongate-oval (*Nepa, Curicta*). Front legs *raptorial.* Terminal abdominal appendages *long, slender,* forming a non-retractable breathing tube. Usually 20–40 mm.

Waterscorpions in the genus *Ranatra* are brownish and resemble walkingsticks. They are common in ponds. Our only species of *Nepa* (*N. apiculata* Uhler) is elongate-oval; it is widely distributed but not very common. Two species of *Curicta* occur in the Southwest; they are somewhat more elongate than *Nepa.* Waterscorpions are predaceous on other insects, and can inflict a painful bite if handled carelessly.

GIANT WATER BUGS Family Belostomatidae
Identification: Brownish, oval, flattened, *about 1–2 in.* Front legs *fitted for grasping prey,* hind legs *somewhat flattened.* Terminal appendages *short,* retractile.

These bugs are fairly common in ponds, where they feed on various insects and small vertebrates. They can inflict a painful bite. Females of *Belostoma* (about 1 in.) lay their eggs on the back of the male, which carries them around until they hatch. Most larger giant water bugs belong to the genus *Lethocerus;* they lay their eggs on aquatic vegetation. Members of this group sometimes leave the water and fly about, and are often attracted to lights.

CREEPING WATER BUGS Family Naucoridae
Identification: Dark brown to brownish black, oval, *5–16 mm.* Membrane of FW *without veins.* Front femora *greatly thickened.*

These bugs are fairly common in ponds, where they feed on various aquatic animals. They can inflict a very painful bite if handled carelessly.

TOAD BUGS Family Gelastocoridae
Identification: Small *toad-shaped* bugs, *10 mm. or less.* Ocelli *present.* Front legs shorter than middle legs.

Toad bugs occur along the shores of ponds and streams, where they feed on smaller insects. They resemble toads in appearance and hopping habits.

VELVETY SHORE BUGS Family Ochteridae   **Not illus.**
Identification: Oval-bodied, 4–5 mm. Velvety bluish or black. Ocelli present. Antennae exposed. Front legs similar to middle legs in form and length. Beak long, extending at least to hind coxae.

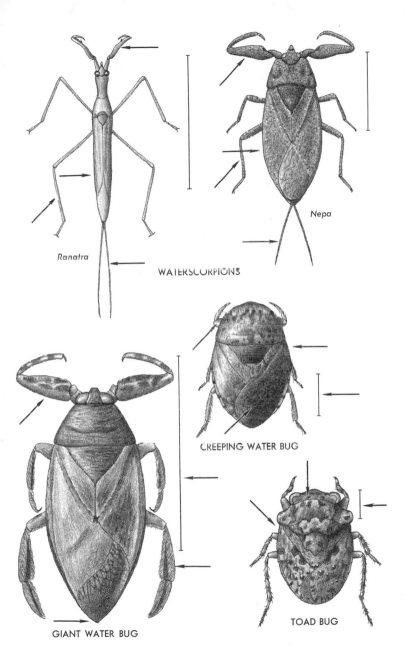

Ranatra

Nepa

WATERSCORPIONS

CREEPING WATER BUG

GIANT WATER BUG

TOAD BUG

These bugs occur along the shores of ponds and slow streams, and are predaceous. They are rather rare.

Long-horned Bugs: Suborder Gymnocerata

Antennae exposed, and longer than head. Mostly terrestrial (aquatic forms are generally surface-inhabiting).

WATER STRIDERS Family Gerridae
Identification: Usually found running about on surface of water. Middle legs *rise closer to hind legs* than to front legs. Tarsi *2-segmented*. Usually over 5 mm.

Gerrids are common on the surface of slow streams and ponds. They are generally slender, elongate, and blackish, the front legs short and the *other legs long and slender*. Some are winged and some are wingless. They feed on various small insects that fall onto the water surface. They do not bite man.

RIPPLE BUGS Family Veliidae
Identification: Small water striders, *usually less than 5 mm.*, found near riffles of streams. Middle legs *rise about equidistant* from front and hind legs (except in *Rhagovelia*, which has 1-segmented front tarsi). Legs short. Tarsi 1- to 3-segmented, claws located *before tip*.

Ripple bugs are brown or black, often with silvery markings. They are fairly common insects generally occurring in swarms. They are sometimes called broad-shouldered water striders because the body is widest near bases of middle or hind legs and the abdomen tapers posteriorly.

JUMPING GROUND BUGS
Families Dipsocoridae and Schizopteridae
Identification: Length 1–2 mm. Antennae 4-segmented, the 2 basal segments short and stout, and the 2 apical segments long and slender. Ocelli present. Tarsi and beak 3-segmented. Dipsocoridae: eyes project outward, not overlapping front edges of pronotum; head and tibiae with strong bristles. Schizopteridae: eyes project outward and backward, overlapping front edges of pronotum; head and tibiae without strong bristles.

These bugs occur on the ground in moist places, usually beneath dead leaves, and jump actively when disturbed. They are chiefly southern in distribution and are quite rare.

BED BUGS Family Cimicidae
Identification: Flat, oval, usually reddish brown. *6 mm. or less.* Wings *vestigial*. Antennae *4-segmented*. Beak and tarsi *3-segmented*. Ocelli *absent*.

Bed bugs feed by sucking blood from man and animals; 1

species, *Cimex lectularius* Linn., is often a serious pest in houses, hotels, and other living quarters, and it also attacks various animals. Other bed bugs attack bats and birds. This group is a small one but widely distributed.

MINUTE PIRATE BUGS Family Anthocoridae

Identification: *Small* (mostly 3–5 mm.), oval, flattened, and black with white markings. FW with *a cuneus*, the membrane with *few or no veins* and no closed cells. Beak and tarsi *3-segmented*. Ocelli present. Antennae *4-segmented*.

Anthocorids are fairly common bugs usually found on flowers (they are sometimes called flower bugs), but some species occur

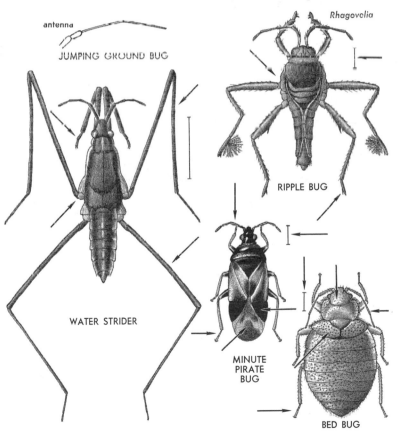

antenna

JUMPING GROUND BUG

Rhagovelia

RIPPLE BUG

WATER STRIDER

MINUTE PIRATE BUG

BED BUG

under bark, in leaf litter, or in fungi. They are predaceous on other insects and insect eggs.

LEAF or PLANT BUGS   See also p. 113 and Pl. 3
Family Miridae
Identification: Small, oval or elongate, soft-bodied, mostly less than 10 mm. FW with *a cuneus*, and membrane with *2 closed cells* (rarely membrane is lacking and cuneus is not distinct, in which case hind femora are enlarged). Ocelli *absent*. Beak 4-segmented. Tarsi *3-segmented*.
This is the largest family of bugs, with several hundred N. American species. They occur on vegetation, and are often abundant. All are rather soft-bodied, many are brightly colored. Nearly all the leaf bugs feed on plants, and some are serious pests of cultivated plants. Fleahoppers (*Halticus*), which are active jumpers, have short, uniformly thickened front wings and enlarged hind femora.

JUMPING TREE BUGS Family Isometopidae   **Not illus.**
Identification: Similar to Miridae but with ocelli. Less than 3 mm.
Five rare species in this group occur in the East. They are usually on bark or twigs and jump when disturbed.

MICROPHYSID BUGS Family Microphysidae   **Not illus.**
Identification: Broadly oval and somewhat flattened, shining black, 1.2 mm. Ocelli present. Antennae 4-segmented. Middle and hind tarsi 2-segmented. FW with a cuneus.
One very rare species of microphysid, *Mallochiola gagates* (McAtee and Malloch), occurs in the U.S. It has been recorded from Maryland and the District of Columbia.

GNAT BUGS Family Enicocephalidae   **Not illus.**
Identification: Slender, gnatlike, about 5 mm. FW entirely membranous. Front femora and tarsi thickened. Middle and hind tarsi 2-segmented. Head elongate, constricted behind eyes. Ocelli present. Antennae and beak 4-segmented.
At least 4 rare species of *Systelloderes* occur in the e. U.S.

AMBUSH BUGS Family Phymatidae   **See also Pl. 3**
Identification: Antennae 4-segmented, *slightly clubbed*. Beak short and 3-segmented. Ocelli present. Front femora *greatly thickened*. Middle and hind tarsi 3-segmented. Abdomen *wider in distal half*, extending laterally beyond wings.
Ambush bugs are common predaceous insects that usually occur on flowers, where they lie in wait for their prey. They often are found on goldenrod, where their greenish-yellow and brownish color provides camouflage. Though small (12 mm. or less), they are able to capture insects as large as bumble bees. They do not bite man.

ASSASSIN BUGS Family Reduviidae   See also Pl. 3
Identification: Body shape *generally oval, sometimes* (Emesinae,
see illus., p. 121) *greatly elongate and resembling a walkingstick.*
Beak *short, 3-segmented,* usually curved and *fitting into a groove
in prosternum.* Head *elongate, a transverse groove between eyes.*
Antennae 4-segmented, sometimes with 1 or more segments
divided into subsegments. Ocelli usually present (absent in
Emesinae). Front femora generally *thickened.* Edges of abdomen
often *extend laterally beyond wings.*

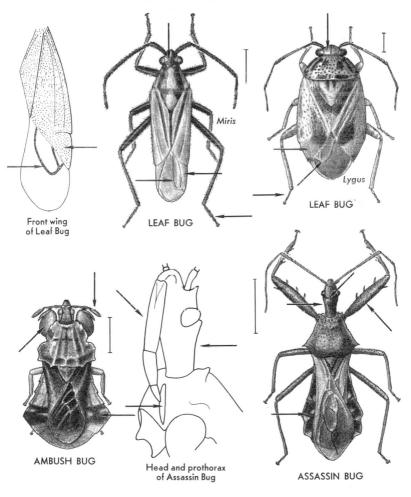

Front wing
of Leaf Bug

LEAF BUG

Miris

LEAF BUG

Lygus

AMBUSH BUG

Head and prothorax
of Assassin Bug

ASSASSIN BUG

Assassin bugs are generally brownish to black and 10–25 mm.
Many are fairly common. Most species occur on foliage, although
a few sometimes enter houses. The majority are predaceous on
other insects, but a small minority are bloodsucking. Many can
inflict a painful bite.

Subfamily Emesinae (considered by some authorities to be a
separate family, the Ploiariidae). These resemble small walkingsticks, and are called thread-legged bugs (see opp.). They are
generally found in old buildings, cellars, and similar places.
Most northern species are 30–35 mm., but many southern
species are 5–10 mm.

DAMSEL BUGS   Family Nabidae
Identification: Elongate-oval. Antennae usually 4-segmented
(rarely, subfamily Prostemminae, 5-segmented). Beak generally 4-segmented. Ocelli *present.* Front femora *slightly thickened.*
Membrane of FW, when developed, with *a number of small cells
around margin.*

Most damsel bugs are yellowish brown and about 8 mm., the
body somewhat narrowed anteriorly and the wings well developed; some are slightly larger and shining black, and a few of
these have very short front wings that lack a membrane.
Nabids are common insects and usually occur on low vegetation.
They are predaceous.

BAT BUGS   Family Polyctenidae   Not illus.
Identification: Wingless, lacking eyes and ocelli, and ectoparasites of bats. 3.5–5.0 mm. Front legs short, femora ′thickened;
middle and hind legs long and slender.

Two rare species of bat bugs occur in the U.S., 1 in Texas and
the other in California.

LACE BUGS   Family Tingidae
Identification: Body and wings with *reticulate sculpturing.*
Pronotum has *a triangular posterior extension* over scutellum.
Antennae and beak 4-segmented. Ocelli absent. Tarsi 1- or
2-segmented. Usually *5 mm. or less.*

Most lace bugs are grayish and somewhat rectangular, with a
hoodlike extension of the pronotum forward over the head and
a lacelike pattern of ridges on the pronotum and wings. Other
(less common) species are narrower, with ridges on dorsal side
of body forming a finer network, and some lack the extension
of the pronotum forward over the head. Lace bugs feed on the
foliage of trees and shrubs, and sometimes cause extensive de-
foliation.

ASH-GRAY LEAF BUGS   Family Piesmatidae
Identification: Small, gray, *oval*, mostly *about 3 mm.* Dorsal side
of body with *numerous small pits.* Pronotum without a posterior
extension over scutellum. Antennae and beak 4-segmented.

Ocelli present. *A pair of fingerlike processes on front of head.*

These bugs feed on various weeds and trees. They are not common.

STILT BUGS Family Berytidae

Identification: *Slender,* usually brownish, 5–9 mm. Legs and antennae *long and slender;* antennae 4-segmented, 1st segment *very long,* 4th segment *short and spindle-shaped.* Beak 4-segmented. Ocelli present. Tarsi 3-segmented.

Stilt bugs are fairly common and usually found on vegetation. They are plant feeders.

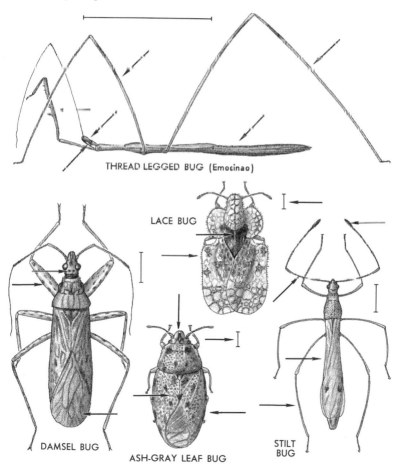

THREAD LEGGED BUG (Emesinae)

LACE BUG

DAMSEL BUG

ASH-GRAY LEAF BUG

STILT BUG

SEED BUGS Family Lygaeidae **See also Pl. 3**
Identification: Small to medium-sized, elongate to oval, and fairly hard-bodied. Antennae and beak *4-segmented*. Tarsi *3-segmented*, with a pad at base of each claw. Front femora sometimes thickened. Membrane of FW with *only 4 or 5 veins*. Ocelli nearly always *present*.

Members of this large group are often common on vegetation. They are 3–15 mm. but most are less than 10 mm. Color varies considerably: majority are brownish, though some are brightly patterned; 2 species that feed on milkweed seeds (for 1, see Pl. 3) are red and black. Most species feed on seeds. The Chinch Bug, *Blissus leucopterus* (Say), which feeds on the sap of the host plant, is a serious pest of wheat and other grains. Members of the subfamily Geocorinae, often called big-eyed bugs, sometimes feed on other insects.

RED BUGS or STAINERS **See also Pl. 3**
Family Pyrrhocoridae
Identification: Similar to Lygaeidae but *without ocelli* and with *more veins in membrane of FW*. Front femora never thickened. 8–18 mm. Usually brightly colored with red and black.

Red bugs occur chiefly in the southern states and are sometimes fairly common. Most are elongate-oval and brightly colored. A few have short front wings and are somewhat antlike in appearance. All are plant feeders. The Cotton Stainer (Pl. 3), *Dysdercus suturellus* (Herrich-Schäffer), is a serious pest of cotton.

LEAF-FOOTED BUGS Family Coreidae **See also Pl. 3**
Identification: Similar to Lygaeidae but with *many veins in membrane of FW*. Usually dark-colored and over 10 mm. Head *narrower* and most often shorter *than pronotum*. Hind tibiae *sometimes dilated and leaflike*.

This is a large group, and most of its members are relatively large bugs. They are widely distributed but are more common in the South. Some are plant feeders and others are predaceous. One of the plant feeders, the Squash Bug (Pl. 3), *Anasa tristis* (De Geer), is a serious pest of cucurbits. Coreids often give off an unpleasant odor when handled.

SCENTLESS PLANT BUGS Family Rhopalidae **Pl. 3**
Identification: Similar to Coreidae but usually pale-colored, less than 14 mm., and lacking scent glands.

Scentless plant bugs occur chiefly on grass and weeds, and are plant feeders; they are common in late summer and early fall. One species, the Boxelder Bug (Pl. 3), *Leptocoris trivittatus* (Say), is blackish with red markings and 11–14 mm.; it feeds on box elder trees, and often enters houses in the fall.

BROAD-HEADED BUGS Family Alydidae
Identification: Similar to Coreidae but head *nearly as wide and as long as pronotum.* Scent glands well developed.

Most alydids are yellowish brown or black and 10–18 mm. They give off an unpleasant odor when disturbed. Some of the black species have a reddish band across middle of dorsal side

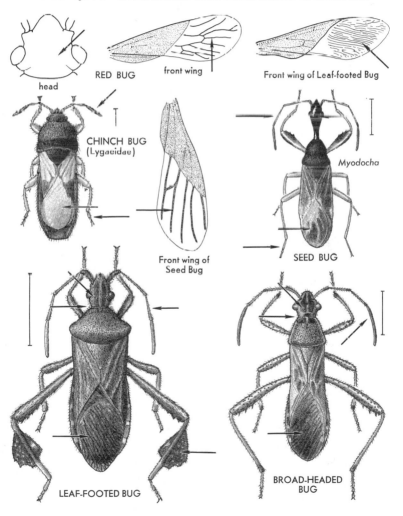

head

RED BUG   front wing

Front wing of Leaf-footed Bug

CHINCH BUG (Lygaeidae)

Myodocha

Front wing of Seed Bug

SEED BUG

LEAF-FOOTED BUG

BROAD-HEADED BUG

of the abdomen, and their nymphs resemble ants. Broad-headed bugs are plant feeders and are usually found on vegetation.

FLAT or FUNGUS BUGS   Family Aradidae
Identification: Small, oval, dark, very flat. Wings *narrow, abdomen extends beyond them.* Antennae and beak 4-segmented, 1st segment of beak very short. Ocelli *absent.* Tarsi *2-segmented.*
Flat bugs are usually found under decaying bark or in woody fungi, and are sometimes fairly common.

TERMATOPHYLID BUGS   Family Termatophylidae **Not illus.**
Identification: Similar to Miridae (p. 118) but with only 1 closed cell in membrane of FW (2 in Miridae), and basal segment of beak scarcely longer than wide (longer than wide in mirids).
One rare species has been reported from New Hampshire and Arizona. It is 4 mm. and dark brown with a white scutellum.

SHORE BUGS   Family Saldidae
Identification: Small, *oval,* flattened, usually brownish. Membrane of FW with *4–5 long closed cells.* Antennae 4-segmented. Beak and tarsi *3-segmented.* Ocelli *close together.*
Saldids are fairly common along grassy shores. When disturbed they fly a short distance and crawl down into the vegetation or into a crevice. They are predaceous.

SPINY SHORE BUGS   Family Leptopodidae   **Not illus.**
Identification: Saldidlike, 3.5 mm., with long spines over most of body, including eyes. Last 2 antennal segments very slender. Beak short. Ocelli close together on a tubercle.
One species in this Old World family, *Patapius spinosus* (Rossi), has been introduced into California.

WATER TREADERS   Family Mesoveliidae
Identification: Small, slender, usually greenish or yellowish, 5 mm. or less. Only anterior basal part of FW thickened, clavus *membranous;* membrane *veinless.* Antennae *long, slender, 4-segmented.* Beak slender, 3-segmented, 3rd segment long. Ocelli *present.* Tarsi 3-segmented, 1st segment very small.
These insects are usually found on aquatic vegetation or running about on the surface of the water. They are sometimes fairly common around ponds. They feed on small aquatic organisms found on surface of the water.

ROYAL PALM BUGS   Family Thaumastocoridae   **Not illus.**
Identification: Minute (2–3 mm.), flattened, oblong-oval. Pale yellow, with red eyes. Antennae 4-segmented. Beak 3-segmented. Tarsi 2-segmented. Ocelli present. Membrane of FW without veins.
This group is represented in the U.S. by a single species,

Xylastodoris luteolus Barber, which occurs in Florida. It feeds on the royal palm.

VELVET WATER BUGS Family Hebridae
Identification: Length *3 mm. or less.* Somewhat oblong, with *a broad-shouldered appearance.* Body densely covered with velvety pubescence. Antennae *4- or 5-segmented.* Membrane and clavus of FW similar in texture, *without veins.*

Hebrids are usually found running or walking on surface of water, and they differ from other bugs found in this situation in being very pubescent. Only 7 species occur in N. America, and they are relatively rare.

WATER MEASURERS Family Hydrometridae
Identification: Grayish, most *about 8 mm.,* the body and legs *very long and slender.* Usually wingless. Head *long and slender,* the eyes bulging and located slightly behind middle of head. Antennae 4-segmented. Tarsi 3-segmented.

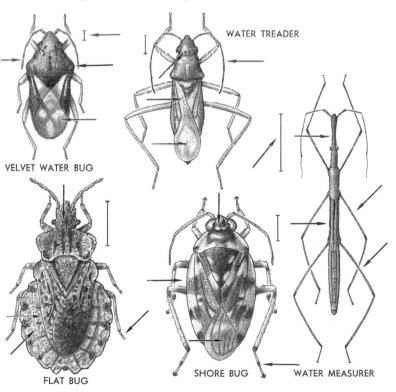

WATER TREADER

VELVET WATER BUG

FLAT BUG

SHORE BUG

WATER MEASURER

Water measurers resemble tiny walkingsticks, and are usually found around the edges of ponds, either on aquatic vegetation or walking slowly over surface of water. They are fairly common insects.

Superfamily Scutelleroidea

This group includes the remaining families of bugs, which have *5-segmented antennae* and are somewhat shield-shaped (with a broad-shouldered appearance).

BURROWER BUGS Family Cydnidae
Identification: Black, generally 7 mm. or less. Similar to Pentatomidae (below) but with *strong spines on tibiae.*
Burrower bugs usually occur under stones and boards or around the roots of grass tufts. They are not common.

NEGRO BUGS Family Corimelaenidae
Identification: Broadly oval, convex, shining black, *3–6 mm.* Scutellum *large, covering most of abdomen and wings.* Tibiae with slender spines or none.
Negro bugs resemble beetles but can be recognized as bugs by the 5-segmented antennae and 4-segmented beak. The wings under the large scutellum are typically hemipterous. These bugs are common on vegetation and flowers.

SHIELD-BACKED BUGS Family Scutelleridae
Identification: Similar to Pentatomidae but scutellum *very large, broadly oval,* and *extending to apex of abdomen.* Pronotum *without a prominent tooth on each side* in front of humeral angle. Usually brownish. 8–10 mm.
These bugs occur on vegetation, and are plant feeders. They are fairly common but not as common as stink bugs.

TERRESTRIAL TURTLE BUGS Family Podopidae
Identification: Similar to Scutelleridae but smaller (*3.5–6.5 mm.*). Scutellum *U-shaped. A prominent tooth on each side of pronotum* just in front of humeral angle.
This is a small group, and its members are quite rare.

STINK BUGS Family Pentatomidae   **See also Pl. 3**
Identification: Broadly oval and somewhat shield-shaped. Scutellum *large and triangular* but not longer than corium and not reaching apex of abdomen. Tibiae with *weak or no spines.* Usually over 7 mm.
This is a large and well-known group, and many of its members are very common bugs. The common name refers to the rather disagreeable odor these bugs produce. Some stink bugs are rather plain-colored, brownish or grayish, but many are brightly colored. Although most of them are plant feeders, some are

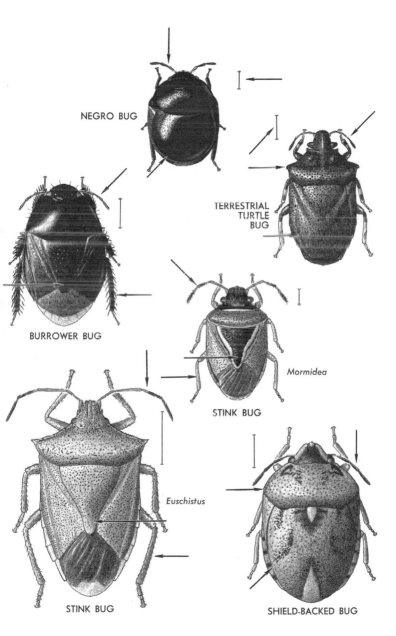

NEGRO BUG

TERRESTRIAL TURTLE BUG

BURROWER BUG

Mormidea

STINK BUG

Euschistus

STINK BUG

SHIELD-BACKED BUG

predaceous. A few of the plant feeders occasionally damage cultivated plants, especially fruits.

Cicadas, Hoppers, Whiteflies, Aphids, and Scale Insects: Order Homoptera

Identification: Mouth parts similar to those in Hemiptera, but beak usually short and rising at back of head, sometimes appearing to rise between front coxae. Winged or wingless. Winged forms with 4 wings (rarely only 2), FW membranous or thickened, HW membranous and a little shorter than FW, wings at rest usually held rooflike over body. Tarsi 1- to 3-segmented. Antennae variable, sometimes short and bristlelike, sometimes long and threadlike, rarely absent. ♀ often with a well-developed ovipositor. Metamorphosis usually simple.

Similar orders: (1) Hemiptera (p. 112): FW nearly always thickened at base and membranous at tip; beak usually rising at front of head. (2, 3) Coleoptera and Orthoptera (pp. 146, 76) mouth parts chewing.

Immature stages: Usually similar to adult but with wings absent or vestigial.

Habits: Homoptera are plant feeders, and each species usually feeds on a particular part of a few species of plants. The feeding results in discoloration, distortion, wilting, or stunting of the plant, and heavily infested plants are sometimes killed. A few homopterans cause the development of plant galls.

Importance: Many members of this order are serious pests of cultivated plants, causing damage by feeding and sometimes by serving as vectors of plant diseases.

Classification and identification: Two suborders, Auchenorrhyncha and Sternorrhyncha, each of which is divided into superfamilies. Most families are separated by easily seen characters, but scale insects can generally be identified to family only from specimens mounted on microscope slides. Many wingless Sternorrhyncha will be difficult or impossible to identify unless one is familiar with their life history.

No. of species: World, 32,000; N. America, 6500.

Cicadas and Hoppers: Suborder Auchenorrhyncha

Tarsi *3-segmented*. Antennae *short, bristlelike*. Active insects.

Cicadas, Treehoppers, Froghoppers, and Leafhoppers: Superfamily Cicadoidea

Antennae rise *in front of or between compound eyes*. Middle coxae short, contiguous. No Y-vein in anal area of FW. Usually 2 ocelli (3 in cicadas). Jumping insects (except cicadas).

CICADAS Family Cicadidae   See also Pl. 1
Identification: Large insects, mostly 1–2 in. FW *membranous*. *3 ocelli*. ♂ usually with sound-producing organs at base of abdomen on ventral side. Nonjumping insects.

Cicadas are common insects, but are more often heard than seen since the majority are arboreal. Song is produced only by males, and is usually a loud (sometimes pulsating) buzz. Most cicadas are large blackish insects, often with greenish markings, that appear each year in July and August; their life cycle lasts 2–5 or more years, but the broods overlap and adults are present each year. The periodical cicadas (*Magicicada*) have a life cycle of 13 or 17 years, and adults are present in a given area only in certain years. These cicadas occur in the East, and adults appear in May and June; they are 19–33 mm, and have the eyes and wing veins reddish. The 17-year cicadas (3 species) are principally northern. The 13-year cicadas (3 species) are principally southern. The species in each life-cycle group differ in size, color, and song; most emerging broods of periodical cicadas contain 2 or 3 species. Eggs are laid in twigs, which usually die and break off, nymphs live in the ground and feed on roots. Nymphs are stout-bodied, brownish, with expanded front tibiae; they usually crawl up on a tree trunk for their molt to the adult. Cicadas are generally of little economic importance,

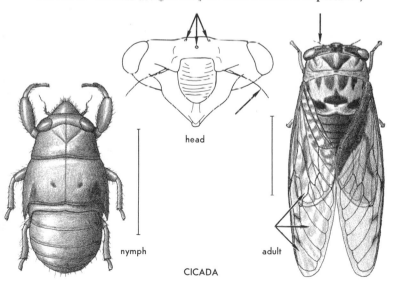

head

nymph   adult

CICADA

but the egg laying of large numbers of periodical cicadas often causes serious damage to young trees.

TREEHOPPERS   Family Membracidae
Identification: Small jumping insects, usually 12 mm. or less. Pronotum *prolonged backward over abdomen.*

Treehoppers are common insects occurring on all types of vegetation. They vary in shape, owing to variations in shape of the pronotum; most of them appear humpbacked and some are shaped like thorns. Adults of most species feed on trees and shrubs, but some feed on weeds and grasses, especially in the nymphal stage. Eggs are usually laid in twigs, and the terminal portion of such twigs generally dies.

FROGHOPPERS and SPITTLEBUGS   Family Cercopidae
Identification: Small jumping insects, generally less than 12 mm. Pronotum does not extend back over abdomen. Hind tibiae with *1 or 2 stout spines*, and usually a circlet of spines at apex.

Adults are called froghoppers because many are somewhat wider posteriorly and are shaped rather like tiny frogs; nymphs produce and become surrounded by a spittlelike mass, and are called spittlebugs. These insects are very common, and spittle masses are often abundant on grass and weeds; a few spittlebugs feed on trees. One of the most common species is the Meadow Spittlebug, *Philaenus spumarius* (Linn.), which varies considerably in color but is usually brownish; it often causes considerable damage to clovers.

LEAFHOPPERS   Family Cicadellidae   See also Pl. 4
Identification: Similar to Cercopidae but body usually tapers posteriorly or is parallel-sided, and hind tibiae have *1 or more rows of small spines.*

This is a very large group; many species are common and abundant insects. Most of them are less than 10 mm., and many are brightly colored. They occur on a diversity of plants but each species is usually rather specific in its selection of a food plant. Many are serious pests of cultivated plants, causing injury by their feeding, and a few serve as vectors of plant diseases. Leafhoppers often discharge from the anus a clear watery fluid called *honeydew*, to which other insects (particularly ants) may be attracted.

Planthoppers: Superfamily Fulgoroidea

Antennae rise *on sides of head beneath eyes.* Middle coxae elongate and separated. Two anal veins in FW usually meet distally to form a Y-vein. Jumping insects, mostly 10 mm. or less. Many species have very short wings that cover only the basal abdominal segments. Both short- and long-winged individuals occur in some species. There is usually a distinct angle between the front (or dorsal) and lateral surfaces of the head; antennae, eyes, and lateral ocelli are on the lateral surface. Planthoppers constitute a large group but are seldom as abundant as other hoppers.

The N. American families of planthoppers may be divided into 2 groups on the basis of the structure of the 2nd segment of the hind tarsus: (1) this segment small to minute, its apex rounded and with a small spine on each side: Tropiduchidae, Acanaloniidae,

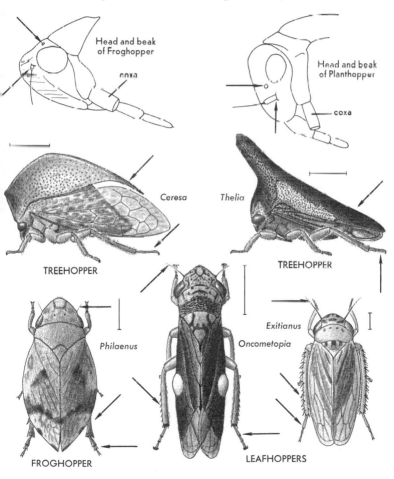

Head and beak of Froghopper

coxa

Head and beak of Planthopper

coxa

Ceresa

Thelia

TREEHOPPER

TREEHOPPER

Philaenus

Oncometopia

Exitianus

FROGHOPPER

LEAFHOPPERS

Flatidae, and Issidae; (2) this segment longer, its apex truncate or emarginate and with a row of small spines: the 7 remaining families.

TROPIDUCHID PLANTHOPPERS   Not illus.
Family Tropiduchidae
Identification: Slender, yellowish to brownish, 7–9 mm. Apical portion of FW set off from rest of wing by a series of cross veins, the apical portion more densely veined than rest of wing. Hind trochanters usually directed backward.

This is a tropical group. There are 3 rather uncommon species in the southeastern states.

ACANALONIID PLANTHOPPERS   Pl. 4
Family Acanaloniidae
Identification: Shape characteristic: FW *broadly oval, held almost vertical at rest.* Costal area of FW reticulate. Hind tibiae with apical spines only.

The members of this small group are usually greenish, sometimes with brown markings. Most of them belong to the genus *Acanalonia*.

FLATID PLANTHOPPERS Family Flatidae   Pl. 4
Identification: Body somewhat wedge-shaped. FW elongate-triangular, held almost vertical at rest, and with *numerous costal and/or apical cross veins.* Hind tibiae with spines on sides in addition to apical ones.

This is a large and widely distributed group, and most of its members are brownish or greenish.

ISSID PLANTHOPPERS   Family Issidae
Identification: Body not particularly wedge-shaped. FW often short, the costal area without numerous cross veins. Hind tibiae with *spines on sides in addition to apical ones.*

Many species in this group are short-winged and some have a weevil-like snout.

FULGORID PLANTHOPPERS Family Fulgoridae   Not illus.
Identification: Anal area of HW with many cross veins (such cross veins are lacking in remaining families).

This group includes some of our largest planthoppers, but they are not very common. Most of them are brownish.

DERBID PLANTHOPPERS   Family Derbidae
Identification: Terminal segment of beak *about as long as wide* (remaining families have this segment longer).

Most derbids have long narrow front wings and are rather delicate in build; many are brightly colored. These planthoppers feed on woody fungi.

DELPHACID PLANTHOPPERS Family Delphacidae
Identification: Hind tibiae with *a large apical spur.*
This is the largest family of planthoppers. Its members are often very common insects. Most of them are quite small and many are short-winged.

ACHILID PLANTHOPPERS Family Achilidae
Identification: Body flattened. FW *overlap at apex,* and the claval suture does not reach wing margin.
Most achilids are brownish, and occur in coniferous forests. Nymphs usually live in dead wood and under bark.

DICTYOPHARID PLANTHOPPERS Family Dictyopharidae
Identification: Head *strongly prolonged anteriorly,* or front with 2 or 3 carinae (keels), or tegulae absent and claval suture obscure. Median ocellus lacking.
Most dictyopharids have the head produced forward into a long, slender, beaklike structure. If the head is not so produced, the body is usually oval and the front femora are broad and flattened.

CIXIID PLANTHOPPERS Family Cixiidae
Identification: Head *little or not at all extended anteriorly;* front with only a median carina (keel) or none. Median ocellus often present. Tegulae present. Claval suture distinct.

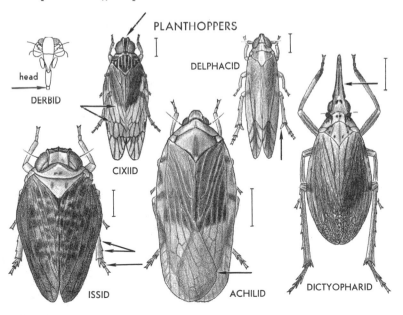

PLANTHOPPERS

head

DERBID

DELPHACID

CIXIID

ISSID

ACHILID

DICTYOPHARID

This is a relatively large group. Its members have *membranous* front wings, often with dark spots along the veins.

KINNARID PLANTHOPPERS Family Kinnaridae   **Not illus.**
  Identification: Pale-colored, 3–4 mm. Similar to cixiids but abdominal terga 6–8 with wax-secreting pores. Vertex narrow and troughlike, narrower anteriorly.
  Six species in the genus *Oeclidius* occur in the West.

Psyllids, Whiteflies, Aphids, and Scale Insects: Suborder Sternorrhyncha

Tarsi 1- or *2-segmented* (legs rarely lacking). Antennae long and threadlike (rarely absent). Mostly rather inactive insects.

Superfamily Psylloidea

PSYLLIDS Family Psyllidae   **See also Pl. 4**
  Identification: FW membranous or thickened. Wings usually *held rooflike over body at rest*. Tarsi *2-segmented*. Antennae *10-segmented*. Active jumping insects, *2–5 mm*.
  Most psyllids are free-living and feed on a variety of plants. A few are gall makers, generally forming galls on hackberry leaves. Nymphs are oval and flat and look very little like the adults; many produce a large amount of waxy filaments, making them look like little blobs of cotton. Some psyllids resemble tiny cicadas. A few species are pests of orchard trees or garden plants. One western species acts as vector of a plant disease known as "psyllid yellows."

Superfamily Aleyrodoidea

WHITEFLIES Family Aleyrodidae
  Identification: *Minute* whitish insects, generally 2–3 mm. Body and wings covered with a white powder. *HW nearly as large as FW*. Wings held horizontal over body at rest. Tarsi *2-segmented*. Antennae *7-segmented*. Compound eyes somewhat elongate vertically and narrowed in middle.
  Whiteflies are chiefly tropical, and our most common species are those attacking citrus or greenhouse plants. The 1st instar is an active insect; subsequent instars become covered with a blackish scalelike covering. Early instars are called larvae and the next to last instar, which is quiescent and does not feed, is called the pupa.

Aphids and Phylloxerans: Superfamily Aphidoidea

Wings, when present, membranous and not covered with a whitish powder. *HW much smaller than FW*. Tarsi 2-segmented. Body oval or pear-shaped, often with *a pair of fingerlike cornicles* near posterior end of abdomen. Many with a complex life cycle.

APHIDS Family Aphididae   **See also Pl. 4**
  Identification: Soft-bodied, usually somewhat pear-shaped, 4–8 mm., *nearly always with a pair of cornicles* near posterior end of abdomen. FW of winged forms with *Rs present and M branched.* Wings at rest usually held vertical above body.

This is a large group, and its members are often found in considerable numbers on the stems, leaves, and flowers of various plants. Many are serious pests of cultivated plants; their feeding causes a curling or wilting of the plant, and some aphids serve as vectors of plant diseases.

Most aphids have a complex life cycle, involving bisexual and parthenogenetic (♀) generations, winged and wingless individuals or generations, and often a regular alternation of food plants. Aphids generally overwinter as eggs, which hatch in the spring into wingless females that reproduce parthenogenetically (without fertilization) and give birth to young (rather than eggs). Two or more generations of such females may be produced; a generation of winged females eventually appears that usually migrate to a different food plant. These winged females also reproduce parthenogenetically, giving birth to young. Late in the season winged forms return to the original food plant, and

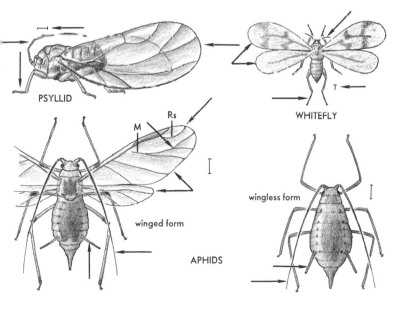

PSYLLID

WHITEFLY

M   Rs

winged form

wingless form

APHIDS

a generation of males and females appears; these mate, and the females lay the eggs that overwinter.

Aphids discharge from the anus a clear watery liquid, called honeydew, to which ants and other insects are attracted. Some ants live closely associated with aphids. They gather aphid eggs and keep them over winter in their nest and transport the aphids to a food plant in spring, tending them during the season and transferring them from one food plant to another. The ants feed on honeydew produced by the aphids.

WOOLLY and GALL-MAKING APHIDS
Family Eriosomatidae

Identification: Most produce *large amounts of a woolly or waxy material* that may nearly or completely cover the body, particularly in nymphal stages. Similar to Aphididae but cornicles very small or absent, and M in FW of winged forms *not branched*.

The life cycle is similar to that in Aphididae, but individuals of the bisexual generation lack mouth parts and a mated female lays only 1 egg. The Woolly Apple Aphid, *Eriosoma lanigerum* (Hausmann), is an important pest of apple and related trees; it usually overwinters on elm, and later in the season migrates to apple; still later some individuals migrate back to the elm, where the bisexual generation is produced, and others migrate to the roots of the apple, where they form gall-like growths. Many members of this family are gall makers, producing galls on leaves, stems, or buds of various plants.

PINE and SPRUCE APHIDS Family Chermidae See also Pl. 4
Identification: Live on needles or twigs of pine or spruce, or in galls on these trees. Cornicles absent. FW with Rs absent, and Cu_1 and Cu_2 *separated at base*. Wings held rooflike over body at rest. Wingless females often covered with waxy threads. Antennae of winged forms 5-segmented. Feed on conifers.

Adults are usually dark-colored. Many species alternate in their life cycle between 2 different conifers, forming galls only on 1. Both bisexual and parthenogenetic generations occur in most species. *Chermes abietis* Linn. is a common member of this group; it forms pineapple-shaped galls on twigs of spruce (see Pl. 4).

PHYLLOXERANS Family Phylloxeridae
Identification: Similar to Chermidae but Cu_1 and Cu_2 in FW *stalked at base*, and wings held horizontal at rest. Wingless individuals sometimes covered with a waxy powder, but never with waxy threads. All forms with 3-segmented antennae. Feed on plants other than conifers.

Members of this group often have a very complex life history. The Grape Phylloxera, *Phylloxera vitifoliae* (Fitch), a serious pest of some varieties of grapes, attacks and forms galls on both the leaves and the roots; different generations occur on these

2 parts of the plant. The chief damage is done by individuals attacking the roots. European varieties of grapes are more susceptible to damage by this insect than American varieties. Where European varieties are grown, damage by the Phylloxera is reduced by grafting European vines to American rootstocks.

Scale Insects: Superfamily Coccoidea

Tarsi usually 1-segmented, with 1 claw, or legs absent. ♂ midge-like, with *1 pair of wings* (rarely wingless) and without a beak. ♀ wingless, often also legless, and usually with a waxy or scalelike covering. Male scale insects differ from Diptera in having 1-segmented tarsi (5-segmented in Diptera) and a long stylelike process (rarely 2) at end of abdomen.

The 1st instar has *legs and antennae* and is an active insect. Subsequent instars are less active and are often sessile. Legs and antennae of sessile scale insects are lost at the 1st molt, and the insect becomes covered with a scalelike covering. The last preadult instar of the male (called a pupa) is quiescent and does not feed.

Many scale insects are serious pests of cultivated plants, particularly orchard trees, shrubs, and many greenhouse plants, and when numerous may kill the plant. A few species are of value because of their secretions: shellac is made from the waxy secretions of a scale insect in India and some cochineal insects have been used as the source of a crimson dye.

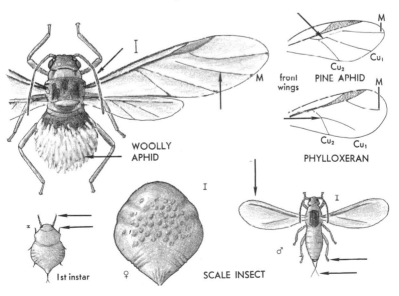

WOOLLY APHID

front wings   PINE APHID

PHYLLOXERAN

1st instar   ♀   SCALE INSECT   ♂

Identification of scale insects to family is based largely on characters of the female that can be seen only when the specimen is mounted on a microscope slide. Since most users of this book will not have facilities for mounting scale insects, structural features that distinguish each family have been omitted in the following accounts.

ENSIGN COCCIDS Family Ortheziidae
Females have well-developed legs and antennae. Body is covered with white waxy plates. These coccids occur on plant roots, and are not often encountered.

GIANT COCCIDS and GROUND PEARLS   Not illus.
Family Margarodidae
Female margarodids are large and rounded; some tropical species are nearly 1 in. long. Wax cysts of female *Margarodes*, some of which are metallic bronze or gold, are sometimes used as beads. The Cottony Cushion Scale, *Icerya purchasi* Maskell, is an important pest of citrus in the West.

ARMORED SCALES Family Diaspididae
Female armored scales are small, flattened, and disclike, lack legs, antennae, and eyes. They live under a scale formed of wax secretions of the insect and cast skins of early instars. Scales vary in shape and color in different species; male scales are usually smaller and more elongate than those of females. This is the largest family of scale insects, and it contains many important pest species. Most armored scales feed on trees and shrubs, and sometimes almost completely incrust twigs or branches. Two important pests of orchard and shade trees are the San Jose Scale, *Aspidiotus perniciosus* Comstock, and the Oystershell Scale, *Lepidosaphes ulmi* (Linn.); see opp. The former has a circular scale, the latter a scale shaped like an oyster shell.

ACLERDID SCALES Family Aclerdidae   Not illus.
Most members of this small group attack grasses and sedges. Generally feed under the leaf sheaths.

WAX and TORTOISE SCALES Family Coccidae
Female coccids are flattened and oval, the body hard and smooth or covered with wax. Antennae reduced or absent. Legs may be present or absent. The Cottony Maple Scale, *Pulvinaria innumerabilis* (Rathvon), about 6 mm., lays its eggs in a cottony mass that protrudes from posterior end of body (see opp.). Some coccids are pests of citrus and greenhouse plants.

LAC INSECTS Family Lacciferidae   Not illus.
This group is chiefly tropical, but a few species occur in the

Southwest on cactus and other desert plants. A species occurring in India, *Laccifer lacca* (Kern), is the source of lac, which is used in making shellac and varnishes.

PIT SCALES Family Asterolecaniidae **Not illus.**
Female pit scales are small and oval. Body covered with a tough waxy film or embedded in a mass of wax. Legs vestigial or lacking. Antennae short and 4- to 6-segmented. The wax of *Cerococcus quercus* Comstock, a pit scale occurring on oak in the Southwest, has been used as chewing gum by the Indians.

MEALYBUGS Families Pseudococcidae and Eriococcidae
These are elongate-oval insects with well-developed legs. Bodies of most of them (Pseudococcidae) are covered with a waxy secretion; some (Eriococcidae) have body bare or only lightly covered with wax. Mealybugs often are pests of citrus and greenhouse plants.

COCHINEAL INSECTS Family Dactylopiidae **Not illus.**
Cochineal insects are similar to mealybugs but are red, and the body is covered with white waxy plates. They occur in the Southwest on cactus. One species has been used by Indians as the source of a crimson dye.

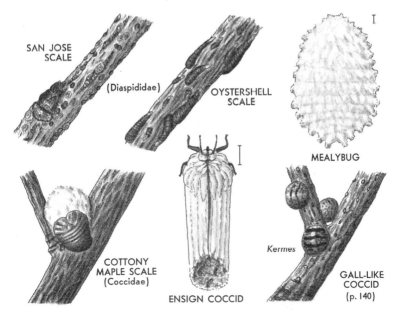

SAN JOSE SCALE

(Diaspididae)

OYSTERSHELL SCALE

MEALYBUG

COTTONY MAPLE SCALE (Coccidae)

ENSIGN COCCID

Kermes

GALL-LIKE COCCID (p. 140)

GALL-LIKE COCCIDS Family Kermidae **p. 139**
Females of this group are somewhat spherical; lack legs but have 6-segmented antennae. Members of the genus *Kermes* live on oak twigs and resemble tiny galls. Females of a species occurring in Israel produce a large amount of honeydew; this honeydew accumulates on the leaves of the host plant to form a sugarlike substance called manna.

Fishflies, Snakeflies, Lacewings, and Antlions: Order Neuroptera

Identification: Four membranous wings: FW and HW about same size or HW a little wider at base; wings usually held rooflike over body at rest; wings generally with many veins, including numerous cross veins in costal area. Antennae long, many-segmented, thread-like, pectinate, or clubbed. Tarsi 5-segmented. Cerci absent. Mouth parts chewing. Metamorphosis complete.
Similar orders: (1) Odonata (p. 68): wings at rest held outstretched or together above body; tarsi 3-segmented; antennae short and bristlelike; wing venation different; harder-bodied. (2) Plecoptera (p. 92): tarsi 3-segmented; cerci present. (3) Mecoptera (p. 208): long-faced; few costal cross veins.
Immature stages: Larvae are campodeiform, and usually have large mandibles; majority terrestrial but a few aquatic. Larvae of most groups are predaceous. Pupation usually occurs in a silken cocoon; silk is spun from the anus.
Habits: Most Neuroptera are relatively poor fliers, and most are predaceous. Many are attracted to lights at night.
Importance: Some species, especially lacewings, are important predators, of value in keeping such pests as aphids under control. Larvae of the aquatic forms are an important item in the food of many freshwater fish, and some (like hellgrammites) are frequently used as fish bait.
Classification: Three suborders, Megaloptera, Raphidiodea, and Planipennia. Families of Planipennia are arranged in 3 super-families. Families are separated chiefly by wing and antennal characters.
No. of species: World, 4670; N. America, 338.

Dobsonflies, Fishflies, and Alderflies: Suborder Megaloptera

HW a little wider at base than FW. Anal area of HW folded fan-wise at rest. Larvae aquatic.

DOBSONFLIES and FISHFLIES Family Corydalidae
Identification: Large soft-bodied insects with a rather fluttery

flight; generally found near streams. Length usually *1 in. or more*. Ocelli *present*. 4th tarsal segment *cylindrical*.

Larvae occur in streams, generally under stones. Dobsonflies (*Corydalus* in the East and Southwest, *Dysmicohermes* in the West) are large insects with front wings over 2 in. long; males have mandibles *about 3 times as long as head*. Dobsonfly larvae, or hellgrammites (see illus.), are often used as fish bait. Fishflies are smaller (FW less than 2 in.), and many have serrate or pectinate antennae; some have clear wings, others extensive black or gray areas in the wings. Corydalid larvae have *a pair of hooked anal prolegs* and *8 pairs of lateral filaments*, and lack a terminal filament.

ALDERFLIES   Family Sialidae

Identification: Similar to Corydalidae but smaller (usually *less than 1 in.*), *without ocelli*, and with 4th tarsal segment *dilated and deeply bilobed*.

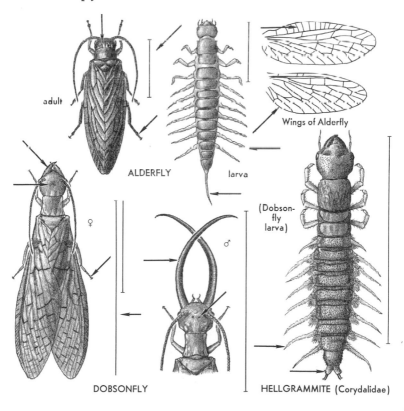

adult

Wings of Alderfly

ALDERFLY   larva

(Dobson-
fly
larva)

♀

♂

DOBSONFLY   HELLGRAMMITE (Corydalidae)

Alderflies are blackish or gray, somewhat lacewinglike in appearance but with the hind wings slightly wider at the base than the front wings. They are usually found near ponds or streams. Larvae are similar to those of Corydalidae but have a *terminal filament* and no hooked anal prolegs, and *only 7 pairs of lateral filaments*.

Snakeflies: Suborder Raphidiodea

Prothorax *elongate*. Front legs rise *from posterior end of prothorax* and are *similar to other legs*.

SNAKEFLIES   Families Raphidiidae and Inocelliidae
   Identification: Raphidiidae: stigma in FW *bordered proximally by a cross vein;* basal m-cu in HW transverse; ocelli present. Inocelliidae: stigma in FW *not bordered proximally by a cross vein;* basal m-cu in HW oblique; ocelli absent.
   Snakeflies occur only in the West, where they are represented by about 17 species of *Agulla* (Raphidiidae) and 2 species of *Inocellia* (Inocelliidae). Adults and larvae are predaceous; larvae are usually found under bark.

Dusty-wings, Lacewings, Antlions, and Owlflies: Suborder Planipennia

FW and HW similar in size and shape, anal area of HW not folded at rest. Either prothorax not lengthened or front legs mantidlike and rising from anterior end of prothorax.

Dusty-wings: Superfamily Coniopterygoidea

Length 3 mm. or less. Covered with a whitish powder. Wings with relatively few veins, Rs with *only 2 branches*.

DUSTY-WINGS   Family Coniopterygidae
   Identification: By the characters of the superfamily.
   Dusty-wings might be confused with psocopterans (p. 102) or whiteflies (p. 134), but can be recognized by their long threadlike antennae, chewing mouth parts, 5-segmented tarsi, and *characteristic wing venation*. They are rather rare.

Superfamily Hemerobioidea

Antennae threadlike, serrate, or pectinate, never clubbed.

MANTIDFLIES   Family Mantispidae
   Identification: Mantidlike in appearance: prothorax *elongate,* front legs *fitted for grasping prey* and rise from *anterior end of prothorax.*
   Mantidflies are widely distributed but are more common in

the South. Adults, less than 1 in., are predaceous. Larvae are parasitic in egg sacs of ground spiders.

PLEASING LACEWINGS Family Dilaridae **Not illus.**
Identification: Superficially resemble small moths, since wings are rather hairy and at rest are often held outspread. Antennae of ♂ pectinate. ♀ with an ovipositor as long as body. Costal cross veins not forked. FW more or less triangular and 3.0–5.5 mm.
   Only 2 species occur in N. America, and they are quite rare.

GIANT LACEWINGS Family Polystoechotidae **Not illus.**
Identification: Wingspread 1½–2½ in. Humeral vein in FW recurved and branched. 1st r-m cross vein in HW longitudinal.
   Only 2 species of giant lacewings occur in N. America, and they are quite rare.

ITHONID LACEWINGS Family Ithonidae **Not illus.**
Identification: Similar to Polystoechotidae, but FW with Sc and R_1 not fused distally and Rs with only a few branches, and 1st r-m in HW short and oblique.

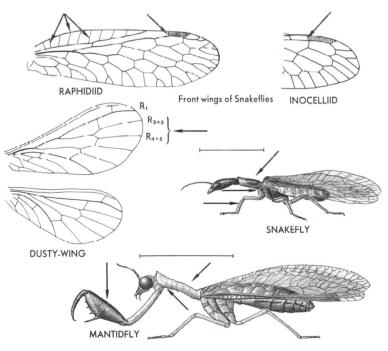

RAPHIDIID

Front wings of Snakeflies

INOCELLIID

R_1
R_{2+3}
R_{4+5}

SNAKEFLY

DUSTY-WING

MANTIDFLY

A single very rare member of this group, *Oliarces clara* Banks, has been reported from s. California. It has a wingspread of about 1½ in.

BEADED LACEWINGS   Family Berothidae   **Not illus.**
Identification: Some costal cross veins forked. Vertex flattened. Outer margin of FW sometimes indented behind apex. Over 8-mm.

Berothids resemble caddisflies, and are often attracted to lights. The group is small, its members quite rare.

GREEN LACEWINGS   Family Chrysopidae
Identification: Wings usually greenish, eyes *golden or copper-colored*. FW with *apparently 1 radial sector*, Sc and R₁ *not fused at wing tip*, and costal cross veins *not forked*.

Green lacewings are very common insects, found on grass, weeds, and shrubs, usually in relatively open areas. They often give off an unpleasant odor when handled. Adults and larvae feed principally on aphids and are important agents in the control of these insects. Eggs are laid *at ends of tiny stalks*, usually on foliage; the larvae, which have *long sickle-shaped mandibles*, pupate in small pea-shaped silken cocoons.

BROWN LACEWINGS   Family Hemerobiidae
Identification: Similar to Chrysopidae but brownish and generally much smaller. FW with *apparently 2 or more radial sectors*. Some costal cross veins *forked*.

Brown lacewings are less common than green lacewings, and are most likely to be encountered in wooded areas. They are widely distributed. Adults and larvae are predaceous.

SPONGILLAFLIES   Family Sisyridae
Identification: Lacewinglike, 6–8 mm. Costal cross veins *not forked*. Sc and R₁ *fused near wing tip*. Vertex convex.

Sisyrids resemble small brownish lacewings, and are found near water. Larvae feed on freshwater sponges.

Antlions and Owlflies: Superfamily Myrmeleontoidea

Large insects resembling damselflies or dragonflies but with longer antennae that are *clubbed*.

ANTLIONS   Family Myrmeleontidae
Identification: Damselflylike, the antennae *about as long as head and thorax together*. *An elongate cell* behind point of fusion of Sc and R₁.

Antlions resemble damselflies but are softer-bodied and have conspicuous knobbed antennae. Larvae (sometimes called doodlebugs) have long sicklelike jaws and usually live at the bottom of a conical pit in dry sandy or dusty areas. They feed on ants and other insects that fall into this pit.

OWLFLIES Family Ascalaphidae

Identification: Large insects resembling dragonflies but with antennae *nearly or quite as long as body.* Cell behind fusion point of Sc and R_1 is *short.*

Owlflies are fairly common in the South and Southwest, relatively rare in the North. Larvae are similar to those of the Myrmeleontidae but do not dig pits; they lie in wait for their

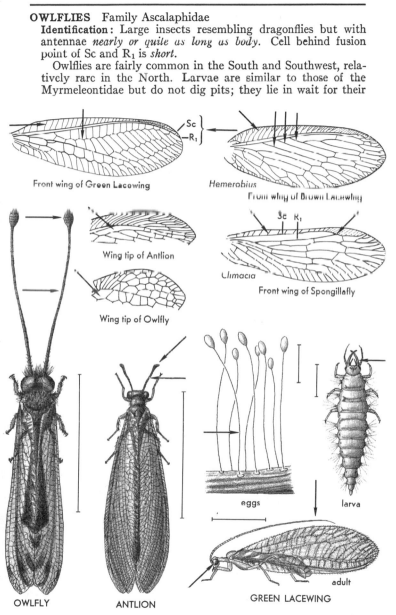

Front wing of Green Lacewing

Hemerobius

From wing of Brown Lacewing

Wing tip of Antlion

Climacia

Front wing of Spongillafly

Wing tip of Owlfly

eggs

larva

OWLFLY

ANTLION

GREEN LACEWING

adult

prey on surface of the ground, often covered with dirt or debris.

Beetles: Order Coleoptera

Identification: FW (*elytra*) horny or leathery, nearly always meeting in a straight line down back and covering HW. HW (the ones used in flight) membranous, usually longer than FW, and folded beneath FW when not in use. FW occasionally short and not covering all of abdomen. 1 or both pairs of wings rarely reduced or absent. Antennae usually with 11 segments, rarely with more, often with 8–10, rarely with as few as 2; antennae variable in form. Mouth parts chewing. Tarsi usually 3- to 5-segmented. Abdomen commonly with 5 segments visible ventrally, sometimes with up to 8. Metamorphosis complete.

Similar orders: (1) Dermaptera (p. 98): abdomen with pincerlike appendages at tip. (2) Hemiptera (p. 112): mouth parts sucking; front wings rarely meeting in a straight line down the back, nearly always overlapping at tip; antennae with 4 or 5 segments. (3) Homoptera (p. 128): mouth parts sucking. (4) Orthoptera (p. 76): front wings with distinct veins (beetle FW lack veins); antennae usually long, threadlike, many-segmented.

Immature stages: Larvae quite variable in form, hardness of body, and development of appendages: campodeiform (like *Campodea;* see illus., p. 63), grublike, or wormlike; some are wirewormlike, and a few are greatly flattened. Feed in the open or burrow into the food material. Occur in a great variety of habitats; many are aquatic.

Habits: This is the largest order of insects. Its members are almost everywhere and feed on all sorts of plant and animal materials. They are abundant on vegetation; they occur under bark, stones, and other objects; many are found on or in the ground, in fungi, rotting vegetation, dung, and carrion. Some are aquatic. A few are parasites of other animals.

Importance: Many plant-feeding beetles are serious pests, and different species attack nearly all parts of plants. Some beetles feed on various stored foods and other materials. Many beetles are of value because they prey upon and help to control injurious insects or act as scavengers.

Classification: Three suborders, Archostemata, Adephaga, and Polyphaga. Archostemata are a primitive group with only 2 rare families. Adephaga have the 1st abdominal sternum interrupted in the middle by the hind coxae. At least the hind part of this sternum extends completely across the abdomen in the Polyphaga. The superfamilies and families of beetles are usually separated by characters of the antennae, legs, head, pronotum, front wings (elytra), thoracic sclerites, and abdomen.

No. of species: World, 290,000; N. America, 28,600.

Identification of beetles: The pictured key (pp. 148–149) to the more common families of beetles should be of help to those not well acquainted with these insects. A comparison of the specimen to be identified with the 1st pair of alternatives will lead to further pairs, and eventually to a group of families in which the specimen belongs. Families marked with a dagger are the most common, and descriptions and illustrations of these should be checked first. Families with an asterisk (somewhat less common) should be checked next, and families without symbols are least common and can be checked last. If, from the descriptions and illustrations

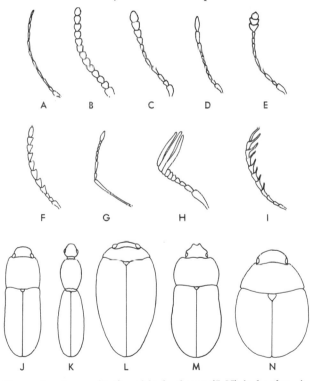

Types of antennae (A–I) and body shapes (J–N) in beetles. A, threadlike; B, beadlike; C, D, and E, clubbed (E, capitate); F, serrate; G, elbowed; H, lamellate; I, pectinate; J, elongate-slender, nearly parallel-sided; K, elongate-slender; L, elongate-oval; M, elongate-robust; N, broadly oval.

Key to the Principal Families of Coleoptera

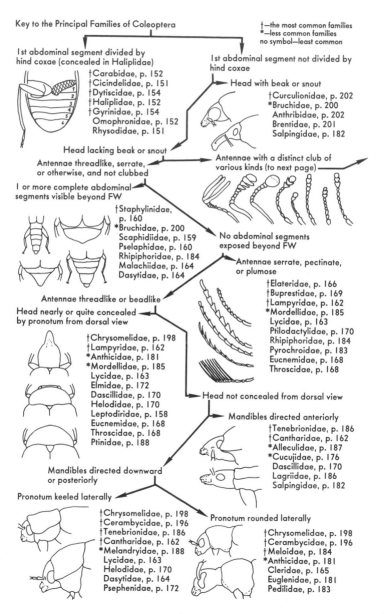

†—the most common families
*—less common families
no symbol—least common

1st abdominal segment divided by hind coxae (concealed in Haliplidae)

†Carabidae, p. 152
†Cicindelidae, p. 151
†Dytiscidae, p. 154
†Haliplidae, p. 152
†Gyrinidae, p. 154
Omophronidae, p. 152
Rhysodidae, p. 151

1st abdominal segment not divided by hind coxae

Head with beak or snout

†Curculionidae, p. 202
*Bruchidae, p. 200
Anthribidae, p. 202
Brentidae, p. 201
Salpingidae, p. 182

Head lacking beak or snout

Antennae threadlike, serrate, or otherwise, and not clubbed

Antennae with a distinct club of various kinds (to next page)

1 or more complete abdominal segments visible beyond FW

†Staphylinidae, p. 160
*Bruchidae, p. 200
Scaphidiidae, p. 159
Pselaphidae, p. 160
Rhipiphoridae, p. 184
Malachiidae, p. 164
Dasytidae, p. 164

No abdominal segments exposed beyond FW

Antennae serrate, pectinate, or plumose

†Elateridae, p. 166
†Buprestidae, p. 169
†Lampyridae, p. 162
*Mordellidae, p. 185
Lycidae, p. 163
Ptilodactylidae, p. 170
Rhipiphoridae, p. 184
Pyrochroidae, p. 183
Eucnemidae, p. 168
Throscidae, p. 168

Antennae threadlike or beadlike

Head nearly or quite concealed by pronotum from dorsal view

†Chrysomelidae, p. 198
†Lampyridae, p. 162
*Anthicidae, p. 181
*Mordellidae, p. 185
Lycidae, p. 163
Elmidae, p. 172
Dascillidae, p. 170
Helodidae, p. 170
Leptodiridae, p. 158
Eucnemidae, p. 168
Throscidae, p. 168
Ptinidae, p. 188

Head not concealed from dorsal view

Mandibles directed anteriorly

†Tenebrionidae, p. 186
†Cantharidae, p. 162
*Alleculidae, p. 187
*Cucujidae, p. 176
Dascillidae, p. 170
Lagriidae, p. 186
Salpingidae, p. 182

Mandibles directed downward or posteriorly

Pronotum keeled laterally

†Chrysomelidae, p. 198
†Cerambycidae, p. 196
†Tenebrionidae, p. 186
†Cantharidae, p. 162
*Melandryidae, p. 188
Lycidae, p. 163
Helodidae, p. 170
Dasytidae, p. 164
Psephenidae, p. 172

Pronotum rounded laterally

†Chrysomelidae, p. 198
†Cerambycidae, p. 196
†Meloidae, p. 184
*Anthicidae, p. 181
Cleridae, p. 165
Euglenidae, p. 181
Pedilidae, p. 183

148

Key to the Principal Families of Coleoptera *(contd.)*

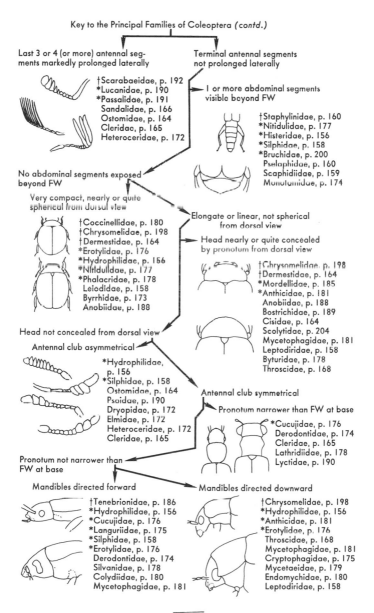

Last 3 or 4 (or more) antennal segments markedly prolonged laterally

†Scarabaeidae, p. 192
*Lucanidae, p. 190
*Passalidae, p. 191
Sandalidae, p. 166
Ostomidae, p. 164
Cleridae, p. 165
Heteroceridae, p. 172

Terminal antennal segments not prolonged laterally

1 or more abdominal segments visible beyond FW

†Staphylinidae, p. 160
*Nitidulidae, p. 177
*Histeridae, p. 156
*Silphidae, p. 158
*Bruchidae, p. 200
Pselaphidae, p. 160
Scaphidiidae, p. 159
Monotomidae, p. 174

No abdominal segments exposed beyond FW

Very compact, nearly or quite spherical from dorsal view

†Coccinellidae, p. 180
†Chrysomelidae, p. 198
†Dermestidae, p. 164
*Erotylidae, p. 176
*Hydrophilidae, p. 156
*Nitidulidae, p. 177
*Phalacridae, p. 178
Leiodidae, p. 158
Byrrhidae, p. 173
Anobiidae, p. 188

Elongate or linear, not spherical from dorsal view

Head nearly or quite concealed by pronotum from dorsal view

†Chrysomelidae, p. 198
†Dermestidae, p. 164
*Mordellidae, p. 185
*Anthicidae, p. 181
Anobiidae, p. 188
Bostrichidae, p. 189
Cisidae, p. 164
Scolytidae, p. 204
Mycetophagidae, p. 181
Leptodiridae, p. 158
Byturidae, p. 178
Throscidae, p. 168

Head not concealed from dorsal view

Antennal club asymmetrical

*Hydrophilidae, p. 156
*Silphidae, p. 158
Ostomidae, p. 164
Psoidae, p. 190
Dryopidae, p. 172
Elmidae, p. 172
Heteroceridae, p. 172
Cleridae, p. 165

Antennal club symmetrical

Pronotum narrower than FW at base

*Cucujidae, p. 176
Derodontidae, p. 174
Cleridae, p. 165
Lathridiidae, p. 178
Lyctidae, p. 190

Pronotum not narrower than FW at base

Mandibles directed forward

†Tenebrionidae, p. 186
*Hydrophilidae, p. 156
*Cucujidae, p. 176
*Languriidae, p. 175
*Silphidae, p. 158
*Erotylidae, p. 176
Derodontidae, p. 174
Silvanidae, p. 178
Colydiidae, p. 180
Mycetophagidae, p. 181

Mandibles directed downward

†Chrysomelidae, p. 198
*Hydrophilidae, p. 156
*Anthicidae, p. 181
*Erotylidae, p. 176
Throscidae, p. 168
Mycetophagidae, p. 181
Cryptophagidae, p. 175
Mycetaeidae, p. 179
Endomychidae, p. 180
Leptodiridae, p. 158

given, the specimen does not appear to belong to any of the families in the group, descriptions and illustrations of related groups should be examined. Should there be doubt at any point in the key as to which way to proceed, try both alternatives. After practice you will recognize the most common groups, and identification of subsequent specimens will be easier. The key is so arranged that some variable families key out in more than one place. In the following text, tarsal segmentation is indicated by a 3-number formula (like 5-5-5), representing the number of segments in the front, middle, and hind tarsi.

Suborder Archostemata

Tarsi 5-5-5. Antennae threadlike or beadlike. This is a primitive group and its members are relatively rare.

RETICULATED BEETLES   Family Cupedidae
   Identification: *Shape distinctive* (resembling leaf-mining leaf beetles, p. 198): body elongate, rather flattened. FW long, parallel-sided, with *rows of square punctures between longitudinal ridges*. Body clothed with broad scales. Antennae *long, threadlike.* 7–20 mm.
   Only 5 species of cupedids occur in N. America, and they are infrequently collected. Adults are found in logs where larvae occur, on vegetation, or flying in the sunlight. Larvae bore in rotting oak, chestnut, and pine logs.

MICROMALTHID BEETLES   Not illus.
Family Micromalthidae
   Identification: Resemble small soldier beetles (p. 162). Antennae short, beadlike, 11-segmented. Pronotum widest anteriorly, head slightly wider than pronotum. FW short, exposing 2 or 3 abdominal segments. 1.5–2.5 mm.
   This family contains only 1 species, *Micromalthus debilis* LeConte, which occurs in the Northeast. This beetle is unique as the only known member of the order in which *paedogenesis* (reproduction by larvae) occurs. Larvae bore in oak that has reached the red-rotten stage of decay, or in rotten, yellowish-brown chestnut. Larvae are common in these situations; adults are very seldom collected, but they can be reared in numbers from infested logs.

Suborder Adephaga

First abdominal segment divided by hind coxae. Notopleural sutures present. Tarsi usually *5-5-5.* Antennae 11-segmented and usually threadlike (rarely beadlike or clubbed). Nearly always predaceous.

Superfamily Rhysodoidea

Antennae beadlike. These beetles are considered to be primitive members of the Adephaga.

WRINKLED BARK BEETLES   Family Rhysodidae
Identification: *Shape distinctive:* elongate-slender, body widest near apex of FW, tapering anteriorly. Antennae *short, beadlike.* Pronotum *grooved.* FW striate. Dorsal surface black or reddish, shiny. 5–8 mm.

Larvae of these beetles bore beneath bark of decaying beech, ash, elm, and pine. Adults are found beneath bark, and often hibernate in groups.

Superfamily Caraboidea

Antennae *threadlike.* Terrestrial, aquatic, or semiaquatic.

TIGER BEETLES   Family Cicindelidae
Identification: Shape distinctive: FW nearly parallel-sided or somewhat wider apically, pronotum *narrower than FW*, head at eyes *as wide as or wider than pronotum.* Antennae inserted above base of mandibles. Legs long, slender. Brownish, black, or green, often patterned, some iridescent and very colorful. 6–40 (mostly 10–20) mm.

Tiger beetles are active, fast-running, fast-flying, and quite a challenge to collect. They are common and occur in bright sunlight in open sandy areas, on sandy beaches, and on open paths or lanes. Their dexterity and strong mandibles make

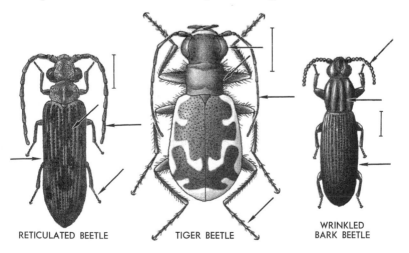

RETICULATED BEETLE   TIGER BEETLE   WRINKLED BARK BEETLE

them well fitted for their predaceous habits. They must be handled carefully when caught, because some can administer a painful bite. The larvae construct vertical tunnels in the ground; they wait at the top of tunnel for passing insects and feed on those they manage to subdue. About ¾ the 130 or so species in N. America belong to the genus *Cicindela*.

GROUND BEETLES   Family Carabidae   **See also Pl. 5**
Identification: Head at eyes nearly always *narrower than pronotum*. Antennae inserted between eyes and base of mandibles. Legs usually long, slender. Generally black and shiny or dark, sometimes brightly colored. 2–35 (mostly 5–15) mm.

This is one of the largest families of beetles, and its members are probably as common and abundant — at least in the East — as any other beetles. Carabids are generally found on the ground beneath objects; some are found on vegetation and flowers. They commonly fly to lights. Most species are nocturnal, and hide during the day. They often run rapidly when disturbed, and seldom fly. Larvae usually occur in the same situation as adults. Adults and larvae are nearly always predaceous and many are very beneficial — they feed on some of our worst pests, including Gypsy Moth larvae, cankerworms, and cutworms. The species of *Calosoma* are called caterpillar hunters; adults and larvae often climb trees or shrubs in search of prey. Many ground beetles give off an unpleasant odor when handled. *Brachinus* species are called bombardier beetles because of their habit of ejecting from the anus a glandular secretion that literally explodes when released, producing a popping sound. This secretion is foul-smelling and irritating, and serves as a means of protection.

ROUND SAND BEETLES   Family Omophronidae
Identification: *Shape and size distinctive:* oval, tapering at each end, very convex, 5–8 mm. Head large. Scutellum *apparently absent*. Brown or black, with light markings. Prosternum enlarged between coxae, concealing mesosternum.

The infrequently collected beetles of this small family occur in burrows in sand or mud along shores of streams and lakes. Throwing water over the banks forces them into the open, where they are easily collected. Larvae occur in similar situations; larvae and adults are predaceous.

CRAWLING WATER BEETLES   Family Haliplidae
Identification: *Shape and size distinctive:* oval, tapering at each end, convex, 2.5–4.5 mm. Hind coxae greatly enlarged and concealing 2 or more abdominal segments. Yellow or brownish, with *black spots*. Head small. Antennae short.

These beetles are fairly common in or around ponds, streams, and lakes, and are usually found creeping or crawling over submerged vegetation. They are not good swimmers.

TROUT-STREAM BEETLES Family Amphizoidae   **Not illus.**
   Identification: Elongate-oval, rather convex dorsally, somewhat flattened ventrally. Dull brownish to dull black. 11–16 mm. Aquatic, occurring in streams in far West.
   Adults and larvae live in the icy waters of swift mountain streams. They cling to driftwood, debris, or to stones in eddies where the water level remains fairly constant. One species lives in relatively warm quiet water in streams near Seattle. Adults

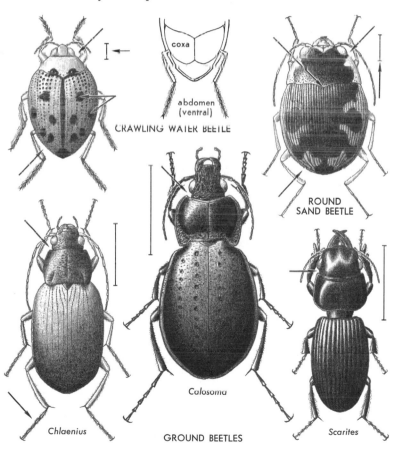

coxa

abdomen
(ventral)

CRAWLING WATER BEETLE

ROUND
SAND BEETLE

Chlaenius

Calosoma

GROUND BEETLES

Scarites

swim very feebly or not at all. Adults and larvae are apparently predaceous.

PREDACEOUS DIVING BEETLES Family Dytiscidae
Identification: Shape often distinctive: elongate-oval, convex, streamlined; hind legs *flattened and fringed* with hairs. Hind tarsi with 1 or 2 claws. Front tibiae lacking spines. Antennae *threadlike*. Scutellum usually visible. Black, brown, or yellowish, often with light markings. 1.4–35.0 mm.

Members of this fairly large group are abundant in ponds, lakes, and streams. They are excellent swimmers, and when swimming move the hind legs in unison; water scavenger beetles (Hydrophilidae, p. 156), with which dytiscids may be confused, move the hind legs alternately when swimming. Predaceous diving beetles frequently fly to lights. Adults and larvae are highly predaceous, and feed on various small aquatic animals, including fish. Larvae (called water tigers) have large sicklelike jaws, and suck the body contents of the prey through channels in the jaws; they do not hesitate to attack an animal larger than themselves.

BURROWING WATER BEETLES Family Noteridae
Identification: Similar to Dytiscidae but scutellum *hidden* and front tibiae *often with a curved spine.* Hind tarsi with 2 claws. Black to reddish brown. 1.2–5.5 mm.

This is a small group whose members are similar in appearance and habits to dytiscids. They are most common in the southeastern states but a few occur in the Northeast. Larvae burrow into the mud near the roots of aquatic plants (hence common name of the group).

Superfamily Gyrinoidea

This group contains a single very distinctive family that differs from nearly all other beetles in having 2 pairs of eyes, and from other Adephaga in having short clubbed antennae.

WHIRLIGIG BEETLES Family Gyrinidae
Identification: Elongate-oval, flattened, 3–15 mm. *2 pairs of compound eyes*, 1 dorsal and 1 ventral. Black, rarely dark metallic green. Front legs *long*, slender, middle and *hind legs very short*, flattened, not fringed with hairs. Antennae *short and clubbed.*

These beetles are often seen swimming in groups in an odd gyrating fashion on the surface of ponds and streams. Their 2 pairs of eyes enable them to watch for enemies and prey both above and below the water surface. They swim rapidly and are as much at home below water as on the surface. Adults and larvae are predaceous. Some adults when handled give off an odor similar to that of pineapples.

Suborder Polyphaga

First abdominal segment not divided by hind coxae, its posterior margin extending completely across abdomen. Prothorax nearly always lacking notopleural sutures. This suborder includes the remaining families of beetles.

Superfamily Lymexylonoidea

Maxillary palps enlarged in ♂. Antennae short, threadlike or serrate. Tarsi 5-5-5. Infrequently collected beetles.

SHIP-TIMBER BEETLES Family Lymexylonidae **Not illus.**
  Identification: Body narrow, elongate, 9.0–13.5 mm. Maxillary palps of ♂ long, flabellate. Antennae short, 11-segmented.
    Two rare species of ship-timber beetles occur in the eastern states, where they are found in decaying wood and under bark. Larvae bore into the heart and sapwood of dead chestnut, poplar, and other trees.

TELEGEUSID BEETLES Family Telegeusidae **Not illus.**
  Identification: Resemble rove beetles (p. 160): body elongate and slender, FW short and exposing much of abdomen; HW extending back over abdomen. Terminal segment of maxillary and labial palps greatly enlarged. 5–6 mm.
    Three very rare species occur in Arizona and California. They are presumed to live under bark. Larvae are unknown.

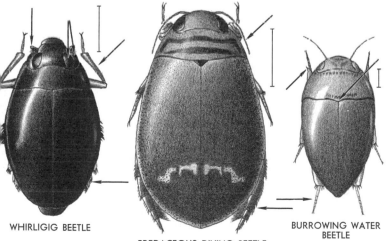

WHIRLIGIG BEETLE

PREDACEOUS DIVING BEETLE

BURROWING WATER BEETLE

Superfamily Hydrophiloidea

Tarsi usually 5-5-5. Mostly aquatic in habit.

HISTER BEETLES   Family Histeridae
  Identification: Antennae *elbowed, clubbed.* FW *short, truncate,*
  exposing 1 or 2 abdominal segments. Body usually oval, sometimes greatly flattened, or elongate and cylindrical. Hard-
  bodied, shiny, black (some with red markings). 1–10 mm.
    These beetles usually occur around decaying organic matter —
  carrion, dung, decaying plants, and oozing sap — and apparently feed on other insects attracted to these materials. Some
  species are very flat and live under loose bark. The elongate
  cylindrical species live in the galleries of wood-boring insects.
  Larvae have much the same habits as adults.

WATER SCAVENGER BEETLES   Family Hydrophilidae
  Identification: Maxillary palps *elongate,* usually longer than
  antennae. Metasternum frequently prolonged posteriorly as a
  sharp spine. Body generally oval or elliptical, convex dorsally.
  Antennae *short, clubbed.* Hind legs *flattened, usually with a
  fringe of hairs.* Black, brown, or yellow, sometimes patterned.
  1–40 mm.
    This is a fairly large group of common insects very similar to
  the Dytiscidae (p. 154). Most species are aquatic, both as
  adults and larvae. Adults are principally scavengers, but larvae
  are predaceous. Larvae feed on a variety of aquatic animals,
  and are very voracious. The members of 1 subfamily (Sphaer-
  idiinae) are terrestrial, and feed in dung, humus, and decaying
  leaves.

MINUTE MOSS BEETLES   Family Limnebiidae   **Not illus.**
  Identification: Similar to Hydrophilidae but with 6 or 7 abdominal segments (5 in Hydrophilidae). Elongate-oval, dark-
  colored. Maxillary palps elongate. 1.2–1.7 mm.
    Members of this small family are found in matted vegetation
  along streams, in decaying moss near the shore, and in swampy
  places. Larvae are predaceous.

SKIFF BEETLES   Family Hydroscaphidae   **Not illus.**
  Identification: Similar to rove beetles (p. 160) in having FW
  short and truncate, exposing 3 abdominal segments, but oval,
  body widest at base of FW, and with notopleural sutures.
  Tarsi 3-3-3. Tan to brown. About 1.5 mm.
    This family contains a single species that occurs in Arizona,
  s. Nevada, and s. California. Adults and larvae occur in streams,
  and are found on filamentous algae growing on rocks, especially
  in shallow water.

Superfamily Staphylinoidea

FW often short, exposing some abdominal segments. Usually small. Most families, except Staphylinidae and Silphidae, are infrequently encountered.

BEAVER PARASITES Family Platypsyllidae **Not illus.**
Identification: Body elongate-oval, flattened. FW short, exposing 4 abdominal segments. Eyes and HW absent. Tarsi 5-5-5. Antennae 11-segmented, not clubbed. About 2.5 mm.
The single species in this family, *Platypsyllus castoris* Ritsema, is an ectoparasite of the American beaver; adults and larvae spend nearly all their life in fur of the host.

HORSESHOE CRAB BEETLES **Not illus.**
Family Limulodidae
Identification: Body finely pubescent and shaped like a horse shoe crab. Antennae clubbed. FW short, exposing 4 abdominal segments. Eyes reduced or absent. 1.7–3.2 mm.
Our 4 species of limulodids are found associated with ants; they feed on body exudates of the ants.

GRASS-ROOT BEETLES Family Brathinidae **Not illus.**
Identification: Shape distinctive: elongate-slender, long-legged, antlike; head as wide as pronotum, narrower posteriorly, necklike. Antennae long, threadlike. Dorsal surface smooth, shiny. Coxae large, contiguous. 3.5–6.0 mm.

WATER SCAVENGER BEETLES

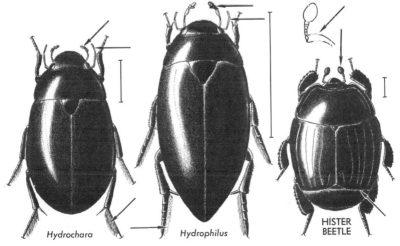

Hydrochara   *Hydrophilus*   HISTER BEETLE

Only 3 species occur in this country, 2 in the East and 1 in California. They are found in roots of grasses growing near water. They are very rare and their habits are not well known.

FEATHER-WINGED BEETLES   Family Ptiliidae   **Not illus.**
Identification: Antennae clubbed, each segment with a whorl of long hair. HW with a fringe of long hair. Oval, often pubescent. 0.25–1.0 mm.

Certain members of this group are not only the smallest beetles known but are among the smallest insects; some species are smaller than many Protozoa. Ptiliids occur in rotten wood, fungus-covered logs, vegetable detritus, and dung. They feed chiefly on fungus spores.

MAMMAL-NEST BEETLES   Family Leptinidae   **Not illus.**
Identification: Brownish, oval, flattened. Eyes reduced or absent. Antennae clubbed. FW with dense golden pubescence. 2.0–2.5 mm.

The few members of this group occurring in the U.S. are found in nests and fur of mice, shrews, moles, and beavers; 1 species has also been found in nests of social Hymenoptera. They probably feed on the eggs and young of mites and other small arthropods.

ROUND FUNGUS BEETLES   Family Leiodidae
Identification: Shape distinctive: oval to nearly spherical, strongly convex, often *capable of rolling into a ball*. Black or brown, shiny. Antennae *clubbed*. 1.5–6.5 mm.

Leiodids occur under the bark of dead trees, in rotten wood, decaying vegetable matter, and rotting fungi. Many species roll into a ball and play dead when disturbed.

SMALL CARRION BEETLES   Family Leptodiridae
Identification: Shape distinctive: elongate-oval, head partly visible dorsally. Antennae clubbed, 8th segment *much smaller than 7th or 9th*. Body with rather dense flattened pubescence. Thorax and FW usually with *cross ridges*. 2–5 mm.

The more common members of this small family feed on carrion. Others are found in fungi, ant nests, or are associated with mammals, and apparently are scavengers.

CARRION BEETLES   Family Silphidae   **See also Pl. 5**
Identification: FW *broad posteriorly*, loosely covering abdomen, or *short, truncate, and exposing 1–3 segments*. Black, often with yellow, orange, or red markings. Body usually soft, flattened. Antennae *clubbed*, last 2 or 3 segments pubescent. 1.5–35.0 (usually over 10.0) mm.

These beetles are commonly found on carrion and decaying

vegetation. They are the largest beetles found in such situations and are often easily recognized by their bright colors. Species of *Silpha* are broadly oval, 10–24 mm., and the elytra (front wings) nearly cover the abdomen. Species of *Nicrophorus* (for 1, see Pl. 5) have short elytra and are called burying or sexton beetles: they burrow under and bury carrion such as dead mice and other small animals. Adults and larvae are scavengers on carrion and decaying vegetation.

SHINING FUNGUS BEETLES Family Scaphidiidae
Identification: Shape distinctive: broadly spindle-shaped, convex. Usually black, shiny, some with red spots. FW *short and*

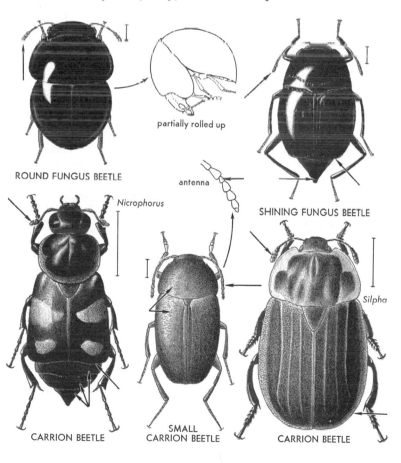

ROUND FUNGUS BEETLE

partially rolled up

Nicrophorus

antenna

SHINING FUNGUS BEETLE

Silpha

CARRION BEETLE

SMALL
CARRION BEETLE

CARRION BEETLE

truncate, exposing *pointed tip of abdomen*. Legs long, slender. Antennae *clubbed*. 2–7 mm.

The members of this small family are found in fungi, dead wood, rotting leaves, and under bark. When disturbed they play dead or run with a characteristic uneven gait. Their habits are not well known.

ROVE BEETLES   Family Staphylinidae

Identification: Form often characteristic: nearly always elongate-slender, parallel-sided. FW *short* and exposing 3 to (often) 5 or 6 abdominal segments. Abdomen flexible, in life bent upward. Antennae threadlike to clubbed. 1–20 (mostly 1–10) mm.

This is a large family, with nearly 3000 species in the U.S., many very common. Rove beetles occur in a variety of habitats: some larger species are found on carrion, others occur on the ground or under objects, along shores of streams and lakes, under bark, in fungi, on flowers, in ant and termite nests, or in decaying vegetable matter. They often run fast, usually with tip of the abdomen bent upward, and are good fliers. The hind wings when not in use are tucked under the short elytra (front wings) with aid of the abdomen. The beginning collector may be rather wary of these beetles because of their habit of holding the abdomen as if they were about to sting. Actually, none can sting but some of the larger species bite readily when handled. Most adults and larvae are predaceous on insects; some feed on decaying organic matter, and a few are parasitic.

SHORT-WINGED MOLD BEETLES   Family Pselaphidae

Identification: Similar to rove beetles, but with abdomen wider than pronotum and head. Chestnut-brown to dark brown. Maxillary palps often with segments enlarged. 0.5–5.5 mm.

This is a fairly large group whose members are found under bark, in or under rotting logs, in moss, on the ground, and in ant nests. They apparently feed on mold.

ANT-LOVING BEETLES   Family Clavigeridae   Not illus.

Identification: Similar to Pselaphidae, but antennae with only 2 or 3 segments. Head and pronotum slender. Eyes present or absent. Tarsi with a single claw. Brownish yellow. 1.8–2.5 mm.

Only 9 species occur in the U.S., and all are found in ant nests. They are fed by the ants, which in turn feed on a substance secreted by the beetles.

FRINGE-WINGED BEETLES   Family Clambidae   Not illus.

Identification: Broadly oval, convex. Capable of partially rolling into a ball. Hind coxae expanded. HW with a fringe of long hairs. Antennae clubbed. Tarsi 4-4-4. About 1 mm.

Most of our 9 species of clambids live in decaying vegetable matter; a few live in ant nests. They are at times fairly numerous, and occasionally fly in numbers at dusk.

ANTLIKE STONE BEETLES Family Scydmaenidae
Identification: Form characteristic: antlike, FW *oval*, head and pronotum *slender*, long-legged. FW shiny, with sparse pubescence. Antennae *slightly clubbed*. Brownish or reddish brown. 1–5 mm.

These beetles occur under bark, in tree holes, under logs, in decaying vegetation, in moss, under stones, and in ant and termite nests. Because of their secretive habits they are seldom observed, but they sometimes fly about in large numbers at twilight. Very little is known about the habits of the larvae.

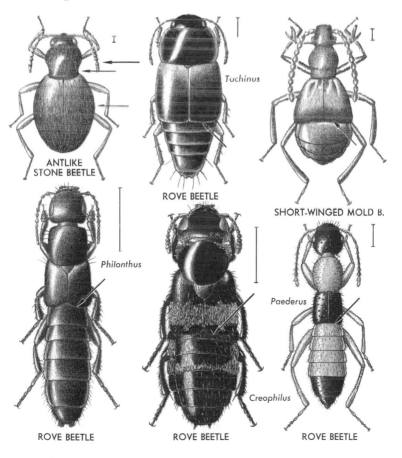

ANTLIKE STONE BEETLE

Tachinus

ROVE BEETLE

SHORT-WINGED MOLD B.

Philonthus

ROVE BEETLE

Creophilus

Paederus

ROVE BEETLE

ROVE BEETLE

FALSE CLOWN BEETLES Family Sphaeritidae **Not illus.**
Identification: Oval, convex. Blackish, with a greenish or bluish luster. FW truncate, exposing last abdominal segment. Antennae clubbed. Similar to hister beetles (p. 156), but FW longer and antennae not elbowed. 4.5–5.5 mm.

The single N. American species occurs along the Pacific Coast (California to Alaska); lives under bark, in decaying fungi, moss, and dung, and is attracted to flowing sap.

MINUTE FUNGUS BEETLES Family Orthoperidae **Not illus.**
Identification: Body rounded or oval. Pronotum usually expanded anteriorly, shelflike, concealing head from above. Black or brownish. Antennae clubbed. HW fringed with hairs. 1 mm. or less.

This is a small group whose members usually occur under decaying bark, in or near rotting fungi, and in decaying vegetable matter. One fairly common species occurs on vegetation and flowers. Adults are predaceous, or feed on fungi.

MINUTE BOG BEETLES Family Sphaeriidae **Not illus.**
Identification: Nearly spherical in shape, convex. Black, sometimes with light markings. Antennae with an abrupt 3-segmented club. Hind coxae large, touching. 0.5–0.75 mm.

This group is represented in the U.S. by 3 species that occur from Texas to California and Washington. These beetles live in mud or under stones near water, in moss, and among the roots of plants in boggy places.

Superfamily Cantharoidea

FW leathery and flexible. Moderate-sized, rather soft-bodied. Tarsi 5-5-5. Abdomen with 7 or 8 segments.

SOLDIER BEETLES Family Cantharidae **See also Pl. 5**
Identification: Body *elongate, parallel-sided* or nearly so. Head *not concealed from above.* Soft-bodied. Black or brown, often with red, yellow, or orange; some predominantly yellow. 1–15 (mostly 5–15) mm.

Soldier beetles are common insects, usually found on flowers or foliage. Many feed on pollen and nectar. Species of *Chauliognathus* are common on goldenrod. Larvae live under bark or on the ground and feed on other insects.

LIGHTNINGBUGS or FIREFLIES **See also Pl. 5**
Family Lampyridae
Identification: Similar to Cantharidae, but head *concealed from above* by pronotum, and last 2 or 3 abdominal segments often luminous. Usually brownish or blackish, frequently with yellow or orange. 5–20 mm.

Lampyrids are unique in that they can flash their lights on

and off; other luminescent insects glow continuously. They are common in spring and early summer and are conspicuous by their blinking lights. Species differ in rhythm of their flashes. Flashing is a recognition signal enabling the sexes to find each other. Some species lack light-producing organs. Larvae live on the ground, under bark, and in moist swampy places. They feed on various invertebrates, including snails.

NET-WINGED BEETLES Family Lycidae **See also Pl. 5**
Identification: FW *reticulate, with longitudinal ridges* and less distinct cross ridges, and often broadest posteriorly. Soft-bodied. Head *concealed from above.* Usually yellow or reddish, with black markings. 5–18 mm.

Members of this small group are fairly common, and live on vegetation, flowers, and foliage of trees and shrubs, usually in wooded areas. Some large species are attractively colored. Adults feed on plant juices or on other insects. Larvae are predaceous and occur under bark.

GLOW-WORMS Family Phengodidae **Not illus.**
Identification: ♂: broad, flat, soft-bodied; antennae plumose; FW short, pointed, HW extending beyond FW and covering abdomen; black or brownish, with red or yellow markings; 10–30 mm. ♀: resemble larvae but with compound eyes.

This is a small group of uncommon beetles. Males occur on foliage or beneath objects, and often fly to lights. Larvae and females are luminescent. Larvae are predaceous and live under bark or beneath objects on the ground.

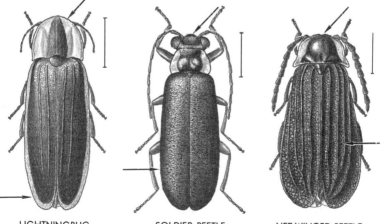

LIGHTNINGBUG   SOLDIER BEETLE   NET-WINGED BEETLE

Superfamily Cleroidea

Size small to moderate. Rather soft-bodied. Tarsi usually 5-5-5.
Abdomen with 5 or 6 segments.

DERMESTID BEETLES Family Dermestidae   **See also Pl. 5**
  Identification: Elongate to broadly oval. Often covered with
  scales or *hair*. Black or brownish, sometimes patterned. Antennae *short, clubbed*, fitting in grooves below sides of pronotum.
  A median ocellus frequently present. 1–12 mm.
    The collector who fails to protect his collection with a repel-
  lant or fumigant will eventually find it infested with these
  beetles. Larvae feed on dried animal or plant materials and can
  completely ruin an insect collection; a pile of powdery material
  below a specimen is evidence of their activity. Dermestids feed
  on a great many things, including cereal products, grains, rugs
  and carpets, various stored foods, upholstery, fur coats, mounted
  birds or mammals, and museum specimens of plants and insects.
  Some species are destructive pests, and most of the damage is
  done by the larvae. Adults of some species occur on flowers,
  where they feed on pollen and nectar. Larvae are subcylindrical,
  brownish, and densely clothed with long hair.

SOFT-WINGED FLOWER BEETLES   **See also Pl. 5**
Families Malachiidae and Dasytidae
  Identification: FW *broadest posteriorly*. Body wedge-shaped,
  soft, often with erect hair. Antennae *serrate* or threadlike, basal
  segments sometimes enlarged. Black, dark blue, or green, often
  with orange, yellow, or red. 1.5–7.0 mm.
    Adults of these 2 families are fairly common on flowers and
  vegetation. Some feed on pollen, others are predaceous. Malachiidae differ from Dasytidae in having protrusible vesicles
  along the sides of the abdomen, and in bearing a lobe beneath
  each tarsal claw.

MINUTE TREE-FUNGUS BEETLES Family Cisidae
  Identification: Elongate, cylindrical, head *concealed from above*.
  Brownish, body with erect pubescence. Tarsi 4-4-4, *1st 3 segments short*. Antennae with a 3-segmented club. Pronotum
  and/or head sometimes with a horn. 1–3 mm.
    Adults and larvae of this small family can be found in numbers
  in shelf fungi, under bark, and in wood. A few species occur in
  galleries of bark beetles (Scolytidae).

BARK-GNAWING BEETLES Family Ostomidae
  Identification: Two body forms: either elongate-narrow, usually
  parallel-sided, head broad, pronotum and FW rather widely
  separated *at sides;* or oval or elliptical, head not broad, pronotum

and FW closely joined, margins of pronotum and FW broadly flattened. Tarsi 5-5-5, *1st segment very short.* Antennae *clubbed,* the club segments often extended laterally. Black, brown, blue, or metallic green. 6–20 mm.

Adults and larvae of this small family normally live under bark, in fungi, and in dry vegetable matter. One species is common in granaries, where larvae and adults feed on other insects or on damaged grain.

CHECKERED BEETLES Family Cleridae   **See also Pl. 5**
Identification: Body *elongate-narrow,* with long erect pubescence.

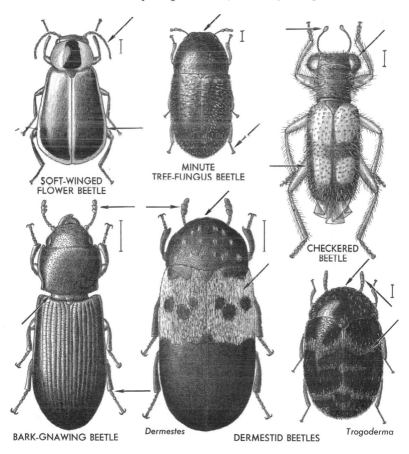

SOFT-WINGED
FLOWER BEETLE

MINUTE
TREE-FUNGUS BEETLE

CHECKERED
BEETLE

BARK-GNAWING BEETLE   *Dermestes*   DERMESTID BEETLES   *Trogoderma*

Pronotum narrower than FW, almost cylindrical. Head usually as wide as or *wider than pronotum*. Often marked with red, orange, yellow, or blue. Antennae variously *clubbed* or thread-like. 3–24 (mostly 3–10) mm.

Clerids are fairly common on trunks of dying or recently killed trees and on flowers and foliage. Most larvae and adults are predaceous on larvae of various wood-boring insects. Some species are beneficial in controlling bark beetles; they track down and devour bark beetle larvae in their burrows. Some adults and a few larvae are pollen feeders. One species, the Red-legged Ham Beetle, *Necrobia rufipes* (De Geer), is destructive to stored meats.

Superfamily Elateroidea

Prosternum often prolonged posteriorly into a lobe that fits into a depression in mesosternum. Abdomen usually 5-segmented. Tarsi 5-5-5. Mostly moderate-sized. Most beetles in this group are plant feeders and some are important pests.

SANDALID BEETLES Family Sandalidae
Identification: Elongate-robust. FW nearly parallel-sided. Pronotum *tapers anteriorly*. Antennae *short, serrate*, pectinate, or flabellate. Mandibles large, prominent. First 4 tarsal segments each with 2 lobes beneath. Brownish to black. 16–24 mm.

The 5 N. American species of sandalids are usually found around elm and ash trees in spring or fall. They are excellent fliers. Larvae are parasites of cicada nymphs.

CEDAR BEETLES Family Rhipiceridae **Not illus.**
Identification: Elongate-oval, black, similar to Sandalidae but smaller (11–15 mm.), tarsal segments not lobed, mandibles small, and antennae serrate.

The single N. American species in this family, *Zenoa picea* (Beauvois), occurs from Florida to Texas, north to Iowa, Ohio, and Pennsylvania, where it is found under logs and bark. This insect is quite rare and little is known of the habits of either adults or larvae.

CLICK BEETLES Family Elateridae
Identification: Shape distinctive: body elongate-narrow, somewhat flattened, usually parallel-sided, and rounded at each end or FW rather pointed at tip; posterior corners of pronotum *prolonged backward into sharp points*. Prosternum with an elongate lobe extending posteriorly into a mesosternal depression (this feature, plus a loose articulation of the prothorax, enables these beetles to "click"). Antennae usually *serrate*, sometimes threadlike or pectinate. Prosternum broadly lobed

anteriorly. Brown or black, sometimes lightly patterned. 3–45 mm.

Click beetles are named for their ability to click and jump. If one is turned onto its back, the head and prothorax are bent backward and then the body is suddenly straightened. This straightening produces an audible click and the beetle is propelled into the air. If it does not land right side up, the performance is continued until it does.

This is a large group, and many species are very common. Adults occur on foliage and flowers, under bark, or in rotting wood. Many apparently do not feed. Larvae (wireworms) are slender, shiny, and hard-bodied; they feed on plant or animal materials, and are found in rotten wood or soil. Some larvae are predaceous but most feed on roots or seeds. A few are very injurious to agricultural crops — they feed on newly planted seeds and the roots of various plants, including vegetables, cereals, and cotton. The Eyed Elater, *Alaus oculatus* (Linn.), is one of the largest (25–45 mm.) and most easily recognized species in this group; it has a salt-and-pepper color, and the pronotum bears 2 eyelike spots. The species of *Pyrophorus*, which occur in the southern states, are dark brown, 12–23 mm., and have 2 light spots at the rear corners of the pronotum that are luminous.

CEROPHYTID BEETLES Family Cerophytidae **Not illus.**
Identification: Elongate-robust. Hind trochanters enlarged, nearly as long as femora. Antennae pectinate in ♂ and serrate

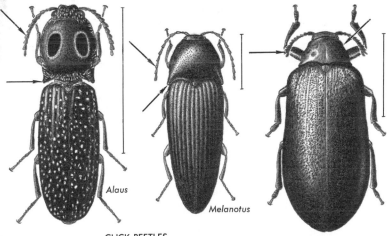

Alaus

Melanotus

CLICK BEETLES

SANDALID BEETLE

in ♀. Hind margin of prosternum with a rounded lobe. Black or brown. 7.5–8.5 mm.

Two very rare species of cerophytids occur in the U.S., one in the East and the other in California; they live under bark and in rotting wood.

CEBRIONID BEETLES Family Cebrionidae

Identification: Body *elongate, broadest at base of FW*. Mandibles *large*, thin, projecting forward in front of head. Abdomen 6-segmented. Brownish, with long, dense, suberect pubescence. 15–25 mm.

Most of the 17 species of cebrionids in the U.S. occur in the Southwest. They are found under bark, on the ground, and beneath objects, and are nocturnal. Males are good fliers, with wasplike flight; females are wingless and live in the ground. Larvae live in or on the ground and feed on roots.

FALSE CLICK BEETLES Family Eucnemidae

Identification: Similar to Elateridae (p. 166), but pronotum *somewhat stouter and more convex above*, anterior margin of prosternum straight across (not lobed), and prothorax more firmly attached to mesothorax, with little capacity for movement. Antennae serrate, pectinate, or flabellate, often fitting into prosternal grooves. Brown or black, sometimes with pale markings. 3–18 mm.

About 70 species of eucnemids occur in N. America, but they are infrequently collected. They are found on foliage, under bark, and in or on rotten wood (especially beech and maple). Adults quiver their antennae continuously (a habit not found in Elateridae). Larvae feed in rotting wood.

PEROTHOPID BEETLES Family Perothopidae **Not illus.**

Identification: Similar to Eucnemidae, but tarsal claws pectinate and prothorax loosely attached to mesothorax. Prosternum narrowly lobed anteriorly. Brownish. 10–18 mm.

Perothopids are rare beetles living on the trunks and branches of old beech trees. The 3 U.S. species occur from Pennsylvania to Florida and in California.

THROSCID BEETLES Family Throscidae

Identification: Similar to Elateridae (p. 166), but body *more compact (broadest near base of FW)*, and prosternal lobe firmly attached to mesothorax and not movable. Antennae serrate or with *a loose 3-segmented club*. Brown or black, sometimes with pale markings. 2–5 mm.

Adults of this small family occur on flowers and vegetation but are not common. Larvae are found in worm-eaten wood, and are probably predaceous.

METALLIC WOOD-BORING BEETLES   See also Pl. 5
Family Buprestidae

Identification: Nearly always metallic or bronzed, especially on ventral surface. Dorsal surface frequently metallic, always shiny, usually lacking pubescence. Hard-bodied; body elongate-slender and parallel-sided to elongate-robust and even strongly oval. Antennae short, serrate or nearly threadlike, rarely pectinate. 3–40 (mostly 5–20) mm.

Buprestid larvae generally bore under bark, in wood, in roots of trees and shrubs, or in leaves, and attack living, dying, or dead plants; these larvae, often called flat-headed borers, are serious pests in orchards and on ornamental plants, and in logs

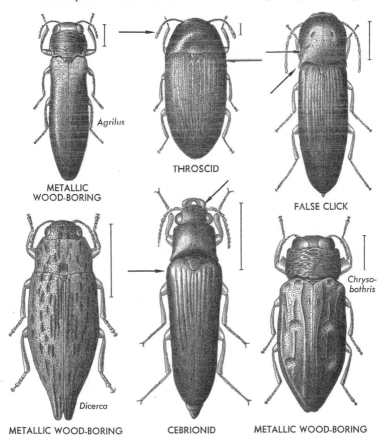

Agrilus

METALLIC
WOOD-BORING

THROSCID

FALSE CLICK

*Chryso-
bothris*

Dicerca

METALLIC WOOD-BORING   CEBRIONID   METALLIC WOOD-BORING

cut for lumber. Many species are stem miners and some produce
galls on certain plants. Adults feed on foliage and bark, or are
attracted to flowers. They are frequently found basking in the
sun on trunks and branches of dying or unhealthy trees. Many
are quite wary and difficult to collect — they may run rapidly
to evade the collector, or take flight if approached too closely.
This group is a large one, and many species are common.
Larvae of *Chrysobothris femorata* (Olivier) attack various trees
and shrubs, and are frequently serious pests in orchards. Adults
of the large genus *Agrilus* are elongate and slender; most are
dark-colored and somewhat metallic; their larvae bore in stems
and roots of trees and shrubs, including raspberries, black-
berries, apple, and pear.

Superfamily Dascilloidea

Usually soft-bodied. Tarsi 5-segmented. Mostly small beetles,
occurring on vegetation in moist areas.

SOFT-BODIED PLANT BEETLES   Family Dascillidae
Identification: Tarsi with 2nd, 3rd, and 4th segments strongly
lobed beneath or hind coxae greatly expanded. Antennae
serrate. FW quite pubescent. 3–9 mm.
   The members of this small group are found on flowers and
foliage in moist shady places, under bark, and in fungi, and are
not common. Larvae live in the ground. Both adults and larvae
are probably predaceous.

PTILODACTYLID BEETLES   Family Ptilodactylidae
Identification: Antennae distinctive: *long, segments 4–10 bearing
a long basal process.* Head nearly or quite concealed from above
by pronotum. Scutellum *heart-shaped*, notched anteriorly. Yellowish brown. 4–6 mm.
   Only 6 species of ptilodactylids occur in N. America. They
are found on vegetation in moist shady places. Larvae live in
decaying logs and leaf mold, or are aquatic.

MARSH BEETLES   Family Helodidae
Identification: Body oval to nearly spherical. Pronotum appears abbreviated, sometimes expanded anteriorly and shelflike,
conceals head. Tarsi with 4th segment lobed beneath. Hind
femora sometimes thickened. Black or brown, sometimes with
orange or red markings. 2–4 mm.
   Marsh beetles are found on vegetation in swampy places or
along lakes and streams. Some live in rotten stumps or tree
holes. Species with enlarged hind femora are active jumpers and
resemble flea beetles; they differ from flea beetles in having the
tarsi distinctly 5-segmented (apparently 4-segmented in flea
beetles). Larvae are aquatic and occur in both stagnant and

flowing water; some live in the water in tree holes or in wet decaying material. The larvae have very long, many-segmented antennae.

Superfamily Dryopoidea

Small (1–8 mm.). Legs frequently long, with strong claws; 5th tarsal segment usually very long. These beetles are aquatic, semi-aquatic, or occur in mud and on foliage beside streams.

MINUTE MUD-LOVING BEETLES   Not illus.
Family Georyssidae
    Identification: Shape distinctive: in dorsal view resembles small snout beetle without the snout, the head concealed from above by pronotum. Antennae clubbed. Body surface roughened, without pubescence. 1.5–3.0 mm.
    This group includes 2 rare species that live in muddy or sandy shores of streams, where they conceal themselves with a coating of mud. One species occurs in Nebraska, the other in Idaho and California.

MINUTE MARSH-LOVING BEETLES   Not illus.
Family Limnichidae
    Identification: Small (1–2 mm.) oval beetles similar to Dryopidae (p. 172), but middle coxae widely separated and hind coxae touching or nearly so. Antennae 10-segmented, short, most segments broader than long. Densely pubescent, golden or gray.
    Adults and larvae of this small family live in the wet sand or

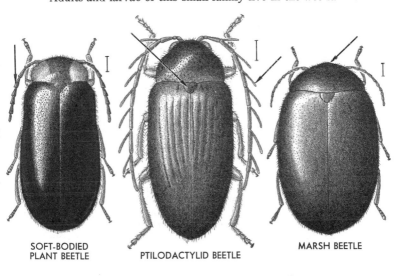

SOFT-BODIED
PLANT BEETLE   PTILODACTYLID BEETLE   MARSH BEETLE

soil along the margins of streams, or are aquatic. Their habits are not well known.

LONG-TOED WATER BEETLES Family Dryopidae
Identification: Body oval or elongate-oval. Legs very long, strongly developed, claws *large*. Antennae *short*, most segments broader than long, usually concealed. Pubescence dense, low-lying, or absent. Front coxae transverse. Black, brown or dull gray. 1–8 mm.

These beetles are usually found on partly submerged sticks and stones in moving water or riffle areas of streams. They hang on firmly with their stout claws or creep slowly over the surface but do not swim. If a stick or stone is removed from the water and placed in the sun to dry, these beetles will be seen to move and can be captured — they are very difficult to see unless they move. Some adults are terrestrial and plant-feeding. Larvae are aquatic.

WATER-PENNY BEETLES Family Psephenidae
Identification: *Oval*, flattened, black or brownish. FW broadest posteriorly, loosely covering abdomen. Dorsal surface sparsely pubescent, ventral surface densely pubescent. Abdomen with 5–7 ventral segments. 4–6 mm.

Water-penny beetles occur on vegetation along streams and on partly submerged rocks in riffles. Larvae are found on the underside of stones in rapidly flowing water; they are brownish, greatly flattened, and nearly circular, hence the common name of the group. Adults and larvae are plant feeders.

RIFFLE BEETLES Family Elmidae
Identification: Form distinctive: body *oval* to cylindrical, legs long, strongly developed, claws *large*. Black to grayish. Antennae short or moderate in length, clubbed or threadlike. 1–8 mm.

Most riffle beetles occur on stones, logs, and other debris in flowing water; a few are terrestrial. The aquatic species spend most of their lives under water. Larvae are aquatic; some are long and slender, but others are flat and oval and resemble larvae of water-penny beetles.

VARIEGATED MUD-LOVING BEETLES
Family Heteroceridae
Identification: Shape and color distinctive: body elongate-robust; black or brownish, FW often with *undulating yellowish bands or spots;* densely pubescent. Front and middle tibiae flattened, *outer margin spiny.* Tarsi *4-4-4.* Mandibles of ♂ *extended forward*, flattened. Antennae short, serrate. 4.0–6.5 mm.

Members of this small family live in galleries in muddy or sandy shores of streams and lakes. They may be driven from their burrows by flooding the shore with water. They often fly to lights at night. Larvae are found in the same habitat as adults, and are very active.

Superfamily Byrrhoidea

Size small to moderate (1.0–9.5 mm.). Oval, convex, or more or less pill-shaped. Uncommon to rare beetles.

PILL BEETLES Family Byrrhidae
Identification: *Shape distinctive:* body oval, strongly convex, pill-shaped, the head deflexed and nearly or quite concealed from above. Tibiae usually flattened, often with spines. Tarsi 4-4-4 or 5-5-5. 1.2–10.0 mm.

Pill beetles are found beneath logs and objects on the ground,

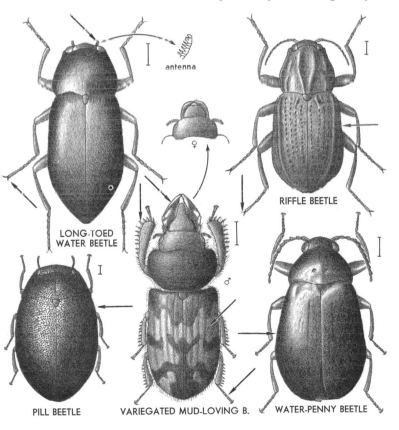

antenna

♀

♂

LONG-TOED
WATER BEETLE

RIFFLE BEETLE

PILL BEETLE   VARIEGATED MUD-LOVING B.   WATER-PENNY BEETLE

often in sandy places or associated with mosses. Larvae live in moist soil and sand, in moss, and under objects. Adults and larvae are plant feeders.

WOUNDED-TREE BEETLES
Not illus.

Family Nosodendridae

Identification: Similar to Byrrhidae (p. 173), but head visible dorsally and FW with rows of short hair tufts. Body oval, convex, black. Antennae with a 3-segmented club. Tibiae broad, with spines. Tarsi 5-5-5. 5–6 mm.

Only 2 species of nosodendrids occur in the U.S., 1 from the Northeast west to Kansas, the other in California; both are quite rare. They are found under bark or in the wounds of trees, where they probably feed on fly larvae. Larvae live in hollow stumps and tree holes.

CHELONARIID BEETLES   Family Chelonariidae   Not illus.

Identification: Elongate-oval. Head withdrawn into prothorax and concealed from above; strongly contractile. FW black, with patches of dense white pubescence. Pronotum reddish. Antennae threadlike. Tarsi 5-5-5. 5–7 mm.

This family contains a single rare species, *Chelonarium lecontei* Thomson, which occurs in N. Carolina and Florida. Adults are found on foliage, and larvae are aquatic.

Superfamily Cucujoidea

This superfamily serves as a sort of dumping ground for families that cannot be placed elsewhere. The included families have little in common, in either morphology or habits.

TOOTH-NECKED FUNGUS BEETLES   Family Derodontidae

Identification: Form of pronotum often distinctive: *lateral margins strongly toothed* or broadly flattened and bent upward. Brownish, often mottled. Head with a pair of ocelli. Antennae with *an elongate 3-segmented club*. FW with rows of punctures. Tarsi 5-5-5 or 4-4-4. 3–6 mm.

The 6 N. American species of derodontids are relatively rare. They occur in shelf fungi, under the bark of willow and tulip trees, and in slime molds. When located, they are sometimes found in large numbers.

SMALL FLATTENED BARK BEETLES
Not illus.

Family Monotomidae

Identification: Form distinctive: elongate, flattened, slender, parallel-sided. FW truncate, exposing tip of abdomen. Antennae with a club of 1 or 2 segments. Front coxae globular. 1.5–3.0 mm.

Members of this group usually occur under bark. They are generally very infrequently collected, but may sometimes be taken in large numbers in molasses traps set in dense woods. Very little is known of the habits of either adults or larvae.

SILKEN FUNGUS BEETLES Family Cryptophagidae
Identification: Usually light yellowish brown, sometimes brown or black, the body nearly always with fine silky pubescence. FW rounded, *broadest near middle.* Pronotum narrower, rounded laterally and broadest near middle; sides of pronotum often toothed or notched. Antennal club *3-segmented.* Tarsi usually 5-5-5, 5-5-4 in some males. 1–5 mm.

Silken fungus beetles are found on flowers and foliage, in fungi and decaying vegetable matter. Some species live in nests of wasps or bumble bees. The group is fair sized, with more than 160 N. American species.

LIZARD BEETLES Family Languriidae   See also Pl. 6
Identification: Shape distinctive: *very elongate-slender, parallel-sided,* the FW, pronotum, and head nearly equal in width. Shining black or blue-black, the *pronotum* and sometimes also the head and FW *reddish, orange, or yellow.* Antennal club *4- to 6-segmented.* Tarsi 5-5-5, appearing 4-4-4. 5–10 mm.

Lizard beetles are common on flowers, leaves, and stems of various plants. Larvae are stem borers, and some are of economic importance. The Clover Stem Borer (Pl. 6), *Languria mozardi* Latreille, often causes considerable damage to clover.

ROOT-EATING BEETLES Family Rhizophagidae   **Not illus.**
Identification: Similar to Monotomidae (p. 174), but with front coxae transverse. FW short, truncate, exposing tip of abdomen. 2–5 mm.

Adults and larvae of these beetles are usually found in rotten,

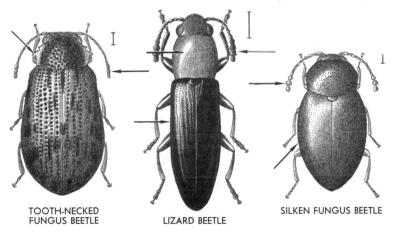

TOOTH-NECKED
FUNGUS BEETLE

LIZARD BEETLE

SILKEN FUNGUS BEETLE

fungus-infested wood. They are predaceous on wood-boring insects and live in the tunnels of these insects. A few species live in ant nests.

DRY-FUNGUS BEETLES Family Sphindidae   **Not illus.**
Identification: Broadly cylindrical, pronotum broad and distinctly convex, head not or only barely visible dorsally. Brownish to black; pubescence sparse, short, suberect. Antennal club 3-segmented. Tarsi 5-5-4. 1.5–3.0 mm.

Sphindids are found in dry fungi and in debris from decaying logs and stumps. They are often difficult to see because of their small size and protective coloration. The 6 U.S. species are relatively rare.

MURMIDIID BEETLES Family Murmidiidae   **Not illus.**
Identification: Small, oval, somewhat flattened. Antennae clubbed, club usually received in cavity of prothorax. Coxae widely separated. Legs retractile. Tarsi 4-4-4. 1st abdominal segment considerably longer than others. 1–2 mm.

The 5 N. American species are relatively rare, and little is known of their habits; they are probably predaceous.

MONOMMID BEETLES Family Monommidae   **Not illus.**
Identification: Shape distinctive: elongate-oval, convex dorsally, flat ventrally. Appendages retractile. Black to brownish. Antennal club 2- or 3-segmented, fitting into grooves on underside of prothorax. Tarsi 5-5-4. 5–12 mm.

Monommids occur on foliage, under rotten wood, and in debris. Our 5 species are in the South, from Florida to s. California. They are not common. Larvae live under rotten wood or bore in the roots of agave.

PLEASING FUNGUS BEETLES   **See also Pl. 6**
Family Erotylidae
Identification: *Elongate to broadly oval.* Black, shiny, lacking pubescence, often marked with red, orange, or yellow. Antennal club *3-segmented.* Tarsi 5-5-5, 4th segment often small. 3–20 mm.

Erotylids are usually found on fungi or in rotten wood; some are fairly common. Adults hibernate under bark, often in groups. Larvae occur in fleshy fungi or in decaying wood; some feed in fungus-infested stored products. Some erotylids are attractively patterned with red or orange and black.

FLAT BARK BEETLES Family Cucujidae   **See also Pl. 6**
Identification: Body often greatly flattened; *elongate, usually narrow, parallel-sided.* Brown, black, or reddish. Antennae threadlike, sometimes clubbed. Tarsi 5-5-5, sometimes appearing 5-5-4. 2.0–13.5 mm.

Most cucujids can be recognized by their greatly flattened body; a few are not so flattened. They are common under loose bark and in dry or decaying plant material, and some are pests of stored grain. Most are predaceous on mites or small insects. A few feed on stored products.

SAP BEETLES Family Nitidulidae   See also Pl. 6

Identification: Antennal club *abrupt, 3-segmented*. Shape variable, usually elongate, robust, sometimes broadly oval; rarely long and slender with short FW. Abdomen often exposed beyond FW. Black or brown, often marked with red or yellow. Tarsi 5-5-5 or 4-4-4. 1.5–12.0 mm.

Many nitidulids are common on decaying fruits, fermenting plant juices, and in fungi; some occur on flowers, some are found in nests of bees and ants, a few breed in carrion, and a few are

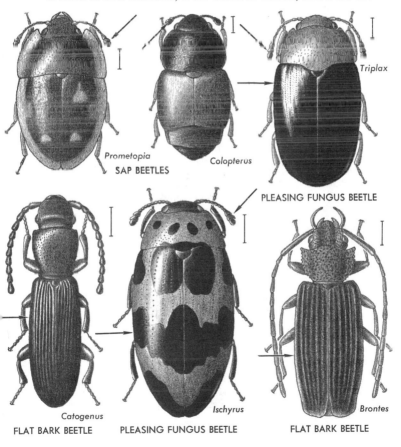

Prometopia   Colopterus   Triplax

SAP BEETLES

PLEASING FUNGUS BEETLE

Catogenus   Ischyrus   Brontes

FLAT BARK BEETLE   PLEASING FUNGUS BEETLE   FLAT BARK BEETLE predaceous on scolytids or scale insects (pp. 204, 137). Species of *Conotelus* (found in flowers of morning glory) resemble rove beetles but have clubbed antennae. Larval habits are similar to those of adults.

FLAT GRAIN BEETLES Family Silvanidae
Identification: *Elongate, rather flattened.* Antennal club *3-segmented*, 3rd segment shorter than 2nd. Tarsi 5-5-5, never lobed. 1–5 mm.
Some silvanids occur under bark, others in stored plant materials (fruit, grain, meal, and flour). The most important grain-infesting species is the Saw-toothed Grain Beetle, *Oryzaephilus surinamensis* (Linn.), named for the teeth along margins of the pronotum.

SHINING FLOWER BEETLES Family Phalacridae
Identification: *Broadly oval to nearly spherical,* dorsal surface very convex. Black or brownish, shining. Antennal club *3-segmented.* Tarsi *5-5-5,* 4th segment small. 1–3 mm.
These beetles are common on the flowers and foliage of various plants, especially goldenrod and other composites. Larvae feed in the heads of these flowers.

MONOEDID BEETLES Family Monoedidae   **Not illus.**
Identification: Reddish yellow, antennae and scutellum black, and 5 long black marks on each FW. FW parallel-sided, a little wider than pronotum. Tarsi 4-4-4, 1st segment broad, flat, oval, 2nd and 3rd minute, 4th large and long with stout claws. Antennal club 1-segmented. 2 mm.
A single rare species is found in s. Florida, where it occurs on a species of milkweed that grows near wet hammocks along the coast.

MINUTE BROWN SCAVENGER BEETLES
Family Lathridiidae
Identification: Shape distinctive: FW *oval, widest at middle,* pronotum and head progressively narrower, pronotum often nearly circular from above. FW with distinct rows of punctures. Antennal club 2- or *3-segmented,* loose. Brown or black. Tarsi *slender, 3-3-3,* 2-3-3, or 2-2-3. *1–3 mm.*
These beetles are fairly common and are usually associated with moldy material. Both adults and larvae occur in rotting vegetation, woodpiles, mammal nests, and sometimes on flowers; they are also found in warehouses, but are thought to feed on fungus and mold and not on stored products.

FRUITWORM BEETLES Family Byturidae
Identification: *Elongate-robust, FW parallel-sided.* Light brown to dark orange, with dense yellowish or grayish hairs. Antennal club *3-segmented.* Tarsi *5-5-5.* 3.5–8.0 mm.
Adults of the eastern species are common on flowers and

foliage. Larvae feed on fruits of raspberry, blackberry, and related plants. Larvae of *Byturus unicolor* Say, the Raspberry Fruitworm, often seriously damage the fruit of raspberry. One western species has been reared from oak galls; adults of this species occur on oak foliage There are 5 species of byturids in the U.S.

MYCETAEID FUNGUS BEETLES Family Mycetaeidae
Identification: Shape distinctive: FW *oval, widest at middle,* margins of pronotum S-shaped and *extending forward, partly enclosing head.* Tarsi *4-4-4.* Black or brown, often with red or orange markings. Shiny, not pubescent. Pronotum often with 2 grooves at base. Antennal club *3-segmented.* 1–4 mm.

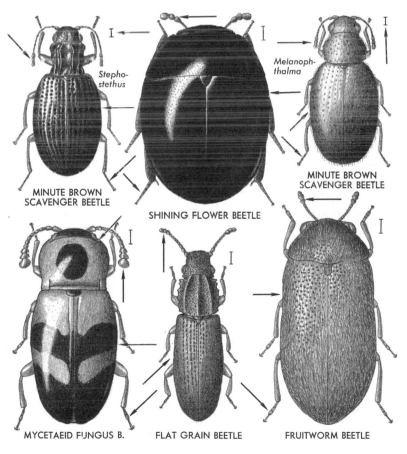

Stephostethus

Melanophthalma

MINUTE BROWN SCAVENGER BEETLE

MINUTE BROWN SCAVENGER BEETLE

SHINING FLOWER BEETLE

MYCETAEID FUNGUS B.

FLAT GRAIN BEETLE

FRUITWORM BEETLE

Members of this small family are found in decaying materials and on flowers, and are not common. One species occurs in granaries, where it feeds on fungus or mold, and may become a pest by spreading mold through the grain.

HANDSOME FUNGUS BEETLES   See also Pl. 6
Family Endomychidae
Identification: Similar to Mycetaeidae (p. 179), but 3rd tarsal segment very small, the tarsi *appearing 3-segmented*. Forward angles of pronotum *prolonged*, partly enclosing head. Shining black or brown, often with red or orange markings. Antennal club *3-segmented*. 3.5–10.0 mm.
Members of this usually uncommon group occur in fungi, under bark, beneath logs, and in rotting wood. They are smooth, shiny, attractively colored, and resemble ladybird beetles. When disturbed they often draw in their appendages and play dead. Larvae occur in the same situations as adults and feed on fungi or rotten wood.

LADYBIRD BEETLES Family Coccinellidae   See also Pl. 6
Identification: Shape often *distinctive:* broadly oval *to nearly spherical*, strongly convex dorsally, nearly flat ventrally. Tarsi *apparently 3-3-3, actually 4-4-4* (3rd segment minute). Head partly or completely concealed by pronotum. Often brightly colored — yellow, orange, or reddish with black markings or black with yellow to reddish markings. Antennae *short, club 3- to 6-segmented*. 0.8–10.0 mm.
This is a large group with many abundant and well-known species. Both adults and larvae of most species are predaceous on aphids, scale insects, mites, and other injurious forms, and are often quite numerous where these pests occur. Some species have been used commercially to combat scale insects injurious in orchards. Adults frequently overwinter in groups, sometimes in tremendous numbers. Adults of the Two-spotted Ladybird Beetle (Pl. 6), *Adalia bipunctata* (Linn.), often overwinter indoors, and may be seen at windows in fall or spring. Two species of *Epilachna* are plant feeders, both as larvae and adults, and are serious garden pests: the Mexican Bean Beetle, *E. varivestis* Mulsant, is yellowish, with 8 small dark spots on each elytron (front wing); the Squash Beetle, *E. borealis* (Fabricius), is yellowish, with 7 large dark spots on each elytron.

CYLINDRICAL BARK BEETLES   Not illus.
Family Colydiidae
Identification: Usually slender and elongate to very elongate, narrow, parallel-sided, rarely oval and flattened. Antennal club 2- or 3-segmented. Black or brown. Pronotum often with ridges or grooves. Tarsi usually 4-4-4, rarely 3-3-3. 1–18 (mostly 1–8) mm.

Many colydiids are predaceous on wood-boring insects and are found under bark and in infested wood. Some are plant feeders and live in shelf fungi, vegetable detritus, or in ant nests. Larvae of a few species are ectoparasites of wood-boring larvae; such parasitic habits are rare among beetles. Few of the 102 N. American species are common.

ANTLIKE FLOWER BEETLES   Family Anthicidae
Identification: Shape distinctive: body antlike, FW elongate-oval, pronotum *distinctly narrower*, often rounded, head oval. Pronotum often with a hornlike process extending forward over head. Antennae threadlike, beadlike, or clubbed. Abdomen 5-segmented. Usually black or brown, often with red or yellow. Tarsi 5-5-4. 2–12 mm.

These beetles are common on flowers and foliage as well as on the ground. Many species possess a distinctive hornlike process on the pronotum. Larvae feed chiefly on vegetable detritus; at least 1 species is predaceous. This group is of moderate size, with 188 N. American species.

ANTLIKE LEAF BEETLES   Family Euglenidae   Not illus.
Identification: Similar to Anthicidae, but abdomen 4-segmented and eyes notched. Black to reddish yellow. Hind femora or antennae often modified. 1.5–3.0 mm.

These beetles are principally eastern in distribution and are found on foliage or flowers. They are not common and their habits are poorly known.

HAIRY FUNGUS BEETLES   Not illus.
Family Mycetophagidae
Identification: Elongate-oblong, hairy or pubescent. Black or

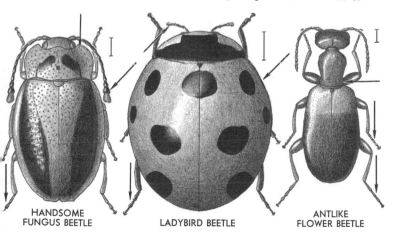

HANDSOME
FUNGUS BEETLE   LADYBIRD BEETLE   ANTLIKE
FLOWER BEETLE

brownish, FW often with orange or reddish. Antennal club 2- to 5-segmented. Tarsi usually 3-4-4 or 4-4-4. 1.5–6.0 mm.

These feed on fungi and are found under fungus-covered bark, in shelf fungi, and in moldy vegetable materials. A few feed on pine pollen.

FALSE LONGHORN BEETLES   Not illus.
Family Cephaloidae
Identification: Elongate, slender, somewhat cylindrical. Similar to Cerambycidae (p. 196), but with tarsi 5-5-4. Tarsal claws pectinate, with a fleshy lobe beneath each claw. Yellowish to brown or black. 8–20 mm.

This is a small, relatively rare group (10 N. American species). The habits are not well known, but adults are usually found on flowers and vegetation in wooded areas.

FALSE BLISTER BEETLES   Family Oedemeridae
Identification: Slender, elongate, soft-bodied, similar to Meloidae (p. 184) or Cerambycidae (p. 196). Tarsi *5-5-4*, next to last tarsal segment wide and densely hairy beneath. Pronotum *broadened anteriorly*, much narrower at base than FW; rounded laterally. Often with yellow, red, or orange. 5–20 mm.

Adults of this group occur on flowers and foliage, and fly to lights at night. They are not common. Adults feed on pollen and nectar, and larvae feed in moist decaying wood, especially conifers and driftwood.

FALSE TIGER BEETLES   Family Othniidae   Not illus.
Identification: Somewhat similar to tiger beetles (p. 151), but antennae with a 3-segmented club and tarsi 5-5-4. Brownish, FW often mottled. 5–9 mm.

Othniids occur on foliage, under bark, on cacti, and in rotting leaves. They are probably predaceous. One species has been reported from Virginia and 4 others from the western states; all are rare.

AEGIALITID BEETLES   Family Aegialitidae   Not illus.
Identification: Subcylindrical, long-legged, black. Tarsi 5-5-4, last tarsal segment longer than others combined. FW meet imperfectly, exposing tip of abdomen. All coxae widely separated. 3–4 mm.

These insects spend their entire life in rock cracks along the seacoast, below the high tidemark. The 3 N. American species occur along the Pacific Coast from California to Alaska and are very rare.

NARROW-WAISTED BARK BEETLES   Not illus.
Family Salpingidae
Identification: Elongate to oval. Pronotum usually narrowed basally (hence the common name) and rounded or faintly mar-

gined laterally. Brownish to dull black. Tarsi 5-5-4. Antennae serrate or slightly clubbed. 2–15 mm.

Adults of this group occur on vegetation, in detritus, under the bark of conifers, and under stones. Some species have the head prolonged anteriorly into a snout. Adults and larvae are predaceous.

PEDILID BEETLES   Family Pedilidae
Identification: Similar to Anthicidae (p. 181), but body more elongate, head *narrowed behind eyes into a neck.* Pronotum oval or rounded. Black or brown, often with pale or red markings. Antennae long, threadlike or serrate. 4.5–15.0 mm.

Adult pedilids usually occur on flowers and foliage, and are attracted to lights at night. They are uncommon. Larvae live in decaying vegetable material, sometimes on the shores of streams, ponds, lakes, or the seashore. Most of our 57 species are western.

FIRE-COLORED BEETLES   See also Pl. 6
Family Pyrochroidae
Identification: Shape and color distinctive: elongate, narrow, elytra (FW) parallel-sided or slightly wider posteriorly, pronotum oval or somewhat square and much narrower than FW, head broad, as wide as pronotum; usually black, pronotum and sometimes head reddish or yellowish, or entire body reddish or yellowish; antennae serrate or *pectinate*, in some males nearly plumose. Tarsi 5-5-4. 6–20 mm.

Adults of this small group are found on flowers and vegetation, are not common. Larvae occur under the bark of logs and stumps, and are predaceous.

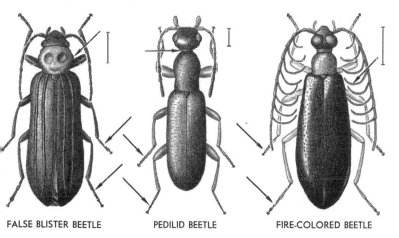

FALSE BLISTER BEETLE   PEDILID BEETLE   FIRE-COLORED BEETLE

HEMIPEPLID BEETLES Family Hemipeplidae **Not illus.**
Identification: Very elongate, slender, parallel-sided, distinctly flattened. Yellow to yellowish brown. 8–12 mm. Similar to Cucujidae (p. 176) but 3rd tarsal segment lobed and front coxal cavities closed behind.

This family is represented in the U.S. by 2 species, found in Florida, Georgia, and California. They are not common.

Superfamily Melooidea

Tarsi 5-5-4. Usually moderate-sized and soft-bodied.

BLISTER BEETLES Family Meloidae **See also Pl. 6**
Identification: Shape distinctive: usually elongate, slender (a few are oval or round), pronotum *narrower than FW*, head *broad*, *usually wider than pronotum*. Body soft, often leathery. FW loosely covering abdomen, rarely shortened. Antennae thread-like or beadlike, intermediate segments sometimes modified. Legs long, slender. Black or brown, sometimes brightly colored, often with light pubescence. 3–20 (usually 10–15) mm.

Blister beetles are common insects occurring on the flowers and foliage of various plants. The name "blister" beetles is based on the fact that the body contains cantharidin, a substance capable of blistering the skin. This chemical is extracted from the body of certain species and used medicinally. Adult blister beetles are plant feeders, and some are serious pests of potatoes, tomatoes, beets, clover, and other plants. They may completely defoliate a plant. Larvae are parasitic and generally beneficial; they usually feed on grasshopper eggs, but some feed on eggs or larvae of bees. Larvae that parasitize bees climb onto flowers and attach themselves to bees visiting the flowers. The bees then carry these larvae to their nest, where the larvae attack the bee eggs. Meloid larvae undergo hypermetamorphosis (in which the various larval instars are quite different in form): the 1st instar is long-legged and active, whereas following instars are grublike or maggotlike. Members of the genus *Meloe*, which are rather large and black or bluish, have very short, overlapping front wings (elytra); they are called oil beetles because they exude an oily substance from the joints of the legs when disturbed; this substance can raise blisters on one's skin.

WEDGE-SHAPED BEETLES Family Rhipiphoridae
Identification: Elongate, humpbacked, wedge-shaped, similar to Mordellidae but abdomen *blunt*, not pointed. Antennae *pectinate* or flabellate in ♂, serrate in ♀. FW entire or *short*, pointed. Usually black and orange. 4–15 mm.

Adults generally occur on flowers but are not common. Larvae are parasitic on wasps, bees, and cockroaches; they undergo hypermetamorphosis (see p. 41). Some females are larviform.

Superfamily Mordelloidea

Unique in shape, and in behavior when captured.

TUMBLING FLOWER BEETLES Family Mordellidae
Identification: Humpbacked, wedge-shaped. Head *bent down, situated ventrally.* Abdomen *pointed* and extending beyond FW. Usually blackish or gray, sometimes with light markings; generally pubescent. Antennae short, threadlike, serrate, or clubbed. Tarsi *5-5-4.* 1.5–15.0 (usually 3–7) mm.

Mordellids are common on flowers and foliage, and when captured tumble about in a comical fashion. They are often difficult to catch, since they run rapidly or take flight when

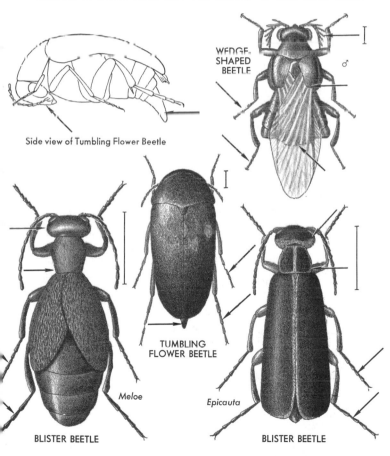

Side view of Tumbling Flower Beetle

WEDGE-SHAPED BEETLE ♂

TUMBLING FLOWER BEETLE

BLISTER BEETLE   *Meloe*

Epicauta   BLISTER BEETLE

alarmed. Larvae occur in rotten wood and plant pith; some are leaf and stem miners, and a few are predaceous.

Superfamily Tenebrionoidea

Tarsi 5-5-4. Front coxal cavities closed behind (except in Melandryidae). Mostly moderate in size.

DARKLING BEETLES Family Tenebrionidae

Identification: Antennae usually 11-segmented, *threadlike, beadlike, or slightly clubbed*. Eyes nearly always notched. Tarsal claws simple. Generally dull black or brown, sometimes with red. 2–35 mm. A varied group usually recognizable by *5-5-4* tarsi, form of eyes, and antennae.

This is a large group whose members are common in a variety of habitats — under bark, in rotten wood, under logs, in fungi, on the ground in desert areas, and in termite and ant nests. Some are pests of stored products. Both adults and larvae are scavengers, feeding on decaying vegetation, fungi, seeds, and other types of organic materials; a few attack living plants.

Darkling beetles are variable in body form and often resemble beetles in other families. Members of the genus *Diaperis*, which inhabit fungi, are similar in appearance and coloration to ladybird beetles. Many species rather closely resemble ground beetles, though they are usually not as shiny. Some fungus-inhabiting tenebrionids have the dorsal surface hard and warty. A species of this type is the Forked Fungus Beetle, *Bolitotherus cornutus* (Panzer), which is 10–12 mm. and common in woody bracket fungi; the pronotum of the male bears 2 hornlike protuberances.

A few are destructive to stored grain and flour. Members of the genus *Tenebrio* (black or dark brown and 13–17 mm.) are pests of stored grain; their larvae are called mealworms. Beetles in the genus *Tribolium* are brown and about 5 mm.; they are pests of flour and other stored products and are known as flour beetles.

Members of the genus *Eleodes* run about with the abdomen raised at an angle of about 45 degrees. They emit a foul-smelling black fluid when disturbed. About 100 species of *Eleodes* occur in the western states.

Darkling beetles are widely distributed but are most abundant in the western states; of the approximately 1400 N. American species, only about 150 occur in the East.

LONG-JOINTED BARK BEETLES Family Lagriidae

Identification: Shape and antennae distinctive: *elongate, narrow*, FW widest apically, pronotum *narrower than FW*, head about as wide as pronotum; antennae threadlike, *last segment elongate*. Dark-colored, often slightly metallic. 6–15 mm.

These beetles occur on flowers, foliage, or under bark, and are generally uncommon. Larvae feed in decaying vegetation, in rotten wood, or under bark.

COMB-CLAWED BEETLES Family Alleculidae

Identification: Tarsal claws *pectinate*. Elongate-robust to elongate-narrow. Brownish to black, with silky pubescence. Antennae usually threadlike or serrate, rarely pectinate. 5–15 mm.

These beetles are common on flowers, under bark, and on vegetation, and probably feed on pollen. Larvae live under

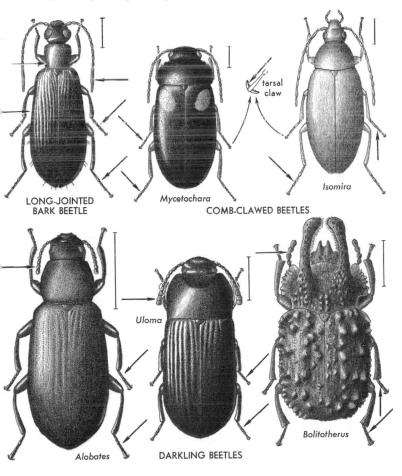

LONG-JOINTED BARK BEETLE

Mycetochara

tarsal claw

Isomira

COMB-CLAWED BEETLES

Uloma

Alobates

DARKLING BEETLES

Bolitotherus

bark, in rotting wood, in fungi, or in vegetable detritus and are similar in appearance to wireworms.

FALSE DARKLING BEETLES   Family Melandryidae
Identification: *Elongate-oval.* Nearly always dark-colored. Pronotum usually with *2 dents at base.* Front coxal cavities open behind. 1st segment of hind tarsi *longer than any other segment.* Antennae generally threadlike. 3–20 mm.

Adults and larvae are found under bark, in dry wood, or in dry fungi; some adults are found on flowers and foliage. Larvae are carnivorous or plant-feeding. The most common eastern melandryids (genus *Penthe*) are elongate-oval, black with black pubescence, and 10–14 mm.

Superfamily Bostrichoidea

Tarsi *5-5-5.* Pronotum usually extending over and partly or *completely concealing head* from above. Generally small.

SPIDER BEETLES   Family Ptinidae
Identification: Shape distinctive: spiderlike, FW elongate-oval, pronotum *narrower than FW* and nearly or completely concealing head from above, legs long and slender. FW usually dull and with erect pubescence, sometimes shiny and not pubescent. Reddish brown to black. Antennae *long, threadlike.* 1–5 mm.

Adults and larvae of this small family are found in various dried animal and plant materials, such as animal carcasses, animal droppings, dry wood, stored products, and in museum specimens of plants and animals. Some species live in ant nests. Spider beetles are not common.

DEATH-WATCH BEETLES   Family Anobiidae
Identification: Pronotum *hoodlike, usually enclosing head* and concealing it from above. Antennae nearly always with last 3 segments *lengthened and expanded,* or simply lengthened, sometimes serrate or *pectinate,* rarely threadlike. Shape variable, usually elongate and cylindrical, sometimes oval to nearly spherical. Appendages often contractile. Hind coxae grooved for reception of femora. Light brown to black. 1.1–9.0 mm.

Nearly all anobiids feed exclusively on plant materials as larvae and adults. Many bore into seasoned wood; some are found indoors after having emerged from furniture, woodwork, flooring, or timbers. Some species produce a ticking sound in their burrows. Superstitious people, thinking this a portent of death, have called these insects "death-watch" beetles. Some live under bark, others in fungi, and some in seeds and galls. Two species, the Drugstore Beetle, *Stegobium paniceum* (Linn.), and the Tobacco Beetle, *Lasioderma serricorne* (Fabricius), are

serious pests of stored products. Both feed on a wide variety of
materials of plant and animal origin, and are among the very
few anobiids that feed on materials of animal origin.

Most anobiids are infrequently collected. The greatest
number and variety, at least in the East, can be taken by
sweeping or beating foliage in wooded areas where the over-
head canopy is dense. Examine the catch closely or you will
overlook them, since many draw in the appendages and play
dead when disturbed.

BRANCH and TWIG BORERS   Family Bostrichidae

Identification: Form distinctive: broadly to narrowly cylindrical, head bent down and appearing on ventral surface of prothorax, *nearly* or completely *concealed from above;* pronotum
usually tuberculate or with *rasplike teeth anteriorly,* not hoodlike

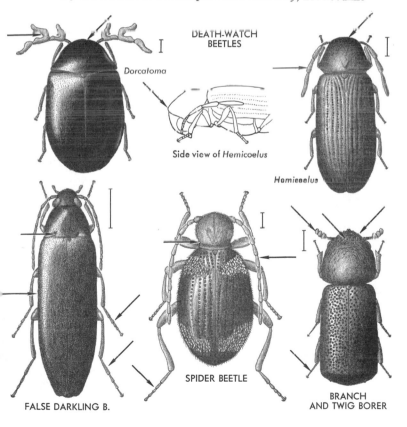

DEATH-WATCH
BEETLES

Dorcatoma

Side view of Hemicoelus

Hemicoelus

FALSE DARKLING B.

SPIDER BEETLE

BRANCH
AND TWIG BORER

or enclosing head. Antennal club *3-* or *4-segmented*, loose. 2–20 mm. (1 western species to 2 in.).

Bostrichids are wood borers, and attack living trees, dead twigs and branches, or dry seasoned timber. The adults of one unusual western species bore into the lead sheathing of telephone cables, allowing moisture to enter the cable and causing a short circuit; this insect, *Scobicia declivis* (LeConte), is called the Short-circuit Beetle (normally lives in wood and does not feed when it bores into cables). The giant of this family is the Palm Borer, *Dinapate wrighti* Horn, which occurs in California; it is about 2 in. long.

PSOID BEETLES   Family Psoidae
Identification: Similar to Bostrichidae (p. 189), but head not bent down, *visible dorsally*. Mandibles large and strong. 6–28 mm.

These beetles occur in the West, where they bore into the heartwood or branches of various trees and shrubs. Some severely prune trees and are very destructive to orchards in Oregon and California.

POWDER-POST BEETLES   Family Lyctidae
Identification: Form distinctive: *narrow, elongate,* flattened, head *visible dorsally*. Antennal club *2-segmented*, abrupt. Reddish brown to black. 1–7 mm.

The common name refers to their habit of boring into seasoned wood and reducing it to powder. Lyctids are very destructive to dried wood and wood products of various kinds, including woodwork, timbers, tool handles, gunstocks, and other manufactured materials.

Superfamily Scarabaeoidea

Terminal antennal segments extended laterally into a club of various types, usually lamellate, sometimes flabellate. Tarsi 5-5-5. Mostly moderate-sized to large.

STAG BEETLES   Family Lucanidae
Identification: Elongate-robust. Antennae *elbowed*, club *3-* or *4-segmented*, segments of club not capable of being held together in a tight ball. Black to reddish brown. Mandibles of ♂ *very large*, sometimes branched. Pronotum without a median groove. FW usually smooth. 8–40 mm.

Most adult stag beetles feed on sap flows; larvae live in decaying logs and stumps and apparently feed on juices of rotting wood (larva of 1 species feeds in sod). The greatly developed, sometimes branched mandibles of the males of a few species give these beetles their common name. Adults of most species are found on the ground in woods, others on sandy beaches. Adults of a large reddish-brown eastern species frequently fly to lights at night.

BESSBUGS Family Passalidae
Identification: Elongate-robust, parallel-sided, black and shiny. Head with *a forward-directed horn*. Pronotum with a distinct median groove, FW with *longitudinal grooves*. Antennae *not elbowed*, segments of club *not capable of being held together in a tight ball*. Pronotum and FW *distinctly separated at sides*. 30–40 mm.

Only 1 of the 3 U.S. species of bessbugs, *Popilius disjunctus*

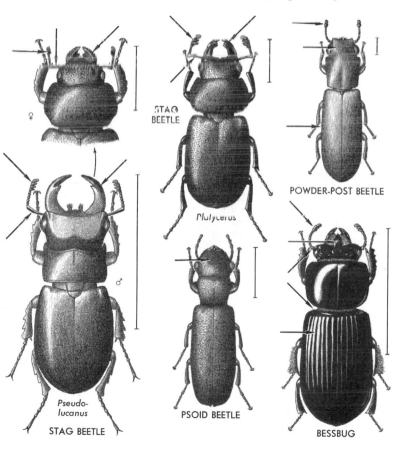

STAG BEETLE

Platycerus

POWDER-POST BEETLE

Pseudo-lucanus

STAG BEETLE

PSOID BEETLE

BESSBUG

(Illiger), occurs in the East; the other 2 occur in s. Texas. The eastern species forms colonies in well-decayed logs and stumps and is fairly common. Both adults and larvae stridulate; the sounds produced probably serve as a means of communication. Bessbugs have a number of common names, including patent-leather beetles, betsy-beetles, and horned passalus beetles. Despite their size and strong mandibles, the passalids do not bite.

SCARAB BEETLES Family Scarabaeidae   See also Pl. 7
Identification: Oval or elongate, usually stout and heavy-bodied, varying greatly in form and size. Antennae 8- to 11-segmented, lamellate (rarely flabellate), segments of club *capable of being held tight together*. 2–50 mm.

This is one of the largest families of beetles, with nearly 1300 N. American species. Many are very important because of damage done by larvae or adults. There are 14 subfamilies in N. America, of which only the most important are discussed here.

Dung Beetles and Tumblebugs, Subfamily Scarabaeinae (see also Pl. 7). Hind legs situated closer to tip of abdomen than to middle legs; hind tibiae usually with *only 1 apical spur*. These beetles occur in or near dung, manure, and carrion. Tumblebugs (*Canthon* and *Deltochilum*) are dull black (some are green), about 20 mm. or less, without horns, and the hind tibiae are rather slender; a pair will form a mass of dung into a ball, roll the ball a distance, dig a hole, and bury it; then the female lays eggs in it. Males of some dung beetles (like *Phanaeus*) bear horns on both head and pronotum, or on either.

Aphodian Dung Beetles, Subfamily Aphodiinae. Members of this group have the hind legs closer to tip of the abdomen than to middle legs, and hind tibiae bear 2 apical spurs. They are usually smaller and more elongate than Scarabaeinae. Apho-diines are quite common in cow dung. Adults are generally black, sometimes with the front wings red or yellowish.

Earth-boring Dung Beetles, Subfamily Geotrupinae (not illus.). The body is very stout, convex, and shiny, and the antennae are 11-segmented. Adults and larvae are found in cow dung, horse manure, carrion, fungi, and under logs. The most common species (*Geotrupes*) are 14–20 mm. and black (often with a purplish luster) or brownish.

Subfamily Acanthocerinae (not illus.). Black, round, 5–6 mm., the middle and hind tibiae dilated and spiny. These beetles can bend head and prothorax downward and roll themselves into a ball. The 3 U.S. species, which are not common, occur from the eastern states to Texas. They are found under bark and in rotting logs and stumps.

Skin Beetles, Subfamily Troginae. These beetles are dull brownish, the dorsal surface is roughened or *tuberculate*, and the 2nd antennal segment rises before tip of the 1st. They are usually found in dry animal carcasses, where they feed on hide, fur, feathers, and dried tissues on the bones. They draw in their legs and play dead when disturbed, and because they are usually covered with debris are easily overlooked. Beneficial scavengers, they represent one of the last stages in the succession of organisms feeding on animal carcasses.

June or May Beetles, Chafers, and Others, Subfamily Melolonthinae. In this and the remaining subfamilies of scarabs the hind legs are situated at about midbody, closer to the middle legs than to tip of the abdomen. Melolonthines have the clypeus expanded laterally, usually concealing the mandibles

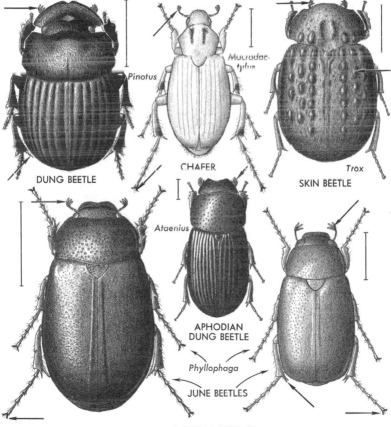

DUNG BEETLE

Pinotus

Macrodactylus

CHAFER

Trox

SKIN BEETLE

Ataenius

APHODIAN
DUNG BEETLE

Phyllophaga

JUNE BEETLES

SCARAB BEETLES

and labrum, and the tarsal claws are of *equal size and usually toothed or bifid*. The best-known members of this group are probably the June beetles, or Junebugs, which are common at lights in early summer. There are over 100 N. American species, most belonging to the genus *Phyllophaga*. June beetle larvae are white grubs that feed in the soil on roots of grasses and other plants; when abundant they cause serious damage to lawns, pastures, and various crops. Adults feed on flowers and foliage of various trees and shrubs and are capable of completely defoliating them.

Shining Leaf Chafers, Subfamily Rutelinae (see also Pl. 7). These beetles have tarsal claws, at least on hind legs, of *unequal size*, and there are 2 apical spurs on the hind tibiae. Adults are foliage and fruit feeders, and larvae usually feed on plant roots. One of the most serious pests in this group is the Japanese Beetle (Pl. 7), *Popillia japonica* Newman, accidentally introduced into the e. U.S. about 1916. It has since spread over much of the country and severely damages lawns, golf courses, pastures, shrubbery, and fruits. Larvae feed in the soil on roots of various plants; adults feed on foliage, fruits, and flowers of more than 200 kinds of plants.

Rhinoceros Beetles, Hercules Beetles, and Elephant Beetles, Subfamily Dynastinae. These beetles have tarsal claws *simple* (and usually equal in size), front coxae transverse, and the dorsal surface is more or less convex; males of many species bear horns on the head and/or pronotum. Many are 1 in. or more long, and a few are over 2 in. The group contains some of the largest N. American beetles. The largest eastern species is the Eastern Hercules Beetle (also called Unicorn Beetle), *Dynastes tityus* (Linn.), which is 2–2½ in. and greenish gray mottled or blotched with black; it is common in the Southeast, and ranges as far north as s. Ohio and Indiana. The Rhinoceros Beetle, *Xyloryctes jamaicensis* (Drury), is dark brown and slightly over 1 in.; males have a large horn on the head (females have only a small tubercle); it occurs in the East. A few small and widely distributed members of this group, without horns on either head or pronotum (such as *Ligyrus*), occasionally damage crops.

Flower Beetles and Others, Subfamily Cetoniinae (see also Pl. 7). These have tarsal claws simple and equal in size, front coxae conical, and the body flattened dorsally; the lateral margins of the elytra (FW) are often *narrowed behind the front corners*, and there are no horns. Our species are small to moderate-sized, but some African species (Goliath beetles) reach a length of 4 or 5 in. Adults of most species are pollen feeders, found on flowers; some feed on the juices of decaying wood and are found under bark. Larvae occur in the soil and often damage roots. Adults of the Green June Beetle, *Cotinis nitida* (Linn.), a dark green beetle about 1 in., feed on grapes, ripening fruit, foliage, and young corn; larvae, which have the habit of crawling on their back, feed in the soil on plant roots. This beetle is common in the southeastern states. Some of the smaller flower beetles (like *Trichiotinus*, 9–12 mm., and *Valgus*, 7.5 mm. or less) have the clytra truncate and not much longer than wide. Members of the genus *Cremastocheilus* (9–15 mm.) occur in ant nests.

Superfamily Cerambycoidea

Tarsi *apparently 4-4-4, actually 5-5-5* (4th segment very small and concealed in bilobed 3rd segment). These are plant feeders. Many of great economic importance.

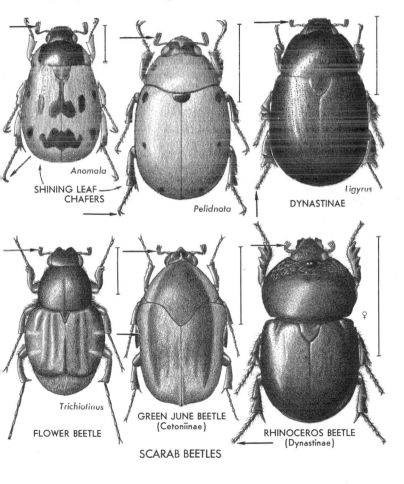

Anomala

SHINING LEAF CHAFERS

Pelidnota

Ligyrus

DYNASTINAE

Trichiotinus

FLOWER BEETLE

GREEN JUNE BEETLE
(Cetoniinae)

RHINOCEROS BEETLE
(Dynastinae)

SCARAB BEETLES

Plate 1

DAMSELFLIES AND DRAGONFLIES (Odonata)

At rest, damselflies hold their wings together above the body or diverging; dragonflies hold them horizontal.

NARROW-WINGED DAMSELFLIES, Family Coenagrionidae   p. 74
Wings *stalked at base* and held together above body at rest; 2 antenodal cross veins; M_3 rises behind nodus.
CIVIL BLUET, *Enallagma civile* (Hagen); N. America.
VIOLET DANCER, *Argia violacea* (Hagen); ne. U.S.
COMMON FORKTAIL, *Ischnura verticalis* (Say), ♀; e. N. America.

BROAD-WINGED DAMSELFLIES, Family Calopterygidae   p. 74
Wings *gradually narrowed at base* and held together above body at rest; wings usually with blackish or reddish areas; 10 or more antenodal cross veins.
BLACK-WINGED DAMSELFLY, *Calopteryx maculata* (Beauvais), ♂; e. N. America.

COMMON SKIMMERS, Family Libellulidae   p. 72
Dragonflies *without* a brace vein; anal loop in HW elongate, with a bisector, and usually *foot-shaped;* hind margin of compound eyes straight or nearly so.
ELISA SKIMMER, *Celithemis elisa* (Hagen); e. N. America.
COMMON AMBERWING, *Perithemis tenera* (Say), ♂; e. N. America.

DARNERS, Family Aeshnidae   p. 70
Dragonflies *with a brace vein*, and with compound eyes *in contact* for a considerable distance on dorsal side of head; ♀ with ovipositor well developed.
GREEN DARNER, *Anax junius* (Drury); N. America.
EASTERN BLUE DARNER, *Aeshna verticalis* Hagen; ne. U.S.

CLUBTAILS, Family Gomphidae   p. 70
Dragonflies *with a brace vein*, and with compound eyes *separated;* terminal abdominal segments sometimes *dilated;* ♀ without an ovipositor.
Gomphus vastus Walsh; e. U.S.

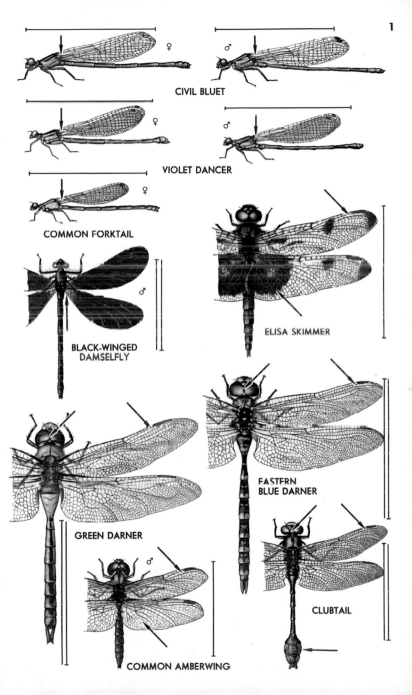

1

CIVIL BLUET

VIOLET DANCER

COMMON FORKTAIL

BLACK-WINGED DAMSELFLY

ELISA SKIMMER

GREEN DARNER

EASTERN BLUE DARNER

COMMON AMBERWING

CLUBTAIL

Plate 2

GRASSHOPPERS, CRICKETS, AND COCKROACHES
(Orthoptera)

Usually large insects; hind legs often (grasshoppers and crickets) enlarged; FW narrow, HW broad and at rest folded fanwise.

LONG-HORNED GRASSHOPPERS, Family Tettigoniidae p. 80
Antennae *long and hairlike;* tarsi *4-segmented;* ovipositor sword-shaped.
BUSH KATYDIDS, Subfamily Phaneropterinae p. 80
Usually green and over 1 in.; FW flat, shorter than HW; prosternal spines absent; vertex rounded.
Scudderia curvicauda (De Geer), ♂; e. N. America.
MEADOW GRASSHOPPERS, Subfamily Conocephalinae p. 82
Slender, greenish, seldom over 1 in.; prosternum usually with a pair of small spines; vertex short.
Conocephalus brevipennis (Scudder), ♂; e. N. America.

SHORT-HORNED GRASSHOPPERS, Family Acrididae p. 78
Antennae relatively *short;* tarsi *3-segmented;* pronotum not prolonged backward over abdomen; wings usually well developed.
SPUR-THROATED GRASSHOPPERS, p. 78
Subfamily Cyrtacanthacridinae
A spine or tubercle on prosternum.
Schistocerca americana (Drury); e. N. America.
BAND-WINGED GRASSHOPPERS, p. 78
Subfamily Oedipodinae
HW usually *brightly colored;* a median longitudinal *keel on pronotum;* hind margin of pronotum *triangularly extended backward;* face vertical or nearly so.
Dissosteira pictipennis Brunner; Calif. to s. Oregon.

CRICKETS, Family Gryllidae p. 82
Body somewhat flattened; antennae *long and hairlike;* tarsi *3-segmented;* ovipositor usually long and cylindrical.
BUSH CRICKETS, Subfamily Trigonidiinae p. 84
2nd tarsal segment heart-shaped and flattened dorsoventrally; no teeth between tibial spines; ovipositor sword-shaped.
Phyllopalpus pulchellus (Uhler); N.Y. to Illinois and Florida.

COCKROACHES, Family Blattidae p. 86
Body flattened and oval, head *concealed from above* by pronotum; antennae *long, hairlike;* tarsi *5-segmented.*
ORIENTAL COCKROACH, *Blatta orientalis* Linn.; N. America (introduced).

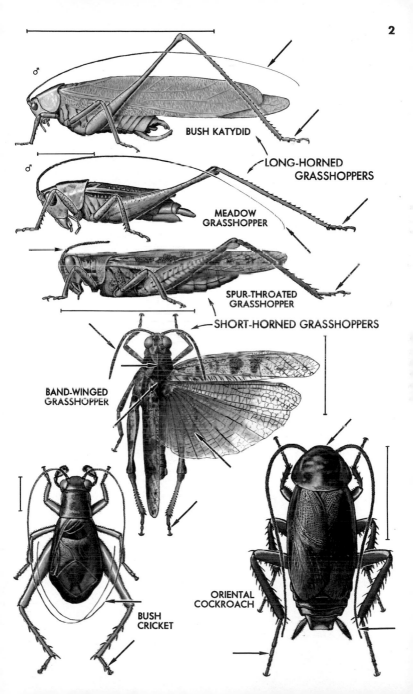

2

♂ BUSH KATYDID

LONG-HORNED
GRASSHOPPERS

♂ MEADOW
GRASSHOPPER

SPUR-THROATED
GRASSHOPPER

SHORT-HORNED GRASSHOPPERS

BAND-WINGED
GRASSHOPPER

BUSH
CRICKET

ORIENTAL
COCKROACH

Plate 3

BUGS (Hemiptera)

FW thickened at base and membranous at tip, the tips overlapping at rest; antennae 4- or 5-segmented; mouth parts in the form of a sucking beak.

SCENTLESS PLANT BUGS, Family Rhopalidae   p. 122
Elongate to oval, fairly hard-bodied; ocelli *present;* membrane of FW with *many veins;* head narrower than pronotum; 14 mm. or less.
BOXELDER BUG, *Leptocoris trivittatus* (Say); U.S.

LEAF or PLANT BUGS, Family Miridae   p. 118
Small, soft-bodied, oval or elongate; FW with a *cuneus* and membrane with *2 closed cells;* ocelli *absent.*
FOUR-LINED PLANT BUG, *Poecilocapsus lineatus* (Fabricius); N. America.

AMBUSH BUGS, Family Phymatidae   p. 118
Abdomen *wider toward rear,* and extending on sides beyond wings; front femora *much thickened;* beak short, 3-segmented; antennae *4-segmented,* last segment *swollen.*
Phymata fasciata (Gray); e. U.S., N.J. to Ariz., Texas.

SEED BUGS, Family Lygaeidae   p. 122
Elongate to oval, fairly hard-bodied; membrane of FW with *4 or 5 veins;* ocelli nearly always *present.*
SMALL MILKWEED BUG, *Lygaeus kalmii* Stål; N. America.

STINK BUGS, Family Pentatomidae   p. 126
Shield-shaped bugs; antennae *5-segmented;* scutellum *large and triangular,* not longer than corium, and not reaching tip of abdomen; tibiae without strong spines.
SOUTHERN GREEN STINK BUG, *Nezara viridula* (Linn); se. U.S.
HARLEQUIN BUG, *Murgantia histrionica* (Hahn); U.S.

RED BUGS or STAINERS, Family Pyrrhocoridae   p. 122
Usually brightly colored with red and black; similar to seed bugs but ocelli *lacking* and *more veins* in FW membrane.
COTTON STAINER, *Dysdercus suturellus* (Herrich-Schäffer); se. U.S., S. Carolina to Alabama.

ASSASSIN BUGS, Family Reduviidae   p. 119
Head *elongate,* with a transverse groove between eyes; beak short, 3-segmented, its tip fitting into a groove in prosternum.
BLOODSUCKING CONENOSE, *Triatoma sanguisuga* (Le-Conte); Md. and N.J. to Florida, Illinois, Texas.

LEAF-FOOTED BUGS, Family Coreidae   p. 122
Similar to seed bugs but with *many veins* in membrane of FW; usually dark-colored and over 10 mm.
SQUASH BUG, *Anasa tristis* (De Geer); N. America.

3

BOXELDER BUG

FOUR-LINED PLANT BUG

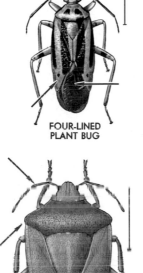

AMBUSH BUG

SMALL MILKWEED BUG

SOUTHERN GREEN STINK BUG

COTTON STAINER

BLOODSUCKING CONENOSE

HARLEQUIN BUG

SQUASH BUG

Plate 4

APHIDS, HOPPERS, CICADAS, AND OTHERS
(Homoptera)

PINE AND SPRUCE APHIDS, Family Chermidae p. 136
Small dark-colored insects, with wings held rooflike over body at rest; cornicles absent; FW with Rs absent, and Cu₁ and Cu₂ separated at base.
EASTERN SPRUCE GALLS, caused by *Chermes abietis* Linn.; e. N. America.

PSYLLIDS, Family Psyllidae p. 134
FW membranous or thickened, usually held rooflike over body at rest; tarsi *2-segmented;* antennae *10-segmented.*
HACKBERRY NIPPLE GALLS, caused by *Pachypsylla celtidis-mamma* Riley; e. U.S.
WOOLLY PSYLLID, *Psylla floccosa* (Patch); e. N. America.

ACANALONIID PLANTHOPPERS, Family Acanaloniidae p. 132
Planthoppers (see p. 130) with FW *broadly oval* and held *almost vertical* at rest; costal area of FW *reticulate* (net-veined); hind tibiae with spines at tip only; 2nd segment of hind tarsi minute, its tip rounded with a small spine on each side.
Acanalonia bivittata (Say); e. N. America.

FLATID PLANTHOPPERS, Family Flatidae p. 132
Similar to acanaloniids (above), but FW *elongate-triangular* and with *numerous costal cross veins*, and hind tibiae with spines on sides in addition to ones at tip.
Anormenis septentrionalis (Spinola); e. U.S.

LEAFHOPPERS, Family Cicadellidae p. 130
Body usually tapers behind; tarsi *3-segmented;* antennae short, bristlelike, rising in front of or between eyes; hind tibiae with 1 or more *rows of small spines.*
Comellus comma (Van Duzee); ne. U.S.
Graphocephala coccinea (Forster); e. U.S.

APHIDS, Family Aphididae p. 135
Small, soft-bodied, usually somewhat pear-shaped, nearly always with *a pair of cornicles* near end of abdomen; wings when present membranous, FW with Rs present and M branched, HW much smaller than FW; tarsi *2-segmented.*
GREEN PEACH APHID, *Myzus persicae* (Sulzer); N. America.

CICADAS, Family Cicadidae p. 129
Large, robust insects, mostly 1–2 in., with *clear wings* and *bulging eyes;* tarsi 3-segmented; antennae short, bristlelike; *3 ocelli;* ♂ usually with sound-producing organs at base of abdomen on underside.
PERIODICAL CICADA, *Magicicada septendecim* (Linn.); ne. U.S.

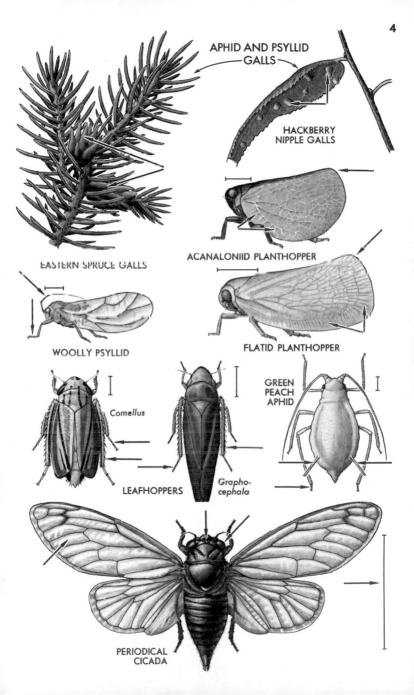

4

APHID AND PSYLLID GALLS

HACKBERRY NIPPLE GALLS

EASTERN SPRUCE GALLS

ACANALONIID PLANTHOPPER

WOOLLY PSYLLID

FLATID PLANTHOPPER

Comellus

LEAFHOPPERS

Grapho-cephala

GREEN PEACH APHID

PERIODICAL CICADA

Plate 5

BEETLES (Coleoptera)

Usually hard-bodied insects, with FW thickened and meeting in a straight line down back; antennae nearly always with 10 or more segments; mouth parts chewing.

CHECKERED BEETLES, Family Cleridae p. 165
Often marked with red, orange, yellow, or blue; pronotum *narrower than FW;* body and legs with *erect pubescence;* tarsi 5-5-5.
Enoclerus ichneumoneus (Fabricius); e. U.S.

DERMESTID BEETLES, Family Dermestidae p. 164
Elongate to broadly oval, often with *scales* or hair forming patterns; antennae *short and clubbed;* tarsi 5-5-5.
CARPET BEETLE, *Anthrenus scrophulariae* (Linn.); N. Amer.

SOFT-WINGED FLOWER BEETLES, Family Malachiidae p. 164
Body soft and wedge-shaped, FW *broad toward rear;* antennae *serrate;* black, blue, or green, often with orange, red, or yellow; tarsi 5-5-5; abdomen with 6 or fewer segments.
Collops vittatus (Say); w. U.S.

LIGHTNINGBUGS or FIREFLIES, Family Lampyridae p. 162
Elongate, *parallel-sided,* rounded at ends; soft-bodied; head *concealed* by pronotum; abdomen often luminescent.
Photuris pennsylvanica (De Geer); e. U.S.

GROUND BEETLES, Family Carabidae p. 152
Generally black and shiny, sometimes brightly colored; legs *long and slender;* head and eyes nearly always *narrower than pronotum;* tarsi 5-5-5; antennae threadlike.
Calosoma scrutator (Fabricius); e. U.S.

SOLDIER BEETLES, Family Cantharidae p. 162
Similar to lightningbugs (above), but head *not concealed* and no abdominal segments luminescent; black or brown, often with red, yellow, or orange, some mostly yellowish.
Chauliognathus pennsylvanicus De Geer; e. U.S.

NET-WINGED BEETLES, Family Lycidae p. 163
FW with a *network of ridges;* body usually *broadest toward rear;* head concealed; black, some with red or yellow.
Calopteron terminale (Say); U.S.

METALLIC WOOD-BORING BEETLES, p. 169
Family Buprestidae
Elongate, usually parallel-sided; body hard, generally metallic, especially below; antennae *short,* threadlike or *serrate;* tarsi 5-5-5.
Acmaeodera pulchella (Herbst); e. U.S.

CARRION BEETLES, Family Silphidae p. 158
Black, often with yellow, orange, or red; antennae *clubbed;* tarsi 5-5-5; body flattened; FW *sometimes short,* exposing 1–3 abdominal segments.
Nicrophorus marginatus Fabricius; N. America.

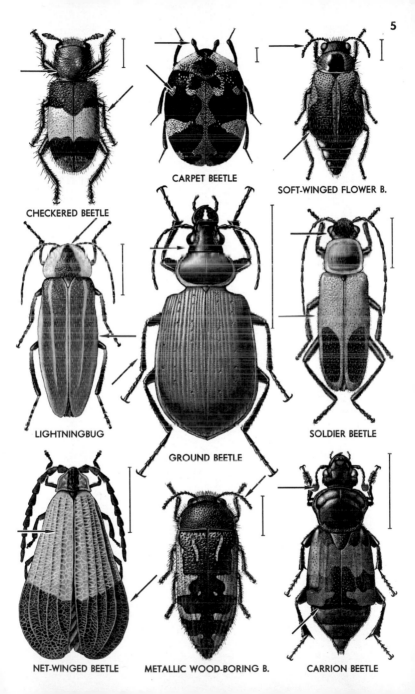

5

CHECKERED BEETLE

CARPET BEETLE

SOFT-WINGED FLOWER B.

LIGHTNINGBUG

GROUND BEETLE

SOLDIER BEETLE

NET-WINGED BEETLE

METALLIC WOOD-BORING B.

CARRION BEETLE

Plate 6

BEETLES (Coleoptera)

SAP BEETLES, Family Nitidulidae p. 177
Antennae with an *abrupt 3-segmented club;* tarsi 5-5-5 or 4-4-4;
tip of abdomen *often exposed* beyond FW; 12 mm. or less.
Glischrochilus quadrisignatus (Say); N. America.

FIRE-COLORED BEETLES, Family Pyrochroidae p. 183
Elongate, narrow, pronotum (sometimes also head and FW)
reddish; pronotum narrower than FW; FW sometimes slightly
wider toward rear; antennae serrate or *pectinate;* tarsi 5-5-4.
Dendroides bicolor Newman; U.S.

LADYBIRD BEETLES, Family Coccinellidae p. 180
Small, broadly oval, convex, often brightly colored; head generally concealed by pronotum; tarsi *appear 3-3-3;* antennae *short
and clubbed.*
a. *Hippodamia convergens* Guérin-Mèneville; N. America.
b. TWO-SPOTTED LADYBIRD BEETLE, *Adalia bipunctata*
(Linn.); N. America.

FLAT BARK BEETLES, Family Cucujidae p. 176
Body very flat, *elongate, parallel-sided;* brown, reddish, or blackish; antennae usually *threadlike,* sometimes clubbed; tarsi 5-5-5,
sometimes appearing 5-5-4.
Cucujus clavipes Fabricius; e. U.S.

HANDSOME FUNGUS BEETLES, Family Endomychidae p. 180
Oval, convex, shiny, often brightly colored; front corners of
pronotum *extend forward; 2 longitudinal grooves* in rear half of
pronotum; tarsi *appear 3-3-3.*
Aphorista vittata (Fabricius); e. U.S.

BLISTER BEETLES, Family Meloidae p. 184
Elongate, slender, soft-bodied; pronotum *narrower than head or
FW;* antennae threadlike; tarsi 5-5-4.
Lytta aenea (Say); e. U.S.

LIZARD BEETLES, Family Languriidae p. 175
Elongate, slender, nearly parallel-sided; dark, shining, pronotum
and sometimes also head reddish or yellowish; antennae *clubbed;*
tarsi 5-5-5.
CLOVER STEM BORER, *Languria mozardi* Latreille; e. U.S.

PLEASING FUNGUS BEETLES, Family Erotylidae p. 176
Elongate to broadly oval; black, shiny, often with red, orange,
or yellow; antennae *clubbed;* tarsi 5-5-5, 4th segment often small.
Megalodacne heros (Say); e. U.S.

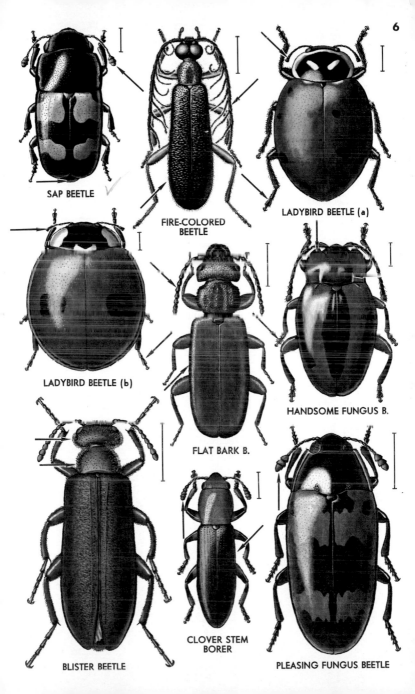

6

SAP BEETLE

FIRE-COLORED BEETLE

LADYBIRD BEETLE (a)

LADYBIRD BEETLE (b)

FLAT BARK B.

HANDSOME FUNGUS B.

BLISTER BEETLE

CLOVER STEM BORER

PLEASING FUNGUS BEETLE

Plate 7

BEETLES (Coleoptera)

SCARAB BEETLES, Family Scarabaeidae p. 192
Stout-bodied, oval or elongate; vary greatly in size and color; antennal club of several *leaflike plates capable of being held tight together;* tarsi 5-5-5.
FLOWER BEETLES, Subfamily Cetoniinae p. 194
Body flattened dorsally; tarsal claws *simple,* equal in size; front coxae conical.
1. *Trichiotinus affinis* (Gory and Percheron); e. U.S.
SHINING LEAF CHAFERS, Subfamily Rutelinae p. 194
Tarsal claws, at least on hind legs, of unequal size; hind tibiae with *2 spurs at tip.*
2. JAPANESE BEETLE, *Popillia japonica* Newman; ne. U.S.
DUNG BEETLES, Subfamily Scarabaeinae p. 192
Robust, with strong, well-clawed front legs; sometimes with rhinoceroslike "horn"; hind tibiae with only *1 spur at tip;* hind legs closer to tip of abdomen than to middle legs.
3. *Phanaeus vindex* MacLachlan; e. U.S. to Rocky Mts.

LONG-HORNED BEETLES, Family Cerambycidae p. 196
Body elongate, usually cylindrical; antennae *at least half as long as body, often longer;* usually over 12 mm.; tarsi *apparently 4-4-4, actually 5-5-5* (4th segment very small).
Subfamily Cerambycinae p. 196
FW parallel-sided; last segment of maxillary palps blunt at tip; eyes usually notched and partly surrounding base of antennae.
4. LOCUST BORER, *Megacyllene robiniae* (Forster); N. America.
5. *Neoclytus scutellaris* (Olivier); e. U.S.
6. *Taranomis bivittata* (Dupont); sw. U.S.
Subfamily Lamiinae p. 196
Similar to Cerambycinae but last segment of maxillary palps cylindrical and pointed at tip.
7. *Saperda cretata* Newman; e. U.S.
8. ELM BORER, *Saperda tridentata* Olivier; e. U.S.
Subfamily Lepturinae p. 196
FW often widest at base, giving the body a broad-shouldered appearance; last segment of maxillary palps blunt at tip; eyes oval or slightly notched.
9. ELDERBERRY LONGHORN, *Desmocerus palliatus* (Forster); e. U.S.

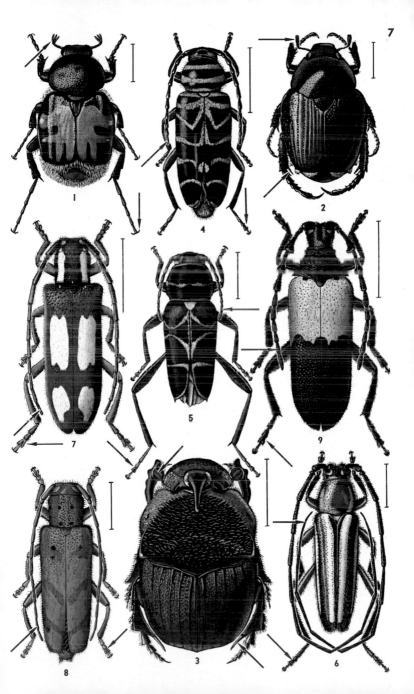

Plate 8

BEETLES (Coleoptera)

SNOUT BEETLES, Family Curculionidae p. 202

Head usually with a well-developed *snout;* antennae *clubbed* and nearly always *elbowed;* palps small and rigid, often concealed within mouth.

BROAD-NOSED WEEVILS, Subfamily Thylacitinae p. 202

Beak *short, squarish*, often widened toward tip; mandibles with a small cusp that breaks off and leaves a scar; grooves in beak for antennae bent down at rear end; basal antennal segment at rest passes below eye.

1. *Scythropus elegans* (Couper); N. America.

Subfamily Cyladinae p. 202

Antlike, 5–6 mm.; pronotum reddish, FW blue-black; antennae *not elbowed, last segment long and swollen.*

2. SWEETPOTATO WEEVIL, *Cylas formicarius elegantulus* (Summers), ♂; Florida to Texas.

LEAF BEETLES, Family Chrysomelidae p. 198

Body usually oval; generally less than 12 mm.; antennae usually less than half as long as body; tarsi *apparently 4-4-4, actually 5-5-5* (4th segment very small).

Subfamily Eumolpinae p. 199

Oval, convex; front coxae rounded; 3rd tarsal segment *bilobed.*

3. *Colaspis oregonensis* (Crotch); n. Calif. to Oregon.

4. DOGBANE BEETLE, *Chrysochus auratus* (Fabricius); e. U.S.

LEAF-MINING LEAF BEETLES, Subfamily Hispinae p. 198

FW with ridges between which are *rows of punctures;* body parallel-sided or wider toward rear.

5. *Microrhopala vittata* (Fabricius); e. U.S.

CASE-BEARING LEAF BEETLES, p. 198

Subfamily Cryptocephalinae

Cylindrical; head *buried in prothorax to eyes;* last abdominal segment *exposed;* FW smooth; antennae *threadlike.*

6. *Cryptocephalus cuneatus* Fall; Georgia, Florida.

TORTOISE BEETLES, Subfamily Cassidinae p. 198

Broadly oval to nearly circular, body expanded and flattened; head largely concealed from above; legs *short.*

7. *Hemisphaerota cyanea* (Say); Georgia, Florida.

Subfamily Criocerinae p. 200

Pronotum rounded on sides, *narrower than FW at base;* punctures of FW in rows; head narrowed behind.

8. *Lema balteata* LeConte; Ariz.

Subfamily Chrysomelinae p. 200

Oval, convex; antennae widely separated at base; front coxae oval or transverse; 3rd tarsal segment *not bilobed;* pronotum with sharp side edges.

9. *Calligrapha serpentina* (Rogers); sw. U.S.

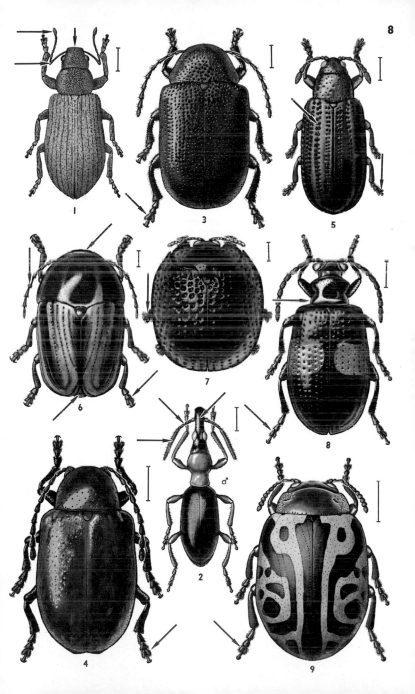

8

Plate 9

BUTTERFLIES
(Lepidoptera, Superfamily Papilionoidea)

Gaudy diurnal insects with scaled wings that are large in proportion to body; antennae knobbed but never hooked at tip (as are
those of skippers, Superfamily Hesperioidea), and close together
at base; wings at rest often held together above body.

SULFURS and Others, Family Pieridae   p. 224
Small to medium-sized, usually white, yellow, or orange, marked
with black; front legs normal or slightly reduced, the tarsal
claws forked; FW with R 3- or 4-branched, and M_1 stalked with
a branch of R beyond discal cell.
ORANGE SULFUR, *Colias eurytheme* Boisduval, ♂; N.
America.

COPPERS, Family Lycaenidae, Subfamily Lycaeninae   p. 224
Small, brownish or reddish with black markings; front legs of ♂
usually reduced; FW with R 4-branched, and M_1 not stalked
with a branch of R beyond discal cell.
AMERICAN COPPER, *Lycaena phleas* Linn.; N. America.

NYMPHS, SATYRS, and ARCTICS, Family Satyridae   p. 226
Small to medium-sized, usually grayish or brownish and often
with eye spots in wings; front legs greatly reduced; some veins
(especially Sc) greatly swollen at base.
WOOD NYMPH, *Cercyonis pegala* Fabricius; N. America.

MILKWEED BUTTERFLIES, Family Danaidae   p. 228
Large, brownish, usually marked with black; front legs greatly
reduced; FW with R 5-branched and 3A present; discal cell in
HW closed by a well-developed vein; antennae without scales.
MONARCH, *Danaus plexippus* (Linn.), ♂; N. America.

BRUSH-FOOTED BUTTERFLIES, Family Nymphalidae   p. 226
Vary greatly in size and color; front legs much reduced; FW
relatively broad and triangular; venation as in milkweed butter-
flies (above), but 3A lacking in FW, and discal cell in HW open
or closed by a weak vein; antennae with scales.
COMMA, *Polygonia comma* Harris; e. N. America.
MOURNINGCLOAK, *Nymphalis antiopa* (Linn.); e. N.
America.
VICEROY, *Limenitis archippus* (Cramer); e. N. America.
GREAT SPANGLED FRITILLARY, *Speyeria cybele* (Fabri-
cius); e. N. America.
RED ADMIRAL, *Vanessa atalanta* (Linn.); e. N. America.

ORANGE SULFUR

AMERICAN COPPER

WOOD NYMPH

MONARCH

COMMA

MOURNING-CLOAK

VICEROY

GREAT SPANGLED FRITILLARY

RED ADMIRAL

Plate 10

MOTHS (Lepidoptera)

Nocturnal (rarely diurnal) Lepidoptera with antennae threadlike or feathery (not clubbed as in butterflies); at rest, wings usually held horizontal or rooflike over body.

GEOMETER MOTHS, Family Geometridae   p. 234
Small to medium-sized, usually slender-bodied; Cu in FW appears 3-branched; Sc in HW abruptly bent into humeral angle of wing, often connected to humeral angle by a cross vein.
NOTCH-WINGED GEOMETER, *Deuteronomous magnarius* (Guenée); n. U.S., Canada.
FALL CANKERWORM, *Alsophila pometaria* (Harris), ♂; N. America.
CHICKWEED GEOMETER, *Haematopis grataria* (Fabricius); e. U.S.

GIANT SILKWORM MOTHS, Family Saturniidae   p. 232
Medium-sized to large, with feathery antennae; wings broad, *usually with eye spots;* Cu in FW appears 3-branched; Sc+R_1 and Rs in HW diverge from base of wing; frenulum small or vestigial; humeral angle of HW not noticeably expanded; HW with only 1 anal vein.
IO MOTH, *Automeris io* (Fabricius), ♂; e. N. America.
POLYPHEMUS MOTH, *Antheraea polyphemus* (Cramer); N. America.
CECROPIA MOTH, *Hyalophora cecropia* (Linn.); e. N. America.
LUNA MOTH, *Actias luna* (Linn.); e. N. America.

SPHINX MOTHS, Family Sphingidae   p. 230
Medium-sized to large; heavy-bodied, with wings relatively small and narrow; FW much longer than HW; antennae *somewhat spindle-shaped;* Sc and Rs in HW parallel to end of discal cell and beyond, connected by an oblique cross vein about opposite middle of discal cell.
WHITE-LINED SPHINX, *Celerio lineata* (Fabricius); N. America.

ROYAL MOTHS, Family Citheroniidae   p. 232
Similar to giant silkworm moths (see above), but wings usually without eye spots; frenulum entirely absent and humeral angle of wing considerably expanded; discal cell in FW closed; 2 anal veins in HW.
REGAL MOTH, *Citheronia regalis* (Fabricius); e. N. America.

NOTCH-WINGED GEOMETER

FALL CANKERWORM ♂

IO MOTH ♂

CHICKWEED GEOMETER

POLYPHEMUS MOTH

REGAL MOTH

WHITE-LINED SPHINX

CECROPIA MOTH

LUNA MOTH

Plate 11

MOTHS (Lepidoptera)

PROMINENTS, Family Notodontidae   p. 234
Medium-sized, heavy-bodied, usually brownish moths, with *threadlike* antennae; wings at rest held rooflike over abdomen; Cu in FW appears 3-branched.
YELLOW-NECKED CATERPILLAR, *Datana ministra* (Drury); e. U.S.

HOOK-TIP MOTHS, Family Drepanidae   p. 236
Small, relatively *small-bodied*, brownish; tip of FW usually *sickle-shaped;* Cu in FW appears 4-branched; frenulum small or absent.
Drepana arcuata (Walker); e. U.S.

NOCTUID MOTHS, Family Noctuidae   p. 238
Small to medium-sized (rarely large), usually dark-colored, with *threadlike* antennae; wings at rest held rooflike over body or outstretched; Cu in FW appears 4-branched; Sc and R in HW fused for only a very short distance at base of wing; ocelli nearly always present.
CABBAGE LOOPER, *Trichoplusia ni* (Hübner); N. America.
ILIA UNDERWING, *Catocala ilia* (Cramer); N. America.

CTENUCHID MOTHS, Family Ctenuchidae   p. 236
Small, day-flying, usually dark-colored, with threadlike antennae; at rest appear triangular, with wings folded back but horizontal; venation as in Noctuidae (see above), but Sc+R$_1$ in HW absent.
YELLOW-COLLARED SCAPE MOTH, *Cisseps fulvicollis* (Hübner); e. N. America.
VIRGINIA CTENUCHA, *Ctenucha virginica* (Charpentier); ne. U.S.

TENT CATERPILLARS, Family Lasiocampidae   p. 236
Small to medium-sized, *heavy-bodied*, brownish, bluish, or blackish, with *somewhat feathery* antennae; frenulum lacking, the humeral angle of HW expanded and with humeral veins.
EASTERN TENT CATERPILLAR, *Malacosoma americanum* (Fabricius); e. N. America.

TIGER MOTHS, Family Arctiidae   p. 238
Medium-sized, *heavy-bodied;* usually light-colored, often brightly spotted or banded; venation as in Noctuidae (see above), but Sc and R in HW usually fused to about middle of discal cell.
BANDED WOOLLYBEAR, *Isia isabella* (Smith); N. America.
SALT-MARSH CATERPILLAR, *Estigmene acraea* (Drury), ♂; N. America.
VIRGIN TIGER MOTH, *Apantesis virgo* (Linn.); ne. U.S., Canada.

11

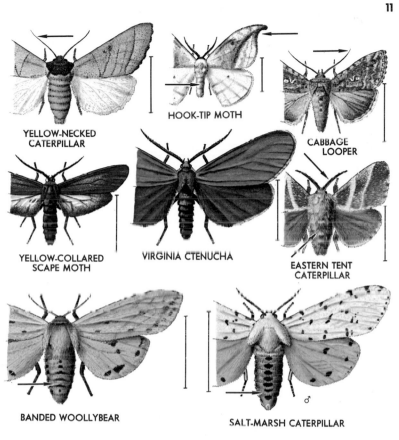

YELLOW-NECKED
CATERPILLAR

HOOK-TIP MOTH

CABBAGE
LOOPER

YELLOW-COLLARED
SCAPE MOTH

VIRGINIA CTENUCHA

EASTERN TENT
CATERPILLAR

BANDED WOOLLYBEAR

SALT-MARSH CATERPILLAR

ILIA UNDERWING

VIRGIN TIGER MOTH

Plate 12

MOTHS (Lepidoptera)

CLEAR-WINGED MOTHS, Family Sesiidae   p. 242
Wasplike, with extensive areas in wings *devoid of scales;* FW
long and narrow, HW broader. Some sphinx moths (p. 230) are
also clear-winged, but are much more robust and beelike.
PEACH TREE BORER, *Sanninoidea exitiosa* (Say), ♀;
e. N. America.

SLUG CATERPILLARS, Family Limacodidae   p. 244
Small, stout, with broadly rounded wings; often with green
markings; 2 anal veins in FW and 3 in HW.
Parasa chloris (Herrich-Schäffer); ne. U.S.

TORTRICID MOTHS, Family Tortricidae   p. 248
Small moths with FW rather *square-tipped;* 2 anal veins in FW
and 3 in HW; Cu_2 in FW rises in basal $\frac{3}{4}$ of discal cell.
FRUIT-TREE LEAF ROLLER, *Archips argyrospilus* (Walker);
n. U.S., Canada.

TUSSOCK MOTHS and Others, Family Liparidae   p. 240
Small moths resembling Noctuidae (see Pl. 11 and p. 238), but
with ♂ antennae *feathery*, and with ocelli absent, cell at base of
HW between Sc and R (basal areole) larger, and Rs and M_1 in
HW sometimes stalked.
TUSSOCK MOTH, *Hemerocampa plagiata* (Walker), ♂; n.
U.S., Canada.
GYPSY MOTH, *Porthetria dispar* (Linn.); ne. U.S.

ERMINE MOTHS, Family Yponomeutidae   p. 250
Small moths with FW usually *brightly patterned;* 2 anal veins in
FW and 3 in HW; Cu_2 in FW rises in distal $\frac{1}{4}$ of discal cell.
Atteva punctella (Cramer); e. U.S.

FORESTER MOTHS, Family Agaristidae   p. 239
Black, with *2 whitish or yellowish spots in each wing.*
EIGHT-SPOTTED FORESTER, *Alypia octomaculata* (Fab-
ricius); ne. U.S.

CARPENTER MOTHS, Family Cossidae   p. 242
Medium-sized, heavy-bodied; wings usually *spotted* or mottled;
2 anal veins in FW and 3 in HW; some branches of R in FW
stalked.
Prionoxystus robiniae (Peck); N. America.

PYRALID MOTHS, Family Pyralidae   p. 246
Small, generally small-bodied; FW narrow and elongate or some-
what triangular, HW broad; HW with 3 anal veins, and Sc + R_1
and Rs fused for a way beyond discal cell.
EUROPEAN CORN BORER, *Ostrinia nubilalis* (Hübner);
N. America.

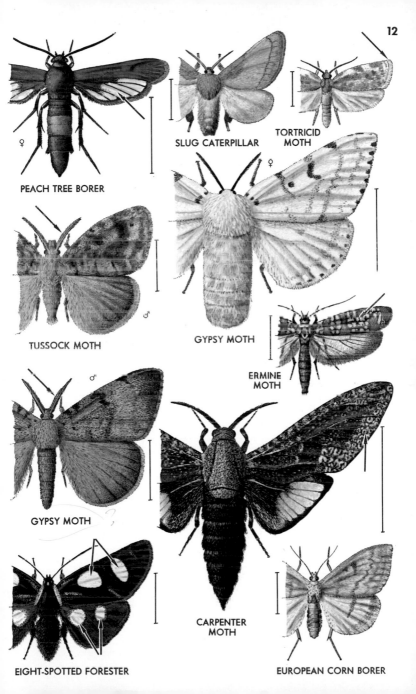

12

PEACH TREE BORER

SLUG CATERPILLAR

TORTRICID MOTH

TUSSOCK MOTH

GYPSY MOTH

ERMINE MOTH

GYPSY MOTH

CARPENTER MOTH

EIGHT-SPOTTED FORESTER

EUROPEAN CORN BORER

Plate 13

FLIES (Diptera)

With only 1 pair of wings (most other insects have 2 pairs).

Suborder Brachycera

Antennae with 5 or fewer (usually 3) segments, and rarely aristate; Rs nearly always 3-branched; anal cell generally long, usually closed near wing margin.

ROBBER FLIES, Family Asilidae p. 276
Long-legged, with thorax stout and abdomen usually long and tapering; top of head *hollowed out between eyes;* tarsi with *2 pads;* 3 ocelli.
a. *Diogmites neoternatus* Bromley; e. U.S.
b. *Laphria sacrator* Walker; se. Canada, ne. U.S.

HORSE and DEER FLIES, Family Tabanidae p. 274
Stout-bodied, often large; 3rd antennal segment *elongate and subdivided;* tarsi with 3 pads; calypters large.
DEER FLY, *Chrysops vittatus* Wiedemann; e. U.S.

BEE FLIES, Family Bombyliidae p. 278
Stout-bodied, *round-headed*, often with patterned wings; 3rd antennal segment *not subdivided;* tarsi with 2 pads; 3 or 4 posterior cells; anal cell often *open.*
Anthrax tigrinus (De Geer); U.S.

MYDAS FLIES, Family Mydidae p. 278
Large flies; black, often with an orange band on abdomen; antennae 4-segmented, last segment *swollen;* 1 ocellus or none; tarsi with *2 pads;* M_1 ends *in front of wing tip.*
Mydas clavatus (Drury); N. America.

LONG-LEGGED FLIES, Family Dolichopodidae p. 278
Small, usually metallic; Rs 2-branched; r-m cross vein in basal $\frac{1}{4}$ of wing or absent; anal cell small or absent; ♂ genitalia often large and folded forward under abdomen.
Dolichopus longipennis Loew; N. America.

Suborder Cyclorrhapha

Antennae 3-segmented, aristate; Rs 2-branched.

SYRPHID FLIES, Family Syrphidae p. 281
Generally beelike or wasplike, often brightly colored; spurious vein *nearly always present;* R_5 cell closed; anal cell long, closed near wing margin; proboscis short, fleshy.
a. *Syrphus torvus* Osten Sacken; N. America.
b. *Allograpta obliqua* (Say); N. America.

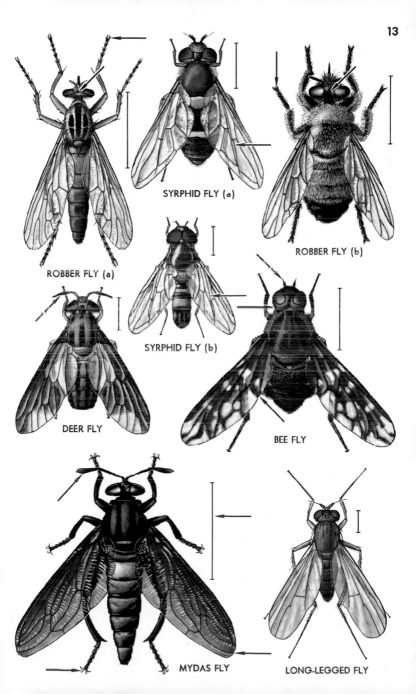

13

ROBBER FLY (a)

SYRPHID FLY (a)

ROBBER FLY (b)

SYRPHID FLY (b)

DEER FLY

BEE FLY

MYDAS FLY

LONG-LEGGED FLY

Plate 14

MUSCOID FLIES (Diptera, Division Schizophora)

Schizophora are flies with a frontal suture.

MUSCID FLIES, Family Muscidae p. 306
Calypters large; hypopleural and/or pteropleural bristles absent; R_5 cell parallel-sided or *narrowed distally;* 2A *does not reach* wing margin.
HOUSE FLY, *Musca domestica* Linn.; N. America.

LAUXANIID FLIES, Family Lauxaniidae p. 292
Calypters very small; Sc complete; postverticals converge; oral vibrissae absent; preapical tibial bristles present.
Minettia lupulina (Fabricius); N. America.

MARSH FLIES, Family Sciomyzidae p. 290
Calypters very small; antennae generally *projecting forward;* Sc complete; postverticals *slightly divergent;* oral vibrissae absent; preapical tibial bristles *present.*
Tetanocera plebeja Loew; N. America.

BLOW FLIES, Family Calliphoridae p. 306
Calypters large; often metallic; hypopleural and pteropleural bristles present; R_5 cell *narrowed distally;* postscutellum not developed; 2 notopleural bristles.
Phaenicia sericata (Meigen); N. America.

PICTURE-WINGED FLIES, Family Otitidae p. 288
Calypters very small; generally black and shiny, the wings often *banded;* Sc complete; anal cell usually with an acute distal projection posteriorly; oral vibrissae and preapical tibial bristles usually absent; postverticals parallel or convergent.
Ceroxys latiusculus (Loew); w. U.S.

FLESH FLIES, Family Sarcophagidae p. 307
Calypters large; similar to blow flies (above), but body not metallic and usually with 4 notopleural bristles.
Sarcophaga haemorrhoidalis (Fallén); N. America.

TACHINID FLIES, Family Tachinidae p. 306
Calypters large; hypopleural and pteropleural bristles present; R_5 cell *narrowed* or closed *distally;* postscutellum developed; arista usually *bare.*
a. *Archytas apicifer* (Walker); N. America.
b. *Bombyliopsis abrupta* (Wiedemann); N. America.

FRUIT FLIES, Family Tephritidae p. 288
Calypters very small; wings generally *banded* or patterned; Sc at apex bent forward at an abrupt angle; anal cell usually with an acute distal projection posteriorly.
APPLE MAGGOT, *Rhagoletis pomonella* (Walsh); e. U.S.

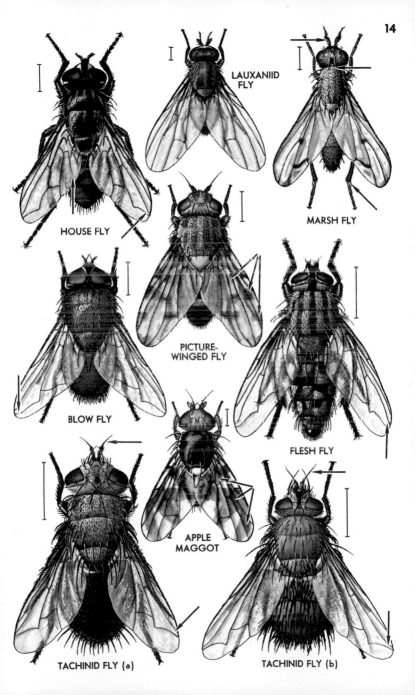

14

HOUSE FLY

LAUXANIID FLY

MARSH FLY

BLOW FLY

PICTURE-WINGED FLY

FLESH FLY

TACHINID FLY (a)

APPLE MAGGOT

TACHINID FLY (b)

Plate 15

SAWFLIES AND PARASITIC HYMENOPTERA
(Hymenoptera)

Wasplike insects.

ICHNEUMONS, Family Ichneumonidae   p. 322
Slender, wasplike; antennae *long*, many-segmented; hind trochanters *2-segmented;* ovipositor often long; costal cell and base
of cubital vein lacking; *2 recurrent veins.*
Lymeon orbum (Say); e. U.S.

BRACONIDS, Family Braconidae   p. 320
Similar to ichneumons but with *1 recurrent vein* or none; base
of cubital vein present or absent.
Rogas terminalis (Cresson); N. America.

COMMON SAWFLIES, Family Tenthredinidae   p. 314
Base of abdomen broadly joined to thorax; front tibiae with
2 spurs at tip; antennae *threadlike*, usually 9-segmented; FW
with 1 or *2 marginal cells*, and without an intercostal vein.
Tenthredo varipictus Norton; w. U.S.

HORNTAILS, Family Siricidae   p. 316
Base of abdomen *broadly joined to thorax;* front tibiae with *1 spur
at tip;* tip of abdomen with a *spear or spine;* pronotum wider
than long, and *shorter along midline* than at sides; usually
25–35 mm.
Tremex columba (Linn.); N. America.

CUCKOO WASPS, Family Chrysididae   p. 338
Metallic blue or green, usually with coarse sculpturing, 6–12
mm.; abdomen with 4 or fewer segments and concave beneath,
the last segment often with *teeth on rear edge;* venation somewhat reduced, with no closed cells in HW.
Chrysis smaragdula Fabricius; e. U.S.

VELVET ANTS Family Mutillidae   p. 344
Very hairy wasps, often brightly colored; ♀ wingless and ant-
like; 6–20 mm.
Dasymutilla occidentalis (Linn.), ♀ ; e. U.S.

CIMBICID SAWFLIES, Family Cimbicidae   p. 316
Base of abdomen *broadly joined to thorax;* antennae with 7 or
fewer segments and *slightly clubbed;* 18–25 mm.
ELM SAWFLY, *Cimbex americana* Leach; e. U.S.

TIPHIID WASPS, Family Tiphiidae, Subfamily Myzininae   p. 342
Black and yellow, long-legged, usually over 20 mm.; meso-
sternum with 2 lobes at rear; 1st discoidal cell *not usually long.*
Myzinum quinquecinctum (Fabricius); N. America.

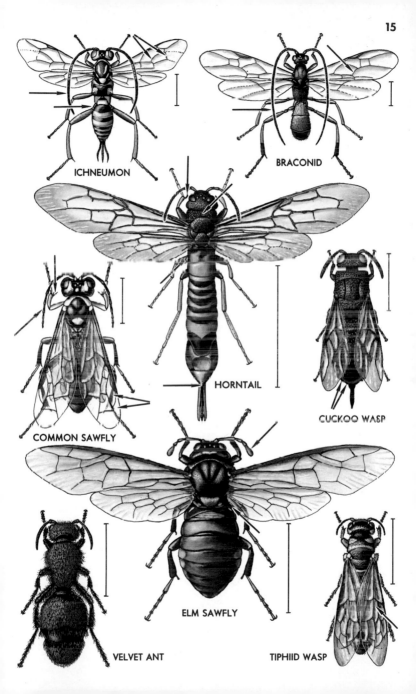

15

ICHNEUMON

BRACONID

HORNTAIL

COMMON SAWFLY

CUCKOO WASP

VELVET ANT

ELM SAWFLY

TIPHIID WASP

Plate 16

WASPS AND BEES (Hymenoptera)

Bees differ from wasps in being more hairy, with the body hairs branched, more robust, and the 1st segment of the hind tarsi is usually elongate and flattened.

SPHECID WASPS, Family Sphecidae   p. 350
 Pronotum *short and collarlike*, with a small rounded lobe on each side that does not reach tegula.
 MUD-DAUBER, *Sceliphron caementarium* (Drury); N. America.
 SAND WASP, *Bembix spinolae* Lepeltier; e. N. America.
 THREAD-WAISTED WASP, *Sphex procerus* (Dahlbom); N. America.

HALICTID BEES, Family Halictidae   p. 356
 Pronotum as in sphecid wasps; jugal lobe in HW as long as or *longer than submedian cell;* basal vein *arched.*
 Augochloropsis metallica (Fabricius); e. N. America.

SPIDER WASPS, Family Pompilidae   p. 346
 Long-legged, wings not folded longitudinally at rest; pronotum in side view more or less triangular, extending to tegula or nearly so; mesopleura with a transverse suture; 1st discoidal cell in FW *not unusually long.*
 Anoplius marginalis (Banks); e. N. America.

LARGE CARPENTER BEES,   p. 360
Family Apidae, Subfamily Xylocopinae
 Large, robust, blackish, resembling bumble bees; 2nd submarginal cell in FW triangular; dorsal surface of abdomen *bare and shining;* HW with a small rounded jugal lobe.
 Xylocopa virginica (Linn.); e. U.S.

BUMBLE BEES,   p. 360
Family Apidae, Subfamily Apinae, Tribe Bombini
 Robust, hairy, generally 15–25 mm., and black with yellow (rarely orange) markings; 2nd submarginal cell in FW *somewhat rectangular*, about as long as 1st; upper surface of abdomen *hairy;* HW without a jugal lobe.
 Megabombus pennsylvanicus (De Geer); N. America.

VESPID WASPS, Family Vespidae   p. 346
 YELLOWJACKETS & HORNETS, Subfamily Vespinae p. 348
 Wings *folded longitudinally* at rest; pronotum in side view triangular, reaching tegula; 1st discoidal in FW *very long;* middle tibiae with 2 spurs at tip; HW without a jugal lobe.
 Vespula maculifrons (Buysson); e. N. America.

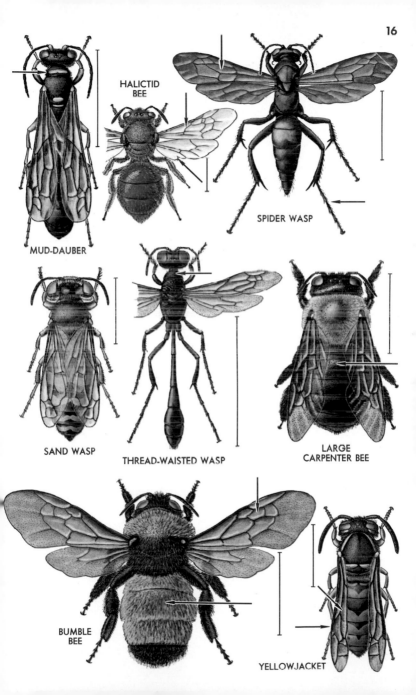

16

MUD-DAUBER

HALICTID BEE

SPIDER WASP

SAND WASP

THREAD-WAISTED WASP

LARGE CARPENTER BEE

BUMBLE BEE

YELLOWJACKET

LONG-HORNED BEETLES   See also Pl. 7
Family Cerambycidae
   Identification: Antennae nearly always *at least half as long as
body, often longer*. Body usually elongate and cylindrical. Eyes
generally notched, antennae often rising in notch. 3–50 mm.
Chrysomelidae (p. 198) are similar and have similar tarsi but
differ as follows: antennae nearly always less than half as long
as body; rarely over 12 mm.; generally oval in shape; eyes
usually oval.
   This is a very large family of about 1200 N. American species
which includes many attractive and brightly colored beetles.
Larvae of most species feed on solid tissues of dead or dying
plants, in trunks and branches of fallen or cut trees and shrubs.
Many species are very destructive to trees and cut logs.
   Subfamily Prioninae. These beetles differ from other cerambycids in having the prothorax *sharply margined laterally*.
Members of the genus *Prionus* are broad and somewhat flattened, brownish or black, and are among the largest of our
cerambycids (some nearly 3 in.). Beetles in the genus *Parandra*
are called aberrant long-horned beetles; they have short
antennae that extend only to about the middle of the pronotum,
and the 4th tarsal segment is large enough to be seen easily;
larvae bore in heartwood.
   Subfamily Lamiinae (see also Pl. 7). The last segment of the
maxillary palps is cylindrical and pointed apically. The genus
Saperda contains important pest species, and some are strikingly
colored. The Roundheaded Apple Tree Borer, *S. candida*
Fabricius, is light brown, with 2 white stripes; the larva bores
in apple and other trees. The Elm Borer (Pl. 7), *S. tridentata*
Olivier, is gray, attractively marked with orange; the larva
bores under bark of dead and dying elms. Sawyer beetles
(*Monochamus*) are usually 1 in. or more and black or mottled
gray; antennae are sometimes twice as long as the body; larvae
bore in freshly cut evergreens.
   Subfamily Lepturinae (see also Pl. 7). The last segment of the
maxillary palps is blunt apically. Lepturinae usually have oval
or only slightly notched eyes, and Cerambycinae usually have
distinctly notched eyes, partly surrounding bases of the antennae. Most Lepturinae have the elytra (FW) broadest at
base and narrowing posteriorly, giving the body a broad-
shouldered appearance. Many species are found on flowers and
are often brightly colored. Members of the genera *Anoplodera*
and *Desmocerus*, which often occur on flowers, frequently bear
yellow, red, or black markings.
   Subfamily Cerambycinae (see also Pl. 7). The last segment
of the maxillary palps is blunt apically. These beetles can be
distinguished from Lepturinae by the eyes — usually notched
and partly surrounding bases of antennae; they do not have the
broad-shouldered appearance of many Lepturinae. A large

group, and many are common. One of the best known is the Locust Borer (Pl. 7), *Megacyllene robiniae* (Forster); adult about 20 mm. and black, with narrow transverse and oblique yellow bands. It is common on goldenrod in fall, and larvae are serious pests of living black locust.

Superfamily Chrysomeloidea

Tarsi as in Cerambycoidea (*apparently 4-4-4, actually 5-5-5*, 4th segment small and concealed). All of these beetles are plant feeders. Many are serious pests.

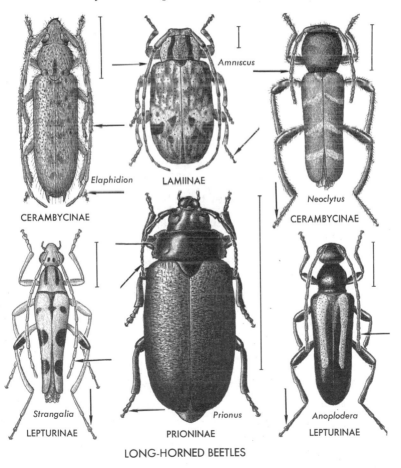

Amniscus

Elaphidion   LAMIINAE

CERAMBYCINAE

Neoclytus

CERAMBYCINAE

Strangalia

LEPTURINAE

Prionus

PRIONINAE

Anoplodera

LEPTURINAE

LONG-HORNED BEETLES

LEAF BEETLES Family Chrysomelidae **See also Pl. 8**
Identification: Antennae nearly always *less than half as long as body.* 1–20 mm., rarely over 12 mm. Body generally oval. Eyes usually not notched. Similar to Cerambycidae and Bruchidae; see pp. 196 and 200 for differences.

This is a very large family (nearly 1400 N. American species), and many species are quite common. Adults occur on flowers and foliage; larvae feed on foliage and roots. Many are serious pests of cultivated plants and a few act as vectors of plant diseases. Numerous subfamilies are recognized, but only the most important are discussed here.

Tortoise Beetles, Subfamily Cassidinae (see also Pl. 8). *Broadly oval* or nearly circular, often with body expanded and flattened, resembling a tiny turtle; the head is largely or *completely concealed from above.* Larvae are oval, flat, spiny, and bear a forked process at posterior end of the body to which are attached cast skins, excrement, and debris; this process is held over the body much like a parasol. A few tortoise beetles resemble ladybird beetles; they differ in the segmentation of tarsi (*apparently 4-4-4* in tortoise beetles, 3-3-3 in ladybirds). Some are very brilliantly colored in life with golden or silvery markings, but these colors often fade after death.

Leaf-mining Leaf Beetles, Subfamily Hispinae (see also Pl. 8). These are 4–7 mm., usually brownish, and the elytra (FW) *bear ridges* between which are rows of punctures; they are parallel-sided or slightly widened apically, and the pronotum is narrower than base of the elytra. Most larvae are leaf miners. One common species, the Locust Leaf Miner, *Xenochalepus dorsalis* (Thunberg), is a serious pest of black locust.

Case-bearing Leaf Beetles, Subfamilies Clytrinae, Cryptocephalinae (see also Pl. 8), and Chlamisinae. Members of these groups are small (usually 6 mm. or less), broadly cylindrical, with head buried in prothorax *nearly to eyes,* and last dorsal abdominal segment (pygidium) not covered by elytra. Clytrinae have smooth elytra, but serrate or pectinate antennae; Cryptocephalinae have smooth elytra, and antennae are *threadlike* or slightly clubbed; Chlamisinae have elytra tuberculate. These beetles are dark-colored, often with red, yellow, or orange markings. They play dead when disturbed. Larvae construct portable cases.

Subfamily Galerucinae. These beetles resemble the Criocerinae (p. 200), but are relatively soft-bodied, the pronotum is usually *margined laterally,* and the head is not narrowed posteriorly into a neck. The Spotted Cucumber Beetle, *Diabrotica undecimpunctata howardi* Barber, and the Striped Cucumber Beetle, *Acalymma vittatum* (Fabricius), feed on cucurbits and other plants. They damage plants by their feeding and act as vectors of cucurbit wilt.

Subfamily Eumolpinae (see also Pl. 8). These beetles are *oval* and convex. They resemble Chrysomelinae (p. 200), but differ in having front coxae rounded and 3rd tarsal segment *bilobed beneath.* Many are metallic or yellow and spotted. The Dogbane Beetle (Pl. 8), *Chrysochus auratus* (Fabricius), is very attractive — iridescent blue-green with a coppery tinge, 8–10 mm., occurs on dogbane and milkweed.

Long-horned Leaf Beetles, Subfamily Donaciinae. The donaciines are elongate, with *long antennae*, and resemble some

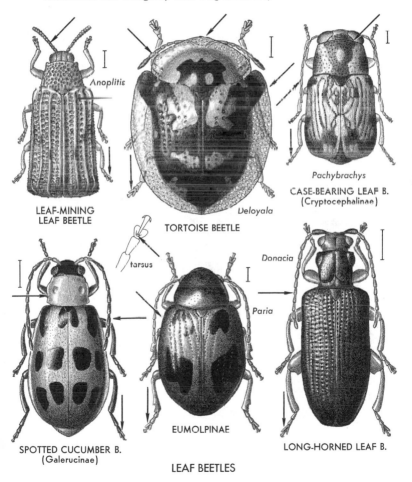

Anoplitis

LEAF-MINING
LEAF BEETLE

Deloyala

TORTOISE BEETLE

tarsus

Pachybrachys

CASE-BEARING LEAF B.
(Cryptocephalinae)

Donacia

Paria

SPOTTED CUCUMBER B.
(Galerucinae)

EUMOLPINAE

LONG-HORNED LEAF B.

LEAF BEETLES

of the cerambycids. They are mostly black or brown, usually with a metallic luster, and range from 5.5 to 12.0 mm. Larvae feed in submerged parts of aquatic plants; adults are found on flowers and foliage of water lilies and other aquatic plants, and are active, fast-flying, often difficult to capture.

Flea Beetles, Subfamily Alticinae. Flea beetles are mostly 2–5 mm., black or bluish (sometimes with light markings) and with *enlarged* hind femora; antennae are close together at the base and front coxae are usually conical. The common name refers to their jumping habits. Some flea beetles are important pests of various cultivated plants; larvae feed in roots of the host plant and adults feed on the leaves. Adult feeding produces holes in the leaves. A heavily infested plant looks as if tiny shot had been fired into the leaves.

Subfamily Criocerinae (see also Pl. 8). These have the pronotum rounded (its base narrower than base of FW, or elytra), the punctures of the elytra are in rows, and the head is prominent and narrowed posteriorly. The group is small, but contains some important crop pests: *Crioceris* (2 species) attacks asparagus, *Lema trilineata* (Olivier) attacks potatoes, and *Oulema melanopus* (Linn.), the Cereal Leaf Beetle, attacks grains.

Subfamily Chrysomelinae (see also Pl. 8). Most members of this group are *oval to nearly circular,* very convex, often brightly colored, and have the head sunk in the prothorax almost to the eyes; antennae are widely separated at base, and pronotum is margined laterally. Most species feed on various weeds and are of little economic importance. The best-known and most important species is the Colorado Potato Beetle, *Leptinotarsa decemlineata* (Say), a serious pest of potatoes; it is large, orange-yellow, and has longitudinal black stripes on the elytra.

SEED BEETLES Family Bruchidae

Identification : Shape distinctive: body oval or egg-shaped, broadest posteriorly; head concealed from above, prolonged into a short broad snout. Antennae *clubbed* or serrate, sometimes pectinate. Black or brown, often mottled or marked with patches of whitish or brownish pubescence. Elytra (FW) short, *exposing tip of abdomen.* 1–10 mm. Similar to Chrysomelidae (p. 198), although differing in body shape and in having a snout.

Seed beetles are usually found on foliage, or in stored peas, beans, or other seeds; larvae of most species feed inside seeds, and some seriously damage beans or peas. The Bean Weevil, *Acanthoscelides obtectus* (Say), attacks beans in storage or in the field and may completely destroy them; in a heavy infestation, as many as a dozen beetles may develop in a single bean. The Pea Weevil, *Bruchus pisorum* (Linn.), attacks peas in the field; larvae consume the central portion of the pea.

Superfamily Curculionoidea

Head prolonged into a more or less distinct snout. Tarsi 5-5-5, usually apparently 4-4-4.

PRIMITIVE WEEVILS Family Brentidae

Identification: FW *elongate and parallel-sided*. Prothorax pear-shaped. Head with *a long or short straight beak*. Femora stout, toothed. Antennae threadlike or *beadlike*. Dark brown to black-ish. FW with orange marks. 7–30 mm.

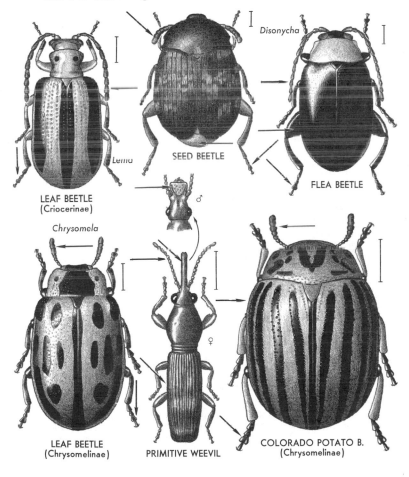

LEAF BEETLE
(Criocerinae)

Lema

SEED BEETLE

Disonycha

FLEA BEETLE

Chrysomela

LEAF BEETLE
(Chrysomelinae)

PRIMITIVE WEEVIL

COLORADO POTATO B.
(Chrysomelinae)

Adults of this group are found under bark and in worm-eaten wood, where they feed on fungi, the sap from tree wounds, or wood-eating insects. Larvae are wood borers.

FUNGUS WEEVILS Family Anthribidae

Identification: Robust, elongate to oval. Usually brownish and often mottled with patches of white, gray, straw-colored, brown, or black pubescence. Beak short, broad. Antennae *clubbed*, not elbowed. Pronotum often with sharp lateral margins posteriorly. 1–11 mm.

Adult anthribids are usually found on dead twigs and branches, under bark, or on fungi (generally woody fungi), and are not common. Most are good fliers and some jump. Larvae live in plant materials on which adults are found. The eastern species most often encountered is *Euparius marmoreus* (Olivier), which is 3.5–8.5 mm. and brown mottled with patches of white, black, ashy, and brown pubescence. Adults and larvae are found on polypore fungi.

SNOUT BEETLES Family Curculionidae   See also Pl. 8

Identification: Snout usually *well developed*. Antennae *clubbed and nearly always elbowed*. Palps small and rigid, often concealed within mouth. Labrum absent. 1–35 mm.

This is one of the largest families of insects (over 2500 N. American species), and its members are common. Practically all are plant feeders; large numbers are serious pests of cultivated plants. Many snout beetles chew holes in fruits, nuts, and other parts of the plant. Adults often bear pubescence or scales, and frequently play dead when disturbed. From N. America 42 subfamilies are recognized, only a few of which can be mentioned here.

Subfamily Cyladinae (Pl. 8). Antlike, 5–6 mm. Pronotum reddish, FW blue-black. Antennae not elbowed, last segment long, swollen. Our only representative of this group is the Sweetpotato Weevil (Pl. 8), *Cylas formicarius elegantulus* (Summers), a pest of sweet potatoes occurring in the South.

Subfamily Apioninae. Black or gray. 1.0–4.5 mm. Somewhat pear-shaped. Antennae *not elbowed*. Trochanters elongate. Larvae bore in seeds, stems, and other parts of various plants (chiefly legumes); adults usually occur on foliage.

Subfamily Baridinae (not illus.). Part of thoracic pleura (mesepimera) visible from above between prothorax and FW. This is the largest subfamily of snout beetles (about 500 N. American species). Several are important pests.

Broad-nosed Weevils, chiefly Subfamilies Leptopiinae, Brachyrhininae, and Thylacitinae (see also Pl. 8). Beak *short, quadrate*, often widened apically, usually with 1 or more longitudinal grooves. Mandibles large, with a small cusp that breaks off and leaves a scar. Antennal scrobe (groove in beak into which basal antennal segment fits) only vaguely defined. Most species are flightless because the elytra are grown together or the hind wings are reduced. White-fringed beetles (*Graphognathus*, Thylacitinae) are important pests in the Southeast, where they feed on many cultivated plants; they reproduce parthenogenetically, and no males are known.

Acorn and Nut Weevils, Subfamily Curculioninae. Snout *slender and very long*, as long as body or longer. Femora with a stout triangular tooth. Eyes often partly covered by prothorax.

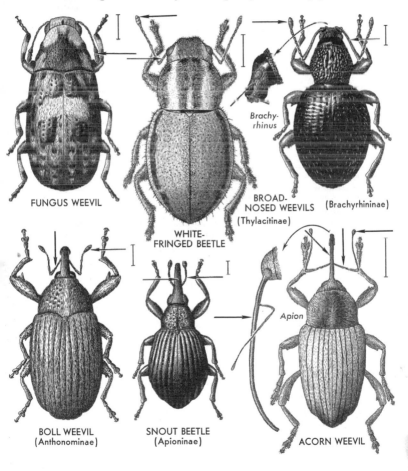

FUNGUS WEEVIL

WHITE-FRINGED BEETLE
(Thylacitinae)

Brachy-rhinus

BROAD-NOSED WEEVILS   (Brachyrhininae)

BOLL WEEVIL
(Anthonominae)

SNOUT BEETLE
(Apioninae)

Apion

ACORN WEEVIL

FW with narrow scales. Larvae develop in the fruit of various nut-producing trees.

Subfamily Anthonominae (see also p. 203). Antennae *elbowed and clubbed*, rising near apex of snout, the basal segment fitting into a well-defined scrobe (groove in beak into which basal antennal segment fits). Front coxae contiguous, located in about middle of prothorax. Snout *fairly long*, not fitting into a prosternal groove at rest. An important pest in this group is the Boll Weevil (illus., p. 203), *Anthonomus grandis* Boheman (about 5 mm., yellowish brown), which does much damage to cotton; adults feed on the seedpod (boll) and lay eggs in the feeding hole; larvae feed inside the bolls and eventually destroy them. This is a large subfamily, with about 200 N. American species.

Subfamily Cryptorhynchinae. Beak at rest fits into a groove in prosternum. Antennae *elbowed and clubbed*. Eyes oval and partly covered by prothorax when beak is in prosternal groove. FW usually wider than prothorax, giving the insect *a broad-shouldered appearance*. The Plum Curculio, *Conotrachelus nenuphar* (Herbst), is a common species and a serious pest of stone fruits; injury results from both adult and larval feeding.

Billbugs and Grain Weevils, Subfamily Rhynchophorinae. Antennae rise *near eyes*, the *basal segment* not fitting into the short scrobe (groove) and *extending past posterior margin of eye*. 6 antennal segments between basal segment and club; 1st segment of antennal club *enlarged and shining*. Pygidium (last dorsal abdominal segment) usually exposed. Members of this group are 3–31 mm. Some of our largest snout beetles belong to the group. Billbugs are large snout beetles that feed on grasses; larvae bore in stems and adults feed on foliage. Grain weevils (*Sitophilus*) are brownish and 3–4 mm.; they attack stored grain and are often serious pests.

PINHOLE BORERS Family Platypodidae

Identification: *Very elongate-slender, parallel-sided*. Brownish. Tarsi long, slender, *1st segment long*. Head visible dorsally, as wide as or slightly wider than pronotum. 2–8 mm.

Pinhole borers attack trees, and their burrows extend deep into the heartwood; adults and larvae feed on a fungus in the burrows. These beetles usually attack weakened or unhealthy trees, and often bore into felled logs. Adults occasionally fly to lights. The group is small, with only 7 species in N. America.

BARK or ENGRAVER BEETLES and AMBROSIA BEETLES Family Scòlytidae

Identification: Elongate, cylindrical. Head visible dorsally or concealed, narrower than pronotum. Tarsi short, 1st segment not elongate. Antennae elbowed and *clubbed*. Brownish to black. 1–9 (mostly 1–3) mm.

Nearly all scolytids bore into bark or wood, both as larvae and as adults. Adults spend most of their lives in their burrows, leaving them only long enough to find a new host. Bark or engraver beetles burrow just under the bark; ambrosia beetles burrow into the heartwood and feed on fungi in the galleries. Bark beetles are very common and usually attack weakened, dying, dead, or recently cut trees; some attack living trees and since they often kill the tree are very serious pests, particularly in the West.

Bark, or engraver, beetles excavate patterns (galleries) under the bark, each species making a characteristic pattern. The adults

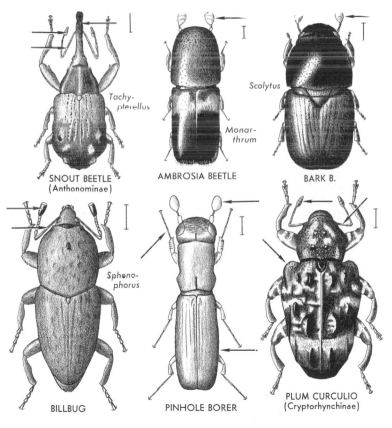

SNOUT BEETLE
(Anthonominae)

Tachypterellus

AMBROSIA BEETLE

Monarthrum

Scolytus

BARK B.

BILLBUG

Sphenophorus

PINHOLE BORER

PLUM CURCULIO
(Cryptorhynchinae)

come in first and excavate 1 or more brood galleries, along the sides of which they lay their eggs; when the larvae hatch they bore away from the brood gallery, their galleries increasing in size as they tunnel, thus forming the typical pattern under the bark. Each species attacks a particular species of tree (or 1 of several related species). The galleries permit the entrance of fungi, which often rot the bark. The Elm Bark Beetle, *Scolytus multistriatus* (Marsham), is the chief vector of Dutch elm disease — a disease that has killed thousands of elms in the e. U.S.

Twisted-winged Parasites:
Order Strepsiptera

Identification: *Minute* insects, 0.5–4.0 mm., the sexes quite different. ♂: blackish; FW *reduced to short clublike structures*, HW *large and fanlike*, with a few weak radiating veins; eyes bulging; antennae 4- to 7-segmented, *some segments with a long lateral process;* tarsi *2- to 5-segmented;* mouth parts chewing but usually reduced. ♀: wingless; usually lacking legs and antennae; mouth parts generally vestigial. Metamorphosis complete, with hypermetamorphosis (see p. 41).

Similar orders: Male strepsipterans are somewhat beetlelike, but can be recognized by the characteristic wings and antennae. Females are generally saclike and without appendages, and must usually be recognized by their location in a host insect.

Immature stages and habits: Strepsipterans are internal parasites of other insects. Adult males on emergence leave the host and fly about; females of most species never leave the host. Each female produces many (a few thousand) tiny larvae, which have well-developed legs and are very active; they leave the host and enter another host, where they molt to a legless stage. Most strepsipterans have bees or various hoppers (Homoptera) as hosts; a few attack insects in other orders. The host is injured but rarely killed; the sex organs may be damaged or the shape or color of the abdomen may be changed. Parasitized hosts often can be recognized by the body of the parasite protruding from between the abdominal segments. If such hosts are caged, male strepsipterans frequently can be reared from them. Adult males are very seldom encountered.

Importance: These insects are of no economic importance.

Classification: Four families, which are separated by the tarsal and antennal characters of males.

No. of species: World, 300; N. America, 60.

MENGEIDS　Family Mengeidae　　　　　　　　**Not illus.**
　　Identification: Tarsi 5-segmented, with 2 claws. Antennae

7-segmented, 3rd and 4th segments with long lateral processes.

Only 1 member of this family, *Triozocera mexicana* Pierce, is known from the U.S.; the female of this species is unknown, but the male has been taken in Texas. Females of some European mengeids are free-living as adults, and are usually found under stones; these mengeids are parasites of bristletails (Thysanura, see p. 60).

STYLOPIDS Family Stylopidae
Identification: Tarsi *4-segmented*, without claws. Antennae 4- to 6-segmented, *only 3rd segment with a lateral process*

This family is the largest in the order, with about 40 species in N. America. Most stylopids are parasites of bees (chiefly Andrenidae and Halictidae); a few parasitize vespid or sphecid wasps.

HALICTOPHAGIDS Family Halictophagidae
Identification: Tarsi *3-segmented*, without claws. Antennae 7-segmented, *3rd to 5th segments with long lateral processes*, last segment elongate.

This family is small (14 N. American species) but widely distributed. Most species are parasites of leafhoppers, tree-hoppers, and spittlebugs; 1 species attacks planthoppers and pygmy mole crickets.

ELENCHIDS Family Elenchidae   **Not illus.**
Identification: Tarsi 2-segmented, without claws. Antennae 4-segmented, 3rd segment with a long lateral process.

The members of this small family are parasites of planthoppers (Fulgoroidea).

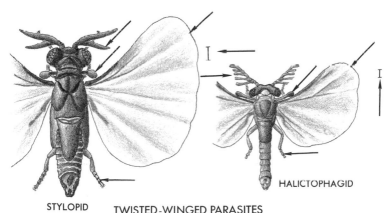

STYLOPID   TWISTED-WINGED PARASITES   HALICTOPHAGID

Scorpionflies and Their Allies: Order Mecoptera

Identification: Small to medium-sized. Body usually slender and relatively soft. *Long-faced*. Mouth parts chewing, at end of a long snoutlike structure. 4 membranous wings (wings rarely absent or vestigial); long and narrow, HW about same size as FW, with rather generalized venation but with extra cross veins; wings often spotted or transversely banded. Legs usually long, slender, the tarsi *5-segmented*, with 1 or 2 claws. Antennae *threadlike, about half body length*. Metamorphosis complete.

Similar orders: (1) Neuroptera (p. 140): without a long-faced appearance; usually with numerous costal cross veins. (2) Diptera (p. 260): only 1 pair of wings. (3) Hymenoptera (p. 312): without a long-faced appearance; HW smaller than FW; wing venation different.

Immature stages: Eggs are generally laid on the ground, and larvae live in or on the surface of the ground or in moss. Larvae are usually caterpillarlike, with 8 pairs of short prolegs. They feed on dead insects and other organic materials.

Habits: Adults are usually found in areas of fairly dense vegetation; some are predaceous on other insects, others are omnivorous or are scavengers. The name "scorpionfly" is derived from the fact that the male genitalia of some species are large and conspicuous and carried curved upward over the back like the sting of a scorpion.

Importance: Scorpionflies are not of economic importance. They do not bite or sting.

Classification: Four families, separated chiefly by wing and leg characters.

No. of species: World, 400; N. America, 85.

SNOW SCORPIONFLIES   Family Boreidae

Identification: Dark-colored, *2–5 mm*. Wings bristlelike or hooklike in ♂, *small and scalelike in ♀*. 10th abdominal segment of ♀ *prolonged posteriorly into an ovipositorlike structure* about half as long as abdomen (other ♀ Mecoptera lack such a structure and have abdomen tapering posteriorly). Tarsi with 2 claws.

   Snow scorpionflies occur in and feed on mosses. Adults appear in winter or early spring, and are usually seen on the snow (hence the common name). The bristlelike or hooklike wings of the male are used in grasping the female at the time of mating. In N. America there are 15 species, 2 in the East and 13 in the West (California to Alaska). They are not often collected.

EARWIGFLIES Family Meropeidae

Identification: Ocelli absent (present in other Mecoptera). Wings relatively broad, Rs and M with *5 or more branches*. Tarsi with 2 claws.

This group is represented in N. America by a single rare species, *Merope tuber* Newman, which occurs in the East Coast states from Georgia to se. Canada. This insect is 10–12 mm., and the male has forcepslike terminal abdominal appendages that resemble the cerci of earwigs (see illus., p. 99).

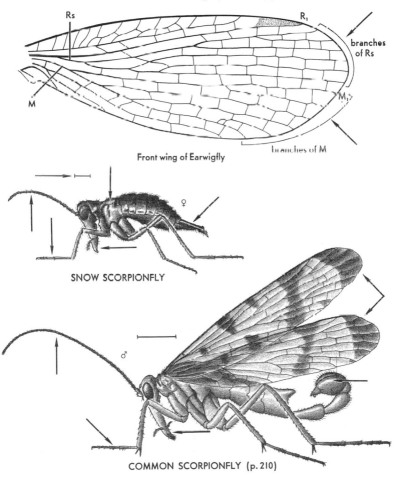

Front wing of Earwigfly

SNOW SCORPIONFLY

COMMON SCORPIONFLY (p. 210)

COMMON SCORPIONFLIES See also p. 209

Family Panorpidae

Identification: Rs 5-branched (R_2 *forked*), M 4-branched. Tarsi *normal*, with 2 claws. ♂ genitalia *large and bulbous, usually curved upward* and forward like the sting of a scorpion.

Most of the 40-odd N. American panorpids are 15–20 mm., brownish, with dark spots or bands on wings. They are widely distributed and fairly common. Adults feed chiefly on dead or dying insects, sometimes on fruits or nectar.

HANGINGFLIES   Family Bittacidae

Identification: Rs and M 4-branched (R_2 *not forked*). Wings *narrower at base* than in Panorpidae. Tarsi with 1 claw; 5th segment *capable of being folded back on 4th.* ♂ genitalia enlarged but not bulbous.

Our 9 species of bittacids are widely distributed and often fairly common. Most are light brown and long-legged, about 20–25 mm., and often resemble large crane flies. Some have brownish areas on certain cross veins, and 1 species has blackish wing tips. This species holds its wings outstretched at rest; others fold the wings back over the abdomen. Bittacids spend most of their time hanging from vegetation by their front legs. They feed on various small insects which they capture with their raptorial hind feet. One California species is wingless.

Caddisflies: Order Trichoptera

Identification: Slender, elongate, mothlike insects. 1.5–25.0 mm. Antennae long and threadlike, usually as long as body or longer. 4 membranous wings (wings vestigial or absent in ♀ of a few species): HW a little shorter than FW, the wings (especially FW) hairy; wings *held rooflike over body at rest;* wing venation rather generalized, with M *usually 4-branched in FW* and 3-branched in HW, and Cu *3-branched;* wings generally with *a small spot* in fork of R_{4+5}. Mouth parts reduced, best described as "sponging," palps well developed. Legs relatively long and slender. Tarsi *5-segmented.* Metamorphosis complete, larvae aquatic.

Similar orders: Lepidoptera (p. 218): wings covered with scales; usually with a coiled proboscis; maxillary palps generally absent or vestigial; M in FW 3-branched; no spot in fork of R_{4+5}.

Immature stages: Larvae are caterpillarlike, with a pair of hooklike appendages at posterior end of the body, and usually with filamentous gills on the abdominal segments. They are fairly active insects, and some move backward more often (and faster) than forward. Many larvae construct portable cases of various small objects fastened together with a gluelike substance or with silk;

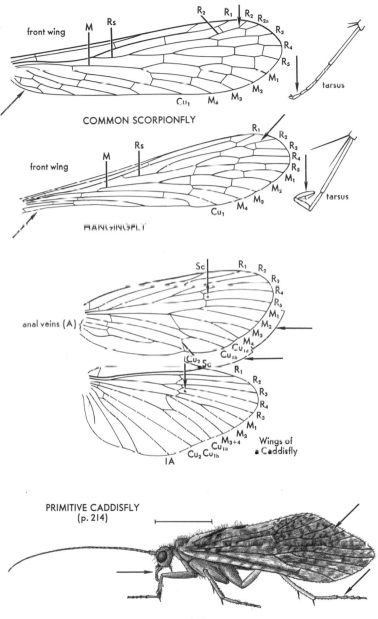

front wing

M   Rs

R₂   R₁   R₂   R₂ₐ
                    R₃
                    R₄
                    R₅
                    M₁
                    M₂
              M₄ M₃
        Cu₁

tarsus

COMMON SCORPIONFLY

front wing

M   Rs

R₁
  R₂
  R₃
  R₄
  R₅
  M₁
  M₂
M₄ M₃
  Cu₁

tarsus

HANGINGFLY

Sc   R₁
        R₂
        R₃
        R₄
        R₅
        M₁
anal veins (A)   M₂
                  M₃
          Cu₁d M₄
        Cu₂ Cu₁b

Sc   R₁
        R₂
        R₃
        R₄
        R₅
        M₁
      M₃₊₄ M₂
  Cu₂ Cu₁a
1A   Wings of a Caddisfly

PRIMITIVE CADDISFLY
(p. 214)

211

as the larva grows the case may be enlarged, or the larva may leave and construct a new and larger case. Larval cases vary considerably, both in shape and in the materials of which they are made; they may be slender or oval, straight or curved, and made of bits of leaves, twigs, sand grains, or small pebbles. Each species makes a characteristic type of case. Larvae of some species construct silken nets and feed on material caught in the nets. Most larvae feed on plant materials, but a few (which usually do not make cases) are predaceous. When a larva is full-grown it attaches its case to some object in the water, then closes the case and pupates inside. The pupa, which usually has well-developed mandibles, cuts or works its way out of the case, crawls out of the water onto a stone or other object, and undergoes its final molt to the adult. Emergence of some species occurs at surface of water.

Habits: Caddisflies are generally rather dull-colored insects, and their flight is jerky and erratic. They are largely nocturnal, spending the day resting in cool dark places; they are strongly attracted to lights at night. Eggs are laid in masses or strings, usually on stones or other objects in the water but occasionally on objects overhanging or near the water.

Importance: Caddisflies, particularly the immature stages, are an important item in the food of many freshwater fish.

Classification: The N. American species are arranged in 17 families; these families are not grouped into superfamilies or suborders.

No. of species: World, 4450; N. America, 975.

Identification of families of Trichoptera: The chief characters used in separating families of caddisflies are those of the ocelli, maxillary palps, thoracic warts, and tibial spurs. Ocelli may be present or absent. The maxillary palps are usually 5-segmented, and the segments may differ in size and shape; the palps sometimes differ in form or segmentation in the two sexes. Thoracic warts are wartlike swellings on the dorsal surface of the thorax; they vary in size, number, and location in different families. Mounting a caddisfly on a pin often damages or destroys these structures; they are more easily studied in specimens preserved in alcohol. The tibial spurs are large, usually brownish, movable structures; there may be 1 or 2 at the apex of the tibia, and 1 or 2 located proximad of the apex. The leg spines are small, usually black structures.

Identification of the families of Trichoptera may be facilitated by dividing them into the following 5 groups:

1. Ocelli present or absent; some hairs on the wings clubbed; very small (1.5–6.0 mm.): Hydroptilidae.
2. Ocelli present; no wing hairs clubbed: Philopotamidae, Rhyacophilidae, Phryganeidae, and Limnephilidae.
3. Ocelli absent; no wing hairs clubbed; terminal segment of maxillary palps much longer than the other segments, and with faint cross striations: Psychomyiidae and Hydropsychidae.

4. Ocelli absent; no wing hairs clubbed; terminal segment of maxillary palps not much longer than other segments, and without cross striations; middle tibiae with preapical spurs, and with or without a row of black spines: Molannidae, Calamoceratidae, Odontoceridae, Goeridae, Lepidostomatatidae, and some Brachycentridae.

5. Similar to Group 4, but middle tibiae without preapical spurs and with a row of black spines: Leptoceridae, Helicopsychidae, Beraeidae, Sericostomatidae, and some Brachycentridae.

MICRO-CADDISFLIES Family Hydroptilidae **See also p. 215**
Identification: Very small (1.5–6.0 mm.), usually with a salt-and-pepper coloration and relatively short antennae. Very hairy, with some wing hairs clubbed. Posterior portion of scutellum forms a flat triangular area with steep sides; meso-

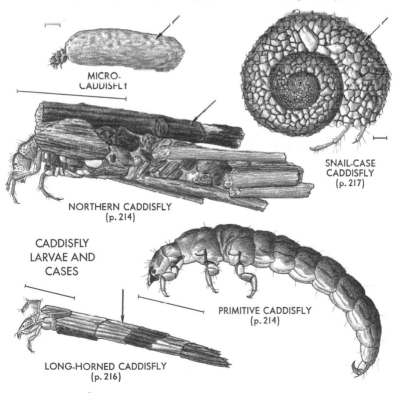

MICRO-
CADDISFLY

SNAIL-CASE
CADDISFLY
(p. 217)

NORTHERN CADDISFLY
(p. 214)

CADDISFLY
LARVAE AND
CASES

PRIMITIVE CADDISFLY
(p. 214)

LONG-HORNED CADDISFLY
(p. 216)

scutum and scutellum *without warts*. Ocelli present or absent. Front tibiae with 1 spur or none.

Members of this group are widely distributed and fairly common. Larvae occur in ponds and streams. *Cases made by the last instar are often purse-shaped*, with both ends open.

FINGER-NET CADDISFLIES Family Philopotamidae
Identification: Usually brownish, with gray or blackish wings, and 6–9 mm. 5th segment of maxillary palps *2 or 3 times as long as 4th*.

Larvae live in rapid streams, where they construct finger-shaped nets attached to stones; they pupate in cases made of pebbles and attached to the underside of stones.

PRIMITIVE CADDISFLIES   See also pp. 211, 213
Family Rhyacophilidae
Identification: Fifth segment of maxillary palps *not much longer than 4th*, and 2nd segment about same length as 1st.

This is a large and widely distributed group. Adults are 3–13 mm. and usually brownish with mottled wings; antennae are relatively short. Larvae live in rapid streams; some make turtle-shaped cases of small pebbles, and others (which are predaceous) do not make cases.

LARGE CADDISFLIES Family Phryganeidae
Identification: Maxillary palps 4-segmented in ♂ and 5-segmented in ♀, in ♀ with 2nd segment much longer than 1st and 5th segment not much longer than 4th. Ocelli *present*. Front tibiae with at least 2 spurs, middle tibiae with 4 spurs.

Adult phryganeids are 14–25 mm. and have the wings mottled with gray and brown. Larvae of most species live in marshes and ponds, and make long slender cases of spirally arranged strips of plant material.

NORTHERN CADDISFLIES Family Limnephilidae   p. 213
Identification: Maxillary palps 3-segmented in ♂ and 5-segmented in ♀; palps of ♀ as in Phryganeidae. Front tibiae with 1 spur or none, middle tibiae with 2 or 3.

This is a large and widely distributed group, but most species are northern or western in distribution. Adults are 7–23 mm. and usually brownish with dark wing markings. Larvae occur chiefly in ponds and slow-moving streams, and *construct various types of cases*.

TRUMPET-NET and TUBE-MAKING CADDISFLIES
Family Psychomyiidae
Identification: Mesoscutum with *a pair of small warts*. 5th segment of maxillary palps with *faint cross striations*.

This is a large and widely distributed group of small caddisflies, mostly brownish and 4–11 mm. Larvae occur in a variety of aquatic habitats. Some larvae construct a trumpet-shaped net, usually in running water; others burrow into the sand at the bottom of streams and cement the walls of the burrow to form a fairly rigid tube.

NET-SPINNING CADDISFLIES   Family Hydropsychidae
Identification: Mesoscutum *without warts*.

Adults of this group are found along streams. They are widely distributed and fairly common. Larvae construct a cup-shaped net, with the open side facing upstream, and spend most of their time in a retreat near the net.

MOLANNIDS   Family Molannidae
Identification: Middle femora with *a row of 6–10 spines*.

Our 5 species of molannids occur east of the Rocky Mts. from s. Canada south to Oklahoma, Illinois, and Pennsylvania. Adults are 10–15 mm. Larvae occur in streams with a sandy bottom; their cases are cylindrical tubes of sand grains, with lateral expansions.

CALAMOCERATIDS   Family Calamoceratidae   p. 217
Identification: Scutellum small, rectangular, *without warts;*

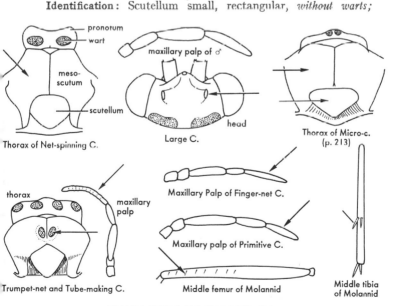

CHARACTERS OF CADDISFLIES

mesoscutal warts represented by *2 rows of small bristly spots* extending length of scutum. Maxillary palps 5- or 6-segmented.

Two of the 5 U.S. species of calamoceratids occur in the eastern states, and 3 occur in the West (Arizona, California, Alaska). They are relatively rare. The larva of 1 eastern species makes its case in a hollowed-out twig.

ODONTOCERIDS Family Odontoceridae
Identification: Scutellum large and domelike, with *a single wart occupying most of sclerite.* Tibial spurs not hairy.

Ten rather rare species of odontocerids occur in N. America, 5 in the East and 5 in the West. The known larvae occur in riffles of streams and make cylindrical, slightly curved cases of sand grains.

GOERIDS Family Goeridae   **Not illus.**
Identification: Scutellum with a single elongate wart occupying only central part of sclerite. Tibial spurs hairy. Maxillary palps 3-segmented in ♂ and 5-segmented in ♀.

This is a small group of rather rare caddisflies, most of them restricted to the East. Three N. American species have been reared; their larvae occur in streams and make cylindrical cases of sand grains and small pebbles, with 1 or 2 larger pebbles glued to each side.

LEPIDOSTOMATIDS Family Lepidostomatidae
Identification: Scutellum with *a pair of warts.* Maxillary palps 1- or 3-segmented in ♂ and 5-segmented in ♀. Preapical tibial spurs long, hairy, located about middle of tibia.

This group is widely distributed but its members are not common. Larvae usually occur in streams or springs; their cases are often square in cross section. Males of some species have rather bizarre characters, such as peculiarly modified maxillary palps, leaflike legs, or widened wings.

BRACHYCENTRIDS Family Brachycentridae
Identification: Scutellum with *a pair of warts.* Maxillary palps 3-segmented in ♂ and 5-segmented in ♀. Middle tibiae with or without apical spurs. Mesoscutum with a pair of small, widely separated warts.

Brachycentrids are 6–11 mm. and occur along streams. Larval cases are elongate, either round or square in cross section, and are made of sand or bits of vegetable material. Young larvae of some species live near shore, whereas older larvae move to midstream and attach their cases to stones.

LONG-HORNED CADDISFLIES   See also p. 213
Family Leptoceridae
Identification: Slender, pale-colored, 5–17 mm. Antennae *long,*

hairlike, often nearly twice as long as body. Pronotum with a pair of warts separated by *a deep notch*. Mesoscutal warts represented by *2 irregular rows of bristly spots*.

This group is widely distributed. Larvae occur in various types of aquatic habitats, and *make cases of different kinds*.

SNAIL-CASE CADDISFLIES See also p. 213
Family Helicopsychidae

Identification: Costal margin of HW with *a row of tiny hooks in basal portion*, and somewhat angulate in middle. Mesoscutum and scutellum each with *a pair of small warts*.

Only 4 species in this group occur in N. America, but 1 is widely distributed. Adults are brownish or mottled and 5–7 mm. Larvae occur in clear, cool, slow-moving streams with a sandy bottom; their cases are *about ¼ in. wide and shaped like a snail shell*.

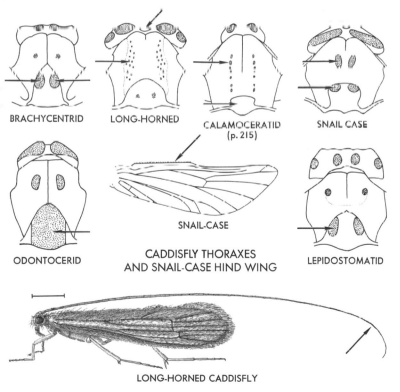

BRACHYCENTRID

LONG-HORNED

CALAMOCERATID
(p. 215)

SNAIL CASE

ODONTOCERID

SNAIL-CASE

CADDISFLY THORAXES
AND SNAIL-CASE HIND WING

LEPIDOSTOMATID

LONG-HORNED CADDISFLY

BERAEIDS Family Beraeidae   **Not illus.**
   Identification: Mesoscutum without warts. Apical spurs of middle tibiae about half as long as basal tarsal segment.
   This group contains only 3 small (about 5 mm.), brownish, and rather rare species, occurring in e. N. America. Their larvae make cylindrical and slightly curved cases of sand grains.

SERICOSTOMATIDS Family Sericostomatidae   **Not illus.**
   Identification: Mesoscutum with a median groove anteriorly and a pair of small warts very close to this groove. Apical spurs of middle tibiae not more than $\frac{1}{3}$ length of basal tarsal segment. FW with a long cross vein between R_1 and R_2.
   The 6 N. American species in this group occur in the mountains of the eastern states, and in the West. One species is fairly common along streams in arid parts of California and Arizona. Larvae live in lakes and streams and make cylindrical cases of sand grains.

Butterflies and Moths: Order Lepidoptera

Identification: With 4 membranous wings (rarely wingless), HW a little smaller than FW, the wings *largely or entirely covered with scales*. Mouth parts sucking, the proboscis *usually in the form of a coiled tube*. Mandibles nearly always vestigial or lacking. Labial palps usually *well developed and conspicuous;* maxillary palps generally vestigial or lacking. Antennae *long, slender*, sometimes plumose, always *knobbed apically in butterflies*. Metamorphosis complete.
Similar orders: (1) Trichoptera (p. 210): few or no scales on wings; no coiled proboscis; maxillary palps well developed; M in FW usually 4-branched (3-branched in Lepidoptera). (2) Hymenoptera (p. 312): no scales on wings; mouth parts chewing, with well-developed mandibles.
Immature stages: Lepidopterous larvae are commonly called caterpillars. They are usually cylindrical, with a well-developed head, *3 pairs of thoracic legs*, and 5 (sometimes fewer) *pairs of abdominal prolegs*. The prolegs are short and fleshy and are provided with a number of tiny hooks (crochets); they are normally present on 4 consecutive segments near the middle of the body (abdominal segments 3–6) and on the last segment. Larvae lack compound eyes but usually have *a group of small ocelli* on each side of the head. Many caterpillars are ornamented with hairs or spines, and although some look very ferocious most are quite harmless to handle. A few give off an unpleasant odor when disturbed, and a few have body hairs that can sting or irritate the skin. Lepidopterous larvae have the salivary glands modified

into silk glands; the silk is spun from the mouth and used principally in making cocoons or shelters. Many larvae pupate in silken cocoons; others make no cocoon. Most butterfly larvae make no cocoon, and their pupae are often called chrysalids (singular, chrysalis); chrysalids are often tuberculate or sculptured, and sometimes brightly colored; moth pupae are usually brownish and smooth. Most caterpillars are external feeders on foliage; a few live inside leaves as leaf miners, a few are gall makers, and a few bore into fruit, stems, and other parts of a plant. A very few caterpillars are predaceous on other insects.

Habits: Adult Lepidoptera feed principally on nectar and other liquid food, and many are common on flowers; a few do not feed as adults. Their flight is usually rather erratic but fairly fast. A few butterflies migrate long distances.

Importance: The larvae of many species are serious pests of cultivated plants; a few are pests of stored foods (grain, flour, and meal), and a few are pests of fabrics.

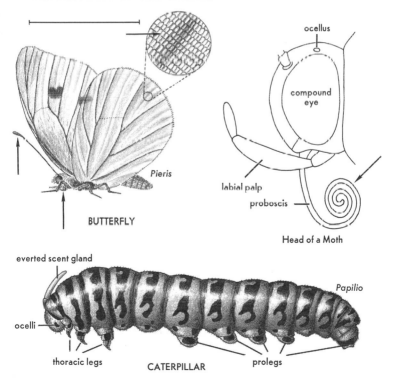

Pieris

BUTTERFLY

ocellus

compound eye

labial palp

proboscis

Head of a Moth

everted scent gland

Papilio

ocelli

thoracic legs

CATERPILLAR

prolegs

Classification: Two suborders, Frenatae and Jugatae, differing chiefly in wing venation and the way the front and hind wings are kept together; Rs in the hind wing is unbranched in frenates, branched in jugates; frenates generally have a frenulum (see opp.), and jugates have a *jugum* (a fingerlike lobe at base of front wing). The suborder Frenatae is divided into a series of superfamilies, which are arranged in 2 groups, Macrolepidoptera (p. 222) and Microlepidoptera (p. 240). Butterflies and skippers, making up the frenate superfamilies Papilionoidea and Hesperioidea, have knobbed antennae and lack a frenulum; moths (rest of order) have antennae of various sorts, usually threadlike or plumose (only rarely slightly clubbed), and they generally have a frenulum.

No. of species: World, 112,000; N. America, 11,000.

Identification of Lepidoptera: The principal characters used in separating families of Lepidoptera are those of the wing venation; other characters include the presence or absence of a frenulum, the presence or absence of ocelli, and characters of the legs, mouth parts, and antennae. Many of these characters are difficult to see, and most beginners try to identify their specimens from pictures. This method may be satisfactory for butterflies if a good illustrated guide such as Klots's *A Field Guide to the Butterflies* is available, but not for most of the moths (which make up the bulk of the order). The identification of most Lepidoptera to family requires a knowledge of wing venation and other characters.

The wing venation in the Lepidoptera is relatively simple, and that of a generalized frenate is shown opposite. Veins indicated by dotted lines are atrophied or lost in many groups. Loss of the basal portion of M results in the formation of a large cell in the central basal part of the wing. This is the *discal cell*, and it provides a starting point in identifying the veins. Sc and the anal veins are always free of the discal cell in the front wing, and the branches of R, M, and Cu come off this cell. Ten veins come off the discal cell in the front wing (if all are present): R_1, R_2, R_3, R_4, R_5, M_1, M_2, M_3, Cu_1, and Cu_2. In the frenate hind wing the anal veins are always free of the discal cell; Sc may or may not be free of this cell; Sc and R_1 always fuse in the hind wing, and Rs is unbranched. The veins in the frenate hind wing (if all are present) are Sc $+ R_1$, Rs, M_1, M_2, M_3, Cu_1, and Cu_2, plus 3 anal veins behind the discal cell. The venational variations encountered involve the number of veins present, how they branch, and where particular veins rise. The *frenulum* (see opp.) is a bristle or group of bristles at the base of the front edge of the hind wing.

The wing venation in most Lepidoptera is obscured by scales, and it is often necessary to bleach or remove the scales to see the venation; if the scales are not too dense, the venation can sometimes be seen if there is a strong light behind the wing. A drop of alcohol on the wing often reveals the venation. When the alcohol evaporates the wing coloration is usually unchanged.

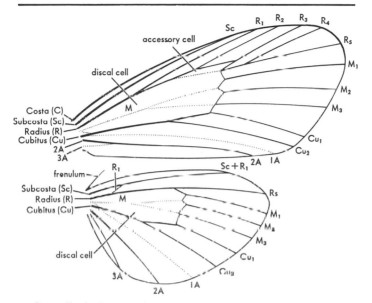

Generalized wing venation of a frenate. Veins shown by dotted lines are often weak or absent.

The best way to see the wing venation in a lepidopteran is to bleach the wings. The procedure for bleaching and mounting lepidopteran wings is as follows:

1. Carefully remove wings from one side of the body (include frenulum if one is present).
2. Dip wings in 95 percent alcohol for a few seconds.
3. Dip wings in a 10 percent solution of hydrochloric acid for a few seconds.
4. Place wings in Clorox; leave them there until color is removed (up to several minutes). If wings are slow in clearing, dip them in acid again and then return them to the Clorox.
5. Rinse wings in water to remove the Clorox.
6. Place wings on a slide (a 2 x 2-in. slide for most wings). This is done by floating wings in a dish and bringing the slide up from underneath. After wings are on the slide, center and orient them.
7. Allow wings to dry, then place another slide (of same size) on top of wings and bind slides together with binding tape. Place the slide label on the outside. The result is a permanent slide mount that can be studied under a microscope, or projected with a slide projector.

Frenate Lepidoptera: Suborder Frenatae

A frenulum present (see illus., p. 221), or humeral angle of HW more or less expanded; Rs in HW unbranched. This suborder includes the vast majority of species in the order.

Macrolepidoptera

Size variable but usually with a wingspread of 1 in. or more. FW more or less triangular, HW rounded. Fringe of hairs on anal margin of HW short. Generally only 1 anal vein (2A) in FW and 1 or 2 (2A, or 2A and 3A) in HW.

Butterflies: Superfamily Papilionoidea

Antennae *knobbed but never hooked at tip*, and close together at base. Some branches of R in FW stalked beyond discal cell, or fewer than 5 branches present. Wings generally large in proportion to body.

SWALLOWTAILS Family Papilionidae
   Identification: FW with R *5-branched* and Cu *appearing 4-branched*. HW with *1 anal vein and with tail-like prolongations*. Front legs normal-sized, not reduced.
   This group includes some of our largest and most strikingly colored butterflies, only a few of which can be mentioned here. The Tiger Swallowtail, *Papilio glaucus* Linn., one of our largest swallowtails, is brightly colored with yellow and black (some individuals have the wings almost entirely dark); its larva (illus., p. 219) feeds on various trees. The Black Swallowtail, *P. polyxenes asterius* Stoll, is black with 2 rows of yellow spots around wing margins; larva feeds on carrots, parsley, and related plants. The Zebra Swallowtail, *Graphium marcellus* (Cramer), is pale greenish with black stripes, and has rather long tails; larva feeds on papaw. The Spicebush Swallowtail, *P. troilus* Linn., is blackish with a row of small yellow spots along outer margin of the front wing and extensive blue-green areas in the hind wing; its larva feeds on spicebush and sassafras.

PARNASSIANS Family Parnassiidae
   Identification: Similar to Papilionidae, but R in FW *4-branched* and HW *without tail-like prolongations*.
   Parnassians are usually gray or white with dark markings. Most species have 2 small reddish eye spots in the hind wing. These butterflies have a wingspread of about 2 in., and are principally western in distribution.

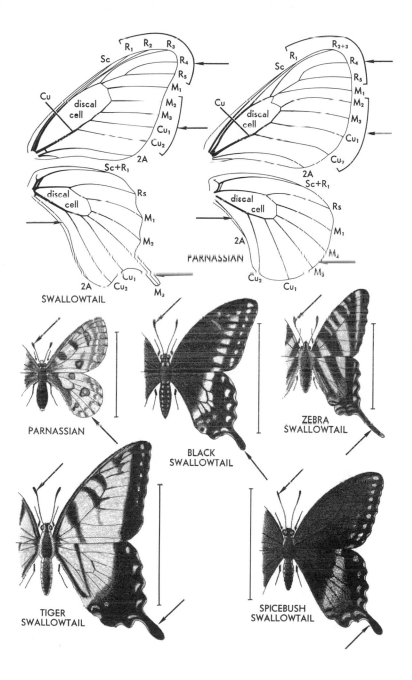

R₁ R₂ R₃
Sc
R₄
R₅
M₁
Cu
discal cell
M₂
M₃
Cu₁
Cu₂
2A

R₂₊₃
R₁
Sc
R₄
R₅
M₁
Cu
discal cell
M₂
M₃
Cu₁
Cu₂
2A

Sc+R₁
discal cell
Rs
M₁
M₂
2A
Cu₁
Cu₂
M₃

SWALLOWTAIL

Sc+R₁
discal cell
2A
Rs
M₁
M₂
Cu₂
Cu₁
M₃

PARNASSIAN

PARNASSIAN

BLACK SWALLOWTAIL

ZEBRA SWALLOWTAIL

TIGER SWALLOWTAIL

SPICEBUSH SWALLOWTAIL

WHITES, SULFURS, and ORANGE-TIPS See also Pl. 9
Family Pieridae

Identification: Small to medium-sized, usually white, yellow, or orange, marked with black. FW with *Cu appearing 3-branched*, R *3- or 4-branched* (5-branched in some orange-tips), and M_1 *stalked with a branch of R* for a distance beyond discal cell. HW with *2 anal veins*, and with or without a humeral vein. Front legs normal or slightly reduced, the tarsal claws forked.

Whites are white with black markings. Our most common species is the Cabbage Butterfly (see also p. 219), *Pieris rapae* (Linn.); its larva feeds on cabbage and related plants, and is a serious pest of cabbage. **Sulfurs** (Pl. 9) are yellow or orange, and our most common species have the wings bordered with black; some sulfurs are very common butterflies; larvae of the common species feed on clovers. **Orange-tips** are small white butterflies that have underside of the wings mottled with greenish, and tips of the front wings are often orange; they are relatively uncommon, and most species are western; larvae feed on shepherd's purse and related plants.

GOSSAMER-WINGED BUTTERFLIES See also Pl. 9
Family Lycaenidae

Identification: Small, delicate, often brightly colored. Wing venation as in Pieridae, but FW with M_1 *not stalked with a branch of R* beyond discal cell (except in harvesters) and R never with more than 4 branches; HW *without a humeral vein and C not thickened*. Front legs of ♂ usually reduced; tarsal claws not forked.

This is a large group, many species being quite common. Larvae are somewhat sluglike, and many secrete a honeydewlike material that attracts ants; some live in ant nests. Adults are rapid fliers.

Harvesters, Subfamily Gerydinae. Differing from other lycaenids in having M_1 in FW *stalked with a branch of R* for a short distance beyond discal cell; R in FW *4-branched*. The single U.S. species in this group, *Feniseca tarquinius* (Fabricius), occurs in the East; it is a small yellowish-brown butterfly with border of the front wing and central basal portion of the hind wing dark brown. Larva feeds on aphids. This butterfly is not common but is most likely to be found near alders growing in swampy places.

Coppers, Subfamily Lycaeninae (Pl. 9). Brownish or reddish, often with a coppery tinge, and with black markings. R in FW 4-branched. Coppers are fast-flying butterflies that generally occur in meadows, marshes, and other open areas. Larvae feed on dock (*Rumex*).

Blues, Subfamily Plebeiinae. Small, delicate, with upper surface of wings usually largely or entirely blue. R in FW usually 4-branched. Females are generally darker than males;

some have little or no blue in the wings. Many larvae secrete a honeydewlike material, and some live in ant nests.

Hairstreaks, Subfamily Theclinae (not illus.). Dark brown or grayish, with delicate striping on underside of wings. Generally 2 or 3 hairlike tails on HW. R in FW 3-branched. Elfins (*Incisalia*) are small, brownish, and the hind wings lack tails and have a somewhat scalloped margin. Hairstreaks usually occur in

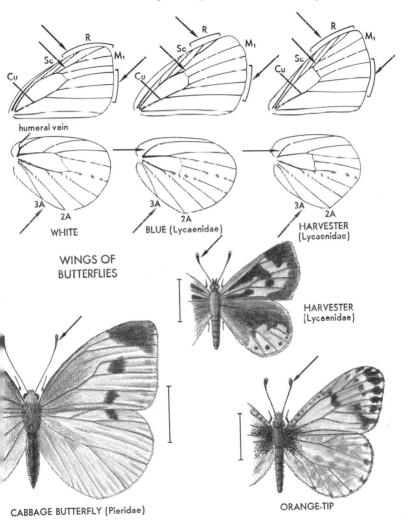

WINGS OF BUTTERFLIES

WHITE

BLUE (Lycaenidae)

HARVESTER (Lycaenidae)

HARVESTER (Lycaenidae)

CABBAGE BUTTERFLY (Pieridae)

ORANGE-TIP

meadows, along roadsides, and in other open areas; they are
often found on flowers.

METALMARKS   Family Riodinidae
Identification: Small dark-colored butterflies. R in FW *4-
branched.* HW with *a short humeral vein* and C *thickened out to
humeral angle of wing.*
    This group is chiefly tropical, and most of our species occur
in the South and Southwest; 3 eastern species range north to
Ohio but are rare in the North. They are brownish, with narrow
lines or spots on wings; most have a wingspread of 1 in. or less;
larvae feed on various weeds.

BRUSH-FOOTED BUTTERFLIES   See also Pl. 9
Family Nymphalidae
Identification: Variable in size and color. Front legs *greatly
reduced.* FW relatively broad and triangular, with R *5-branched,*
Cu appearing *3-branched,* and 3A *lacking.* HW with *2 anal veins,*
humeral vein *straight* or bent toward wing tip, and discal cell
open or closed by a weak vein. No veins greatly swollen at base.
    This is the largest family of Papilionoidea, and includes many
common species; only a few can be mentioned here. Fritillaries
(*Speyeria* and *Boloria*) are medium-sized to large, brownish,
with numerous black spots or narrow bands on the wings, and
usually with silvery spots on underside of the hind wings; larvae
feed on violets. Crescent-spots (*Phyciodes*) are small and brownish, with numerous black markings on the wings; larvae feed
chiefly on asters. Anglewings (*Polygonia*) are small to medium-
sized, brownish, with dark markings on the wings; wing margins
are irregular, and underside of the wings resembles a dead leaf;
larvae feed mainly on nettles and elm. The Mourningcloak
(Pl. 9), *Nymphalis antiopa* (Linn.), is common and widely distributed; its wings are blackish, margined with yellow; larva is
gregarious and feeds on willow and elm. The Red Admiral
(Pl. 9), *Vanessa atalanta* (Linn.), is common and widely distributed; its larva feeds on nettles. The Viceroy (Pl. 9), *Limenitis archippus* (Cramer), looks much like a Monarch but is somewhat smaller and has a black line across the hind wing; its larva
feeds on willow and poplar. Mimicry like that of the Monarch
and Viceroy occurs in many butterflies and is believed to offer
1 species (Viceroy in this case) some protection from predators.
Body fluids of the Monarch are apparently distasteful to predators, so they avoid it; the Viceroy's body fluids are not distasteful, but its resemblance to the Monarch may cause predators
to avoid it.

NYMPHS, SATYRS, and ARCTICS   See also Pl. 9
Family Satyridae
Identification: Small to medium-sized butterflies, usually gray-

ish or brownish and often with eye spots in the wings. Venation as in Nymphalidae, but some veins (*especially Sc*) *greatly swollen at base.* Front legs much reduced.

One of the most common and strikingly marked species in this group is the Wood Nymph (see also Pl. 9), *Cercyonis pegala* (Fabricius), a medium-sized, dark brown butterfly having a broad yellowish band across the front wings with 2 small eye spots. Wood satyrs (*Euptychia*) are small grayish butterflies, about 1 in. in wingspread, with small black eye spots in the wings. The Pearly Eye, *Lethe portlandica* Fabricius, is a woodland species that often alights on tree trunks; it is brownish, with a row of black spots along outer edge of the hind wings. Arctics (*Oeneis*) occur chiefly in the arctic region and on mountaintops; 1 species, *O. jutta* Hübner, may be found in sphagnum bogs in New England. Satyrid larvae feed on grasses.

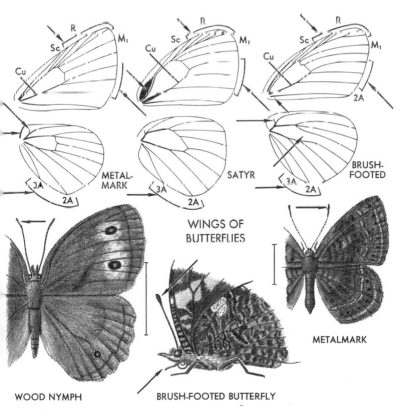

WINGS OF
BUTTERFLIES

METALMARK

SATYR

BRUSH-FOOTED

METALMARK

WOOD NYMPH   BRUSH-FOOTED BUTTERFLY

SNOUT BUTTERFLIES   Family Libytheidae   **Not illus.**
   Identification: Similar to Nymphalidae (p. 226), but labial palps
   longer than thorax, thickly hairy, and projecting forward.
      Our only species in this group has distinctively shaped front
   wings and a wingspread of about 1½ in. It occurs throughout
   the East and Southwest, but is uncommon in the Northeast.
   Larva feeds on hackberry.

HELICONIANS   Family Heliconiidae
   Identification: Similar to Nymphalidae (p. 226), but FW narrow,
   elongate, and HW humeral vein *bent toward base of wing.*
      Our few species of heliconians occur in the South. One of the
   most strikingly colored is the Zebra Butterfly, *Heliconius
   charitonius* (Linn.), which is black with yellow stripes. Its larva
   feeds on passion flowers.

MILKWEED BUTTERFLIES   Family Danaidae   **Pl. 9**
   Identification: Large brownish butterflies, usually marked with
   black. Similar to Nymphalidae (p. 226) but 3A present in FW,
   discal cell in HW closed by a well-developed vein, and antennae
   without scales.
      Our most common danaid is the Monarch (Pl. 9), *Danaus
   plexippus* (Linn.), which occurs throughout the U.S. and s.
   Canada. The Viceroy (Nymphalidae) is very similar, but is
   slightly smaller and has a black line across the hind wing. The
   Monarch is one of the few butterflies in this country that
   migrate; it migrates south in fall, often in immense aggregations,
   and reappears in the North the following spring. The Monarchs
   appearing in the North are usually not the same individuals that
   migrated south the season before, but their offspring; the insect
   reproduces on its wintering ground or after a short northward
   flight in spring. Larvae of danaids feed on milkweed.

Skippers: Superfamily Hesperioidea

Antennae *clubbed and usually also hooked at tip,* and widely separated at base. R in FW *5-branched,* all branches coming off discal
cell. Relatively stout-bodied. Strong fliers. Larvae generally
pupate in a cocoon formed of leaves and silk.

COMMON SKIPPERS   Family Hesperiidae
   Identification: Head about as wide as or wider than thorax.
   Hind tibiae usually with 2 pairs of spurs. Wingspread generally
   less than 30 mm. Widely distributed.
      This is a large group and many species are quite common.
   The front and hind wings at rest are often held at a slightly
   different angle. Larvae feed on leaves, and usually live in a
   shelter formed of a rolled-up leaf or several leaves tied together;
   they are smooth-bodied, with a small and necklike prothorax.

GIANT SKIPPERS Family Megathymidae   **Not illus.**
Identification: Head narrower than thorax. Antennae not hooked at tip, but with a large club. Hind tibiae with only 1 pair of spurs. Wingspread 40 mm. or more.

Giant skippers are fast-flying insects that hold their wings vertical at rest. They occur in the South and West. Larvae bore in stems and roots of yucca and related plants. Larvae are

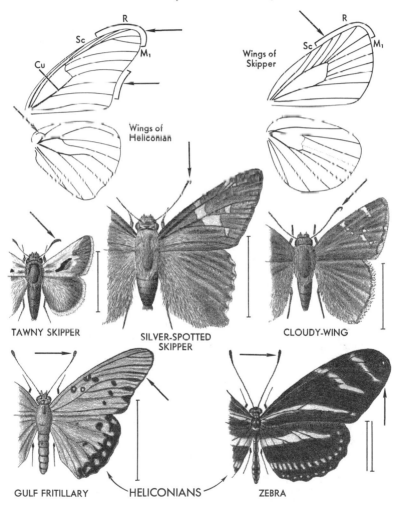

TAWNY SKIPPER   SILVER-SPOTTED SKIPPER   CLOUDY-WING

GULF FRITILLARY   HELICONIANS   ZEBRA

edible, and in Mexico are considered a delicacy; they are cooked by frying in deep fat, and are canned and sold under the name "Gusanos de Maguey."

Macro-Moths: Macrolepidoptera, in part

Antennae threadlike or plumose. Frenulum usually present. 1A nearly always lacking in FW and HW. Fringe on anal margin of HW short. Tibial spurs short. Wings never lanceolate. Wing-spread usually over 20 mm. The 21 families of macro-moths may be divided into 3 groups: (1) Sphingidae — fairly distinct in appearance and venation; (2) Dioptidae through Zanolidae — Cu in FW appears 3-branched; and (3) Ctenuchidae through Drepanidae — Cu in FW appears 4-branched.

Macro-Moths, Group 1

SPHINX or HAWK MOTHS   See also Pl. 10
Family Sphingidae

Identification: Medium-sized to large, *heavy-bodied*, with wings relatively small. Antennae *thickened, somewhat spindle-shaped apically*. Sc and Rs in HW parallel to end of discal cell and beyond, connected by *an oblique cross vein* about opposite middle of discal cell. Frenulum *present but sometimes small*.

Sphinx moths are strong fliers, with a very rapid wingbeat. They feed on flowers, much like hummingbirds. Most species feed at dusk or at night; a few feed during the day. Some species have large areas of the wings devoid of scales and resemble bumble bees. Larvae of most species have a soft spinelike process near the posterior end of the body and are often called hornworms; some are pests of tomatoes, tobacco, and other plants. They usually pupate in the ground.

Macro-Moths, Group 2

Cu in FW appears 3-branched. The families in this group may be divided into 4 groups on the basis of venation in anterior part of hind wing.

1. Sc + R_1 and Rs diverging at base of wing: Dioptidae, Saturniidae, Citheroniidae, Lacosomidae, Epiplemidae.
2. Sc rather abruptly angled into humeral angle of wing, often connected to humeral angle by a cross vein; beyond this bend Sc and R are either fused or are closely parallel for a short distance along anterior side of discal cell: Geometridae, Manidiidae.
3. Sc and R approximately parallel along basal half of discal cell, then connected by a distinct cross vein, diverging beyond the cross vein: Bombycidae.
4. Sc + R_1 and Rs close and parallel along at least basal half of anterior side of discal cell, often farther, then diverging: Thyatiridae, Notodontidae, Zanolidae.

Arctiidae are discussed under Group 3 of the Macro-Moths because Cu in the front wing appears 4-branched in most of them; in the Lithosiinae, which have Cu in the front wing appearing 3-branched, Sc and Rs in the hind wing are fused to beyond the middle of the discal cell, then they diverge.

OAK MOTHS Family Dioptidae.   **Not illus.**
Identification: Slender, pale brown moths. Wingspread 25–35 mm. Frenulum well developed. M_3 and Cu_1 in FW and HW stalked for a short distance beyond discal cell.

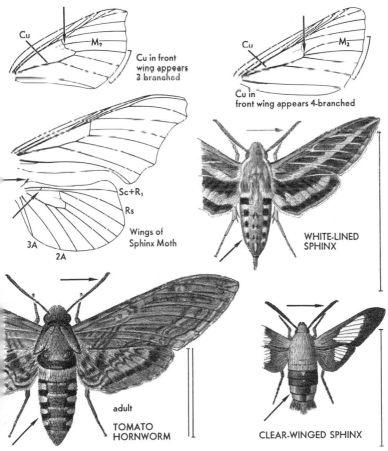

Cu in front wing appears 3-branched

Cu in front wing appears 4-branched

Wings of Sphinx Moth

WHITE-LINED SPHINX

adult
TOMATO HORNWORM

CLEAR-WINGED SPHINX

A single species of oak moth occurs in California. Its larva feeds on various oaks (and occasionally other trees), and sometimes does considerable damage.

GIANT SILKWORM MOTHS   See also Pl. 10
Family Saturniidae

Identification: Medium-sized to large moths, with broad wings. Frenulum small or *vestigial*. Humeral angle of HW *not noticeably expanded;* HW with *only 1 anal vein*. Wings usually with eye spots. Discal cell in FW *generally open*.

This group includes our largest moths, some of which have a wingspread of about 6 in.; many are brightly colored. The antennae are somewhat plumose, especially in males. Mouth parts are reduced, and adults do not feed. Larvae are large caterpillars and many have tubercles or spines on the body; they pupate in silken cocoons. Our largest saturniid is the Cecropia Moth (Pl. 10), *Hyalophora cecropia* (Linn.), with a wingspread of 5 or 6 in. The Luna Moth (Pl. 10), *Actias luna* (Linn.), is light green, with a long tail on each hind wing. The Polyphemus Moth (Pl. 10), *Antheraea polyphemus* (Cramer), is large, yellowish brown, with a round windowlike spot near the center of each wing. The Io Moth (Pl. 10), *Automeris io* (Fabricius), has a wingspread of about 2–2½ in.; males are bright yellow, with a large eye spot in each hind wing; females are similar but have dark brown front wings. The larva of the Io is a spiny green caterpillar with a narrow lateral stripe of red above and white below; the stinging spines make handling an Io larva like handling nettles.

ROYAL MOTHS   Family Citheroniidae   See also Pl. 10
Identification: Medium-sized to large, usually yellowish or brownish. Frenulum *absent*. Humeral angle of HW *considerably expanded*. Venation as in Saturniidae, but *with 2 anal veins in HW* and discal cells *closed*. Usually no eye spot in wings. Antennae plumose only in basal half.

Royal moth larvae usually have horns or spines on the anterior segments. They feed on various trees and pupate in the ground. The largest species in this group is the Regal Moth (Pl. 10), *Citheronia regalis* (Fabricius), which has a wingspread of 5 or 6 in.; its larva feeds chiefly on walnut and hickory. The Imperial Moth, *Eacles imperialis* (Drury), is large, yellow, marked with pinkish purple; its larva feeds on various trees. Most of the moths in the genus *Anisota* are brownish, with a wingspread of about 1½ in.; they resemble tent caterpillar moths (Lasiocampidae) but have Cu in the front wing appearing 3-branched and there are no humeral veins in the hind wing.

SACK-BEARERS   Family Lacosomidae
Identification: Medium-sized, stout-bodied moths, usually yel-

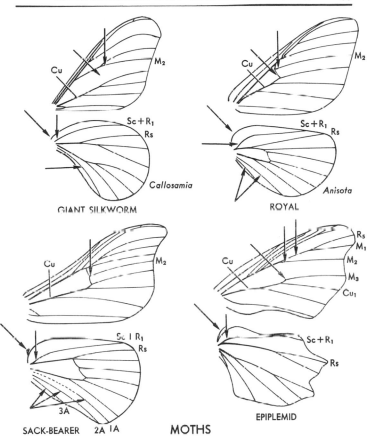

Callosamia
GIANT SILKWORM

Anisota
ROYAL

SACK-BEARER   2A 1A   MOTHS

EPIPLEMID

lowish or brownish with dark lines or spots. Frenulum *small* or vestigial; *2 (rarely 3) anal veins in HW*. Sc and Rs in HW not connected by a cross vein.

Larvae of these moths construct portable cases of leaves. Three species (none common) occur in the U.S. — 2 in the East and 1 in Arizona. One eastern species has apex of the front wing somewhat sickle-shaped, and the other has distal margin of the front wing toothed or scalloped.

EPIPLEMID MOTHS   Family Epiplemidae
Identification: Grayish or yellowish, with a wingspread of about 20 mm. Frenulum *well developed*. M_1 and R_5 in FW *stalked*. M_3 and Cu_1 *not stalked*.

Only 5 species of epiplemids occur in the U.S. They are relatively rare.

GEOMETER MOTHS Family Geometridae **See also Pl. 10**
Identification: Small to medium-sized, usually slender-bodied. Sc in HW with *a rather abrupt angle basally*, and often connected to humeral angle of wing by a cross vein. Antennae not dilated apically, or if so then eyes are bare.

This is a large group, with some 1200 N. American species, occurring almost everywhere. Larvae are small slender caterpillars with only 2 or 3 pairs of prolegs; they move in a looping fashion, and are called measuringworms or inchworms. The moths usually hold wings outstretched at rest rather than back over the body. Females of a few species (like cankerworms) are wingless. Larvae feed on many different plants; some tree feeders occasionally damage orchard and shade trees.

MANIDIID MOTHS Family Manidiidae **Not illus.**
Identification: Similar to Geometridae, but antennae dilated apically and eyes hairy.

A single rare species in this group occurs in Arizona.

SILKWORM MOTHS Family Bombycidae
Identification: Heavy-bodied white moths. Wingspread 35–40 mm. Sc and R in HW *connected by a cross vein* opposite middle of discal cell, then diverging. Frenulum *very small.*

Silkworm moths are not native to N. America, but 1 species, *Bombyx mori* (Linn.), which is the source of natural silk, is sometimes reared here. This insect, the larva of which feeds on mulberry, is not a wild species in this country.

THYATIRID MOTHS Family Thyatiridae **Not illus.**
Identification: Similar to Noctuidae (p. 238), but Cu in FW appears 3-branched, and Sc + R_1 and Rs in HW are approximately parallel along anterior side of discal cell. Differ from Notodontidae in that Cu in HW appears 4-branched (3-branched in Notodontidae), and Rs and M_1 in HW are not stalked.

Moths are medium-sized, and usually brownish with wavy or zigzag lines on the front wings. They are not common.

PROMINENTS Family Notodontidae **See also Pl. 11**
Identification: Medium-sized, usually brownish moths. Sc + R_1 and Rs in HW *close together and parallel* along discal cell; Rs and M_1 in HW *stalked a short distance* beyond discal cell.

Larvae of these moths are usually gregarious. When disturbed they often freeze with ends of the body elevated. Larvae of most species feed on trees and shrubs, and some attack orchard trees. Most larvae are striped. Notodontids are fairly common moths.

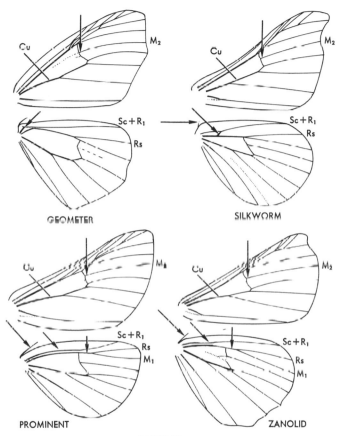

GEOMETER

SILKWORM

PROMINENT

ZANOLID

MOTHS

ZANOLID MOTHS Family Zanolidae

Identification: Similar to Notodontidae, but Sc + R₁ and Rs in HW *diverge about middle* of discal cell. Proboscis lacking. FW with 1 or 2 small clear spots near tip. Tarsal claws without a blunt tooth near base.

This group is represented in the U.S. by 3 species of *Apatelodes*, which have a wingspread of 1½–2 in., and are gray with window-like spots near tip of the front wings. One species has the distal margin of the wings somewhat toothed. Larvae feed on various trees and shrubs.

Macro-Moths, Group 3

Cu in FW appears 4-branched (except in a few Arctiidae). Frenulum well developed except in Lasiocampidae and Drepanidae.

TENT CATERPILLARS and LAPPET MOTHS See also Pl. 11
Family Lasiocampidae
Identification: Medium-sized, stout-bodied, rather hairy moths. HW *without a frenulum* but with humeral angle *expanded* and with *humeral veins.* Cu_2 in FW rises *in basal* ½ *or* ⅓ of discal cell. Antennae somewhat plumose.

A common eastern species in this group is the Eastern Tent Caterpillar (Pl. 11), *Malacosoma americanum* (Fabricius). Its larvae are gregarious and construct a silken tent, usually in the fork of a branch, and use it as a shelter. These larvae feed on apple, cherry, and related trees, and may completely defoliate a tree. Overwintering eggs hatch early in the season, and larvae become full-grown by late May or early June. Larvae spin silken cocoons in various protected places and moths appear about 3 weeks later. A similar species, the Forest Tent Caterpillar, *Malacosoma disstria* Hübner, does not construct a tent. Larva of *M. americanum* has a yellow stripe down the back; that of *M. disstria* has a dorsal row of keyhole-shaped spots. Adults of both species have a wingspread of about 1–1½ in. (females are slightly larger than males) and are yellowish brown with 2 narrow light lines across front wings; adults of *M. americanum* are somewhat darker than those of *M. disstria*. Lappet moths (*Tolype*), about the same size as tent caterpillar moths, are bluish gray with white markings; their larvae feed on various trees. *Gloveria arizonensis* Packard, a common southwestern species, is dark gray, with a wingspread of about 2½ in.

HOOK-TIP MOTHS Family Drepanidae   See also Pl. 11
Identification: Small, brownish, slender-bodied. Apex of FW usually *sickle-shaped.* HW with Sc + R_1 and Rs *separated along anterior side* of discal cell. Frenulum small or *absent.*

These moths have a wingspread of about 1 in. or less, and can usually be recognized by the sickle-shaped apex of the front wing. The 6 U.S. species are not common.

CTENUCHID MOTHS Family Ctenuchidae   See also Pl. 11
Identification: Sc + R_1 in HW *absent.*

Ctenuchids are common moths that are active during the day and feed on flowers. Larvae are quite hairy, and cocoons are formed principally of larval body hairs; most larvae feed on grasses. Scape moths (*Cisseps*) have narrow wings and are slate-colored, the central part of the hind wings being lighter.

The Virginia Ctenucha (Pl. 11), *Ctenucha virginica* (Charpentier), is larger and broader-winged; its wings are blackish, with a narrow white margin along rear edge of the hind wings, and the body is metallic blue; larva is a hairy yellowish caterpillar. The Lichen Moth, *Lycomorpha pholus* (Drury), is small, narrow-winged, and blackish, with the base of the wings yellow; adults occur commonly on goldenrod and larvae feed on lichens.

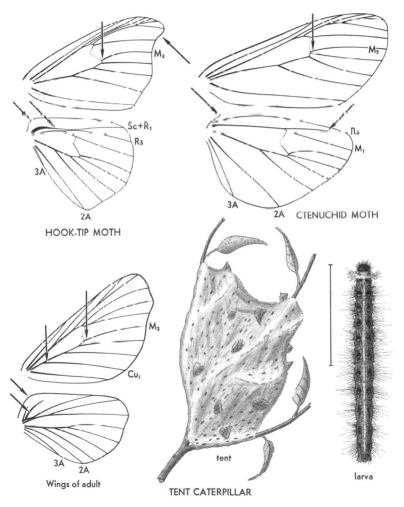

HOOK-TIP MOTH

CTENUCHID MOTH

Wings of adult

TENT CATERPILLAR

larva

TIGER MOTHS Family Arctiidae **See also Pl. 11**
Identification: Small to medium-sized, usually light-colored, often brightly spotted or banded. Cu in HW appears 4-branched; Sc and R in HW usually *fused to about middle* of discal cell, or for a short distance beyond basal areole, and Sc swollen at base. Ocelli present (except in Lithosiinae). If Cu appears 3-branched (some Lithosiinae), Sc and R in HW are fused to middle of discal cell or beyond, and M_2 and M_3 in FW are absent. Noctuids are very similar but generally dark-colored, the palps usually longer (extending beyond middle of face), and Sc and R in HW fuse for only a short distance beyond the basal areole.

This is a large group, many species of which are common moths. Larvae are usually very hairy, and some are called woollybears; most of them feed on grasses but a few feed on trees and shrubs. Adults are generally heavy-bodied and hold wings rooflike over the body at rest; many are beautifully colored and some are largely white. Larvae of the Fall Webworm, *Hyphantria cunea* (Drury), often seriously damage trees and shrubs; they spin silk over the foliage, skeletonizing the leaves as they feed, and may form a web over foliage of entire branches; adults are small white moths.

NOLID MOTHS Family Nolidae **Not illus.**
Identification: Wing venation as in Arctiidae. FW with tufts of raised scales. Ocelli absent.

This is a small group and its members are not common. Most nolids have a wingspread of 1 in. or less. Some larvae feed on lichens, others feed on various trees; larva of the Sorghum Webworm, *Celama sorghiella* (Riley), feeds on sorghum.

NOCTUID MOTHS Family Noctuidae **See also Pl. 11**
Identification: Sc and R in HW *fused for a short distance* beyond a small basal areole, then separating; Cu in HW appears 3- or 4-branched (M_2 in HW *often weak or absent*). Ocelli nearly always present. Antennae slender and threadlike, never plumose. Palps extend to middle of face or beyond.

This is the largest family in the order, with some 2700 N. American species, and many are common moths. Most are nocturnal. Noctuids vary considerably in size and color but most have a wingspread of 20–40 mm. and are dark-colored. The wings at rest may be held flat or rooflike over the body. Underwings (*Catocala*), most of which have a wingspread of $1\frac{1}{2}$–$2\frac{1}{2}$ in., are strikingly colored; the front wings are generally a mottled brownish or gray, but the hind wings have concentric bands of red, yellow, or orange. Noctuid larvae are smooth and dull-colored and most have 5 pairs of prolegs; a few, called loopers, have only 3 pairs, and move like inchworms. Larvae of some species (cutworms) feed on roots and shoots of various plants and often cut off the stem just above the ground. Larva of the Corn Earworm, *Heliothis zea* (Boddie), feeds on the growing ears of corn, and also burrows into tomatoes and the bolls (seedpods) of cotton. Larvae of other species bore into stems and fruits.

FORESTER MOTHS   Family Agaristidae   **Pl. 12**
Identification: Black, with 2 whitish or yellowish spots in each wing. Wingspread about 1 in. Venation as in Noctuidae. Antennae swollen apically. Frenulum well developed.

The Eight-spotted Forester (Pl. 12), *Alypia octomaculata* (Fabricius), is a common and widely distributed species; larva feeds on grape and Virginia creeper. Most of the other 27 U.S. species occur in the West or in the Gulf states.

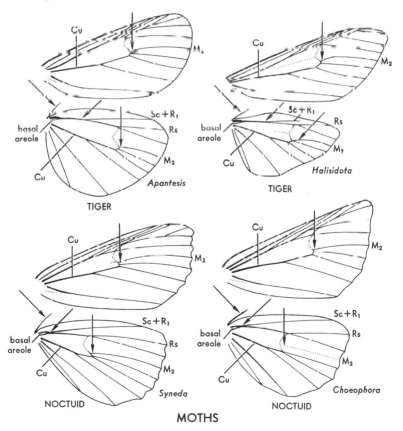

MOTHS

PERICOPID MOTHS Family Pericopidae   **Not illus.**
Identification: Medium-sized, black or bluish with extensive light areas in wings. 2 large rounded prominences on dorsal surface of 1st abdominal segment, separated by about ⅛ width of abdomen. Venation as in Noctuidae (p. 238).

Most of the 6 U.S. species of pericopids occur in the West. *Composia fidelissima* Herrich-Schäffer, which is found in s. Florida, is dark blue marked with red and white.

TUSSOCK MOTHS and Others   **See also Pl. 12**
Family Liparidae
Identification: Similar to Noctuidae (p. 238), but with *a much larger basal areole* in HW and without ocelli. Rs and M_1 in HW *sometimes stalked.* Antennae of ♂ plumose.

Tussock moth larvae are very distinctive caterpillars. They have a pair of pencil-like hair tufts at the anterior end, a single similar tuft at the posterior end, and 4 short thick hair tufts on the back. Adult male tussock moths are grayish brown, with broad wings and a wingspread of 20–25 mm.; females are wingless. The Gypsy Moth (Pl. 12), *Porthetria dispar* (Linn.), a species introduced from Europe, is most common in the East; its larva often does a great deal of damage to forest trees. Male Gypsy Moths are brownish, with plumose antennae, and are good fliers; females, white with black markings, do not fly.

Microlepidoptera

Size variable but usually with a wingspread of 20 mm. or less. Wing shape variable, sometimes relatively broad, sometimes narrow and pointed apically. Tibial spurs generally long. Many, especially narrow-winged forms, with a fringe of long hairs on anal margin of HW. Wing venation variable but broad-winged forms usually with 2 anal veins in FW and 3 in HW. This is a large group making up about half of the order; identification of many is often difficult. The families of Microlepidoptera are discussed below in 3 groups:

1. With a distinctive wing shape or scaling (p. 240).
2. Wings *relatively broad* and more or less rounded apically, FW *somewhat triangular*, HW *usually as broad as or broader* than FW, and with 3 anal veins (p. 242).
3. Wings *narrow and more or less pointed apically*, HW *usually narrower* than FW and with anal area reduced; a long fringe on anal margin of HW (p. 252).

A few families of Microlepidoptera contain species that fall into 2 of the above groups, but in most families all species in the family can be placed in just 1 of these groups.

Microlepidoptera, Group 1

Wings with distinctive shape or scaling: lobed (Pterophoridae and

Alucitidae), distal margin of HW excavated behind tip, with tip produced (Gelechiidae), or wings with extensive areas devoid of scales (Sesiidae).

PLUME MOTHS Family Pterophoridae

Identification: Small, slender, long-legged, brownish or gray. Wings at rest held horizontal, at right angles to body. FW with 2–4 apical lobes, *HW with 3*.

Plume moths are common insects, easily recognized by the wing position at rest and the lobed wings; the lobes of the hind wing, with their long fringe, are somewhat plumelike. Plume moth larvae are chiefly leaf rollers or stem borers, and some occasionally damage cultivated plants. The Grape Plume Moth, *Pterophorus periscelidactylus* Fitch, is a pest of grape.

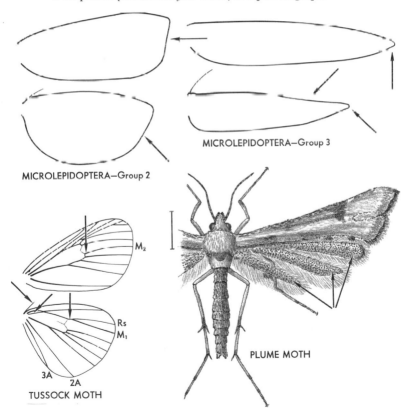

MICROLEPIDOPTERA—Group 3

MICROLEPIDOPTERA—Group 2

TUSSOCK MOTH

PLUME MOTH

MANY-PLUME MOTHS Family Alucitidae **Not illus.**
Identification: Similar to Pterophoridae, but each wing divided into 6 plumelike lobes.

A single rare species of many-plume moth occurs in the northeastern states. It has a wingspread of about 12 mm.

GELECHIID MOTHS Family Gelechiidae
Identification: Small to minute moths. FW narrowly rounded or *pointed at apex*. HW somewhat trapezoidal, the apex *usually prolonged* and margin behind apex *concave*. Head smooth-scaled. Palps long and upcurved, *3rd segment long and tapering*. If HW is narrow and pointed apically, then R_5 in FW is stalked with R_4 and extends to costal margin of wing.

This is one of the largest families of Microlepidoptera, with some 600 N. American species. Many are common moths. Larvae vary considerably in habits; some are leaf rollers or leaf tiers, some are leaf miners, a few are gall makers, and 1 species attacks stored grain. Larvae of *Gnorimoschema* live in stem galls on goldenrod; the galls are elongate, spindle-shaped, and thin-walled. Larva of the Angoumois Grain Moth, *Sitotroga cerealella* (Olivier), feeds in the kernels of corn and other grains and often causes serious damage to grain in storage. Larva of the Pink Bollworm, *Pectinophora gossypiella* (Saunders), attacks the bolls of cotton.

CLEAR-WINGED MOTHS Family Sesiidae **See also Pl. 12**
Identification: Wasplike moths, with extensive areas in wings (especially HW) devoid of scales. FW *long, narrow*, rounded apically, HW broader; posterior margin of FW and costal margin of HW with a series of interlocking spines.

Aegeriids are day-flying and often brightly colored; many strongly resemble wasps. The sexes usually differ in color, and the male often has more clear area in the wings than the female. Larvae bore in roots, stems, and trunks of various plants and trees. Some species are serious pests of garden crops, orchard trees, or forest trees.

Microlepidoptera, Group 2

Size variable, but wingspread usually 20 mm. or more. Similar to macro-moths (pp. 230 ff.), but HW generally with *3 anal veins* and FW with *1A often preserved*, at least near wing margin; HW usually as wide as or wider than FW; FW more or less triangular, apically rounded or somewhat square-tipped; tibial spurs usually long.

CARPENTER and LEOPARD MOTHS **See also Pl. 12**
Family Cossidae
Identification: Medium-sized, heavy-bodied moths. Wings

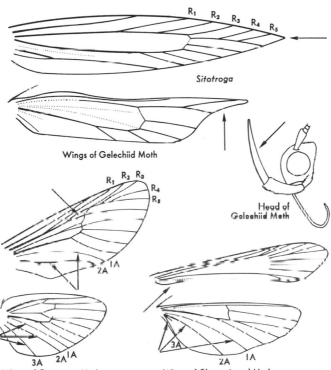

Sitotroga

Wings of Gelechiid Moth

Head of
Gelechiid Moth

Wings of Carpenter Moth

Wings of Clear-winged Moth

usually spotted or mottled; *2 complete anal veins* in FW; FW with *an accessory cell*, and some branches of R stalked. Abdomen extends beyond HW.

Cossid larvae are wood-boring, and sometimes seriously damage trees. Adults of the Carpenterworm (Pl. 12), *Prionoxystus robiniae* (Peck), a common species of carpenter moth, have a wingspread of about 2 in. The Leopard Moth, *Zeuzera pyrina* (Linn.), is slightly smaller and whitish, with black spots on the wings.

DALCERID MOTHS Family Dalceridae   **Not illus.**
Identification: Superficially resemble some megalopygids (p. 244), but venation as in the Cossidae.

Two rare species of dalcerids occur in Arizona. They are orange-yellow without dark markings, and have a wingspread of about 1 in.

SLUG CATERPILLARS   Family Limacodidae   **See also Pl. 12**
Identification: Small to medium-sized, stout-bodied, with broad
rounded wings. Mostly brownish, marked with green, silver, or
some other color. FW has *2 complete anal veins;* 3A in FW
short, meeting 2A near base of wing; M_2 in FW rises *closer to
M_3 than to M_1.* Sc and R in HW *separate at base, fused for a short
distance* near middle of discal cell. No accessory cell in FW.

Limacodid larvae are rather fleshy, with short thoracic legs
and no prolegs; they move about a little like slugs. Some larvae
have stinging hairs. A common species of this type is the
Saddleback Caterpillar, *Sibine stimulea* (Clemens), which feeds
on various trees.

FLANNEL MOTHS   Family Megalopygidae
Identification: Stout-bodied, very hairy, generally brownish or
cream-colored, with a wingspread of 25–35 mm. Sc and R in
HW *fused to middle of discal cell or beyond.* Some branches of R in
FW stalked beyond discal cell. M_2 in FW *rises near M_3.*

Larvae of these moths are stout and hairy, with some of the
hairs forming a crest down middle of the back; some body hairs
are stinging. This is a small group and its members are not
common.

PLANTHOPPER PARASITES   Family Epipyropidae   **Not illus.**
Identification: Small, with broad wings and plumose antennae.
FW with an accessory cell; no branches of R stalked.

Larvae of these moths live, probably as parasites, on the
bodies of planthoppers (Fulgoroidea). Two very rare species
occur in the U.S.

SMOKY MOTHS   Family Pyromorphidae
Identification: Small gray or black moths, with wings thinly
scaled. HW with 2 or *3 anal veins* (if with 2, then 2 complete
anal veins in FW), and Sc and R *fused to near end of discal cell.*
All branches of R in FW rise from discal cell, or R_3 and R_4 *short-
stalked.*

The more common smoky moths resemble scape moths
(Ctenuchidae, p. 236), but can be recognized by the wing venation. Larvae of most species feed on grape or Virginia creeper.

WINDOW-WINGED MOTHS   Family Thyrididae
Identification: Small, dark-colored, with light translucent spots
in wings. M_2 in FW *rises near M_3;* all branches of R and M in
FW rise from the usually *open discal cell* (R_3 and R_4 are stalked
in 1 genus in the Gulf states).

The front wings of these moths are somewhat triangular, the
hind wings are rounded or irregularly scalloped. The group is
small and its members are not common.

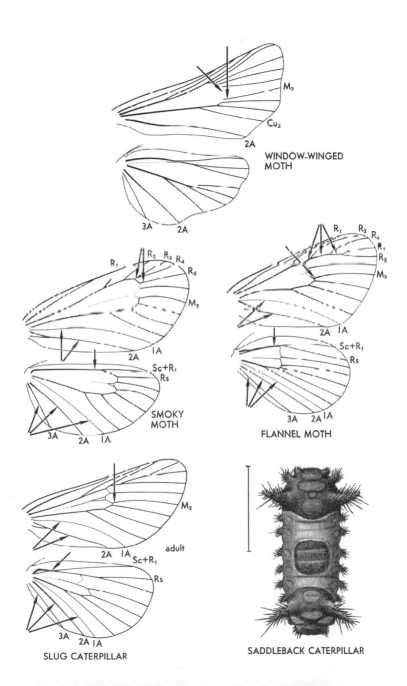

WINDOW-WINGED
MOTH

M_2

Cu_2

2A

3A 2A

R_1 R_2 R_3 R_4

R_5

M_2

2A 1A

Sc+R_1,
Rs

3A 2A 1A

SMOKY
MOTH

R_1 R_2 R_3

R_4

R_5

M_2

2A 1A

Sc+R_1

Rs

3A 2A 1A

FLANNEL MOTH

M_2

2A 1A

adult

Sc+R_1

Rs

3A 2A 1A

SLUG CATERPILLAR

SADDLEBACK CATERPILLAR

PYRALID MOTHS Family Pyralidae **See also Pl. 12**
Identification: Small moths. FW usually elongate-triangular, HW broad and rounded. M_2 in FW rises *near M_3*. Sc + R_1 and Rs in HW *fused* or very close together for a distance beyond discal cell, then separating. Palps often large and projecting forward.

This is the largest family of Microlepidoptera, with over 1100 N. American species. Its members occur almost everywhere and many are very common. Larvae vary in habits: many feed on foliage in the early instars and bore into stems in later instars; many feed about roots of grasses and other plants; a few feed on stored grain or meal, and a few are aquatic. Many are pests of cultivated plants. One of the most important pest species is the European Corn Borer (Pl. 12), *Ostrinia nubilalis* (Hübner). Members of the genus *Crambus*, often called closewings, are common in meadows; they are whitish or pale yellowish brown, and the wings at rest are held close about the body. A species in this group that feeds on cactus has been introduced into Australia, where it helps control the prickly pear cactus.

BAGWORM MOTHS Family Psychidae
Identification: Small, mostly stout-bodied moths. ♀ usually wingless and with or without legs, antennae, and eyes. Mouth parts vestigial. Most ♂ (Psychinae) with wings thinly scaled or almost devoid of scales, HW with 2 anal veins, FW with 1A and 2A *fused at tip* or connected by a cross vein, and HW *about as wide as long*. Species with more elongate wings and 3 anal veins in HW can be distinguished from other similar Microlepidoptera by the vestigial mouth parts.

Psychid larvae construct portable bags, or cases, of bits of leaves and twigs, and eventually pupate in this bag. Wingless females lay their eggs in the bag, and usually never leave it until the eggs are laid. The Evergreen Bagworm, *Thyridopteryx ephemeraeformis* (Haworth), is a common species with wingless females; larva feeds on cedars, and the male is black with almost clear wings. Other species are somewhat smoky in color and have the wings thinly scaled.

BURROWING WEBWORMS Family Acrolophidae
Identification: Noctuidlike moths with a wingspread of 12 mm. or more. Eyes usually hairy. 1st segment of labial palps *as large as 2nd or larger*, the palps upturned and in ♂ reaching back over thorax. Wing venation complete, no veins stalked.

Larvae of these moths live in the ground and feed on roots of grasses. They usually construct a tubular web leading from the surface down into the ground, and retreat into this tube when disturbed. These insects are sometimes destructive to young corn plants.

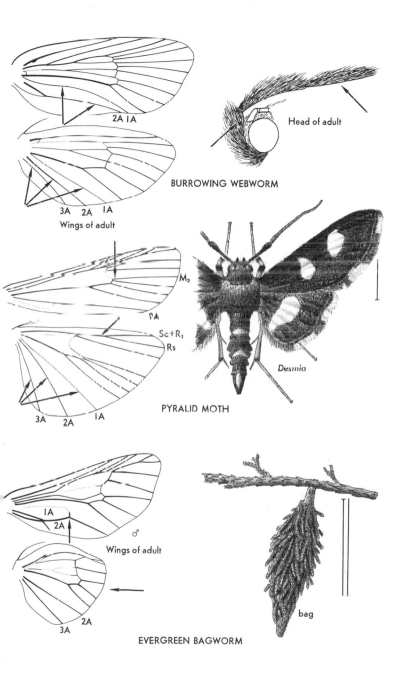

2A 1A

3A 2A 1A

Wings of adult

BURROWING WEBWORM

Head of adult

M_2

2A

$Sc+R_1$

Rs

3A 2A 1A

PYRALID MOTH

Desmia

1A

2A

♂

Wings of adult

2A

3A

EVERGREEN BAGWORM

bag

OLETHREUTID MOTHS   Family Olethreutidae

Identification: Small brownish or gray moths, FW *rather square-tipped*. Cu_2 in FW rises *in basal ¾ of discal cell*. FW with R_4 and R_5 separate, or M_2, M_3, and Cu_1 strongly converge distally. Upper side of Cu in HW usually with a fringe of long hairs.

This is a large family, with over 700 N. American species. Many are serious pests of cultivated plants; the larvae usually bore into the stems or fruits of the plant. One of the most important pests in this group is the Codling Moth, *Carpocapsa pomonella* (Linn.), which attacks apple and other fruits; a small caterpillar found inside an apple is very probably the larva of this moth.

TORTRICID MOTHS   Family Tortricidae   See also Pl. 12

Identification: Similar to Olethreutidae, but Cu in HW lacking fringe of long hairs, R_4 and R_5 in FW usually stalked or fused, and M_2, M_3, and Cu_1 in FW parallel or divergent.

Most tortricid larvae are leaf rollers or leaf tiers. Some tie a number of leaves together with silk and feed inside the shelter so formed. An important pest of the group is the Spruce Budworm, *Choristoneura fumiferana* (Clemens), which may defoliate and kill spruce or other evergreens; this species is important in n. New England and the Maritime Provinces of Canada. Tortricids are common moths.

PHALONIID MOTHS   Family Phaloniidae   Not illus.

Identification: Small moths, HW broad and rounded. Cu_2 in FW rises in distal ¼ of discal cell. 1A absent in FW. R_5 in FW usually not stalked with R_4, and extends to outer margin of wing. M_1 in HW usually stalked with Rs. 3rd segment of labial palps short, blunt, the palps beaklike.

Some phaloniid larvae tie leaves together to form a shelter. They are mostly seed or stem borers and are uncommon.

CARPOSINID MOTHS   Family Carposinidae   Not illus.

Identification: Small moths. HW with only 1 branch of M. 1A completely lacking in FW.

This is a small, relatively rare group. The larva of 1 species bores into the fruit of currants.

OECOPHORID MOTHS   Family Oecophoridae

Identification: Small, somewhat flattened. Wings relatively broad, rounded apically, HW as wide as FW or nearly so. Cu_2 in FW rises *in distal ¼ of discal cell*, R_4 and R_5 in FW *stalked;* Rs and M_1 in HW *not stalked*. Head usually smooth-scaled. Palps long, upcurved, usually extending beyond vertex.

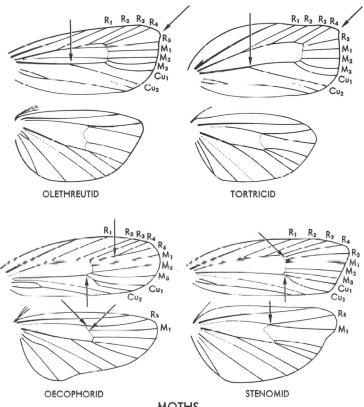

OLETHREUTID

TORTRICID

OECOPHORID

STENOMID

MOTHS

Oecophorids are fairly common, usually brownish moths. *Depressaria heracliana* (Linn.) attacks wild carrot, parsley, and related plants; the larva ties up the flower heads with silk and feeds inside them, then burrows into the stem.

STENOMID MOTHS Family Stenomidae
Identification: Cu_2 in FW rises *at end of discal cell.* Rs and M_1 in HW *stalked;* R_4 and R_5 in FW *usually not stalked,* or if stalked then R_5 extends to outer margin of wing. Head smooth-scaled.

Stenomid larvae usually feed in webs on leaves of oak. The moths are relatively large and plain-colored, whitish or gray, sometimes with dark markings. This group is small (24 N. American species), its members not common.

ETHMIID MOTHS   Family Ethmiidae   Not illus.
Identification: Similar to Oecophoridae (p. 248), but M_2 in HW rises closer to M_1 than to M_3.

Most ethmiids have the front wings marked with black and white. Larvae live in webs and feed chiefly on bindweed.

GLYPHIPTERYGID MOTHS   Family Glyphipterygidae
Identification: Small, FW rather square-tipped. Ocelli *large and conspicuous.* Head *smooth-scaled.* Cu_2 in FW rises in *distal ¼ of discal cell.* Rs and M_1 in HW *separated at origin.* R_5 in FW free or stalked with R_4, *extending to* apex or *outer margin* of wing. 2A in HW *forked at base.*

About 50 species occur in N. America but they are not common. Larvae of most species are leaf tiers.

ERMINE MOTHS   Family Yponomeutidae   See also Pl. 12
Identification: Small moths, FW usually brightly patterned. Head smooth-scaled. Cu_2 in FW rises *in distal ¼ of discal cell.* Rs and M_1 in HW *separated at their origin.* R_4 and R_5 in FW *separate,* R_5 extending to apex or outer margin of wing. M_1 and M_2 in HW *not stalked* (except in *Argyresthia,* which has the wings pointed apically).

Ermine moths are so called because some species (*Yponomeuta*) have the front wings white with black spots; other species are differently colored, but most have the front wings brightly patterned. Larvae of most species feed in webs spun over the leaves; a few are leaf miners, and some bore into fruit. Ermine moths are fairly common.

CLOTHES MOTHS and Others   Family Tineidae
Identification: Wings usually somewhat rounded apically, the HW about as wide as FW, sometimes narrowly rounded or *pointed apically* and HW narrower than FW. Maxillary palps usually present, folded at rest. Head rough-scaled or bristly. Antennae with a whorl of erect scales on each segment.

Most of the more than 130 species of N. American tineids are small and plain-colored. Many larvae are scavengers or feed on fungi, and some feed on fabrics; relatively few feed on foliage. Many larvae are casemakers. Three species in this group that feed on clothes and various woolen materials are often called clothes moths. The Webbing Clothes Moth, *Tineola bisselliella* (Hummel), is straw-colored and has a wingspread of 12–16 mm.; larva does not form cases. The Casemaking Clothes Moth, *Tinea pellionella* (Linn.), is about the same size but is more brownish and has 3 small dark spots in each front wing; its larva is a casemaker. The Carpet Moth, *Trichophaga tapetzella* (Linn.), is 12–14 mm. in wingspread and base of the front wings

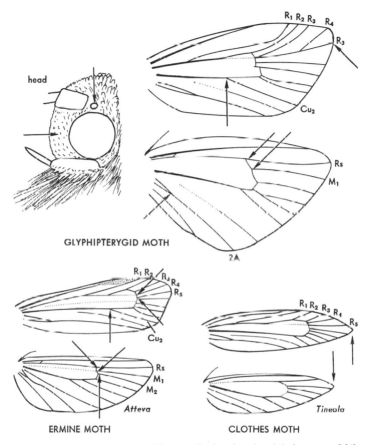

head

GLYPHIPTERYGID MOTH

2A

$R_1 R_2 R_3 R_4$
R_5
Cu_2

Rs
M_1
M_2

Atteva

ERMINE MOTH

$R_1 R_2 R_3 R_4$
R_5

Tineola

CLOTHES MOTH

is dark; larva forms silken galleries in the fabrics on which it feeds.

YUCCA MOTHS and Others   Family Incurvariidae   **p. 253**
Identification: Small moths. Wing surface with minute spines under the scales. Wings narrowly rounded apically, HW about as wide as FW. Venation usually complete, with R_4 and R_5 in FW generally *stalked*, and R_5 *extending to costal margin* of wing. Maxillary palps usually *well developed, folded*, the folded part $\frac{1}{2}$ to $\frac{2}{3}$ as long as width of head (sometimes vestigial or short and not folded).

Subfamily Incurvariinae (not illus.). The maxillary palps are folded, the folded part about half as long as width of head. Most are black or metallic bluish. Larva of the Maple Leaf Cutter, *Paraclemensia acerifoliella* (Fitch), is a leaf miner when young and a casemaker when older; cases of older larvae are made from 2 circular pieces of the leaf and are somewhat turtle-like. The members of this group are not common.

Yucca Moths, Subfamily Prodoxinae. These are similar to the Incurvariinae but white. Yucca moths in the genus *Tegeticula* pollinate the yucca; the female lays her eggs in the ovary of a yucca flower, and the larvae feed on the seeds; after ovipositing, the female thrusts a mass of yucca pollen into the stigma of the flower in which the eggs are laid. Pollen is collected by means of long curled tentacles on the maxillae. Yucca moths in the genus *Prodoxus* lack maxillary tentacles and do not pollinate yucca, but their larvae feed on its flower stem and the fleshy part of the fruit.

Fairy Moths, Subfamily Adelinae (not illus.). These moths have antennae at least as long as the front wings (in males to several times as long). Larvae usually live in flowers or seeds, and pupate inside 2 oval pieces of a leaf. There are about a dozen N. American species, and they are not common.

Microlepidoptera, Group 3

Small to minute, wingspread 3–20 mm. Wings narrow and more or less pointed apically; HW usually narrower than FW, with anal area reduced and with a long fringe on anal margin of wing.

Separation of families in this group will be a problem for the beginner. It is based principally on wing venation, mouth parts, head scaling, and occasionally other characters. Wing venation often is difficult to make out, even after the wings are cleared. Maxillary palps are vestigial or absent in most of these moths but in a few are well developed and at rest held in a folded position on either side of the proboscis. Labial palps are usually well developed, the basal segment small and the other 2 elongate; they curve up in front of the head, often to middle of the face or beyond. Some of these moths have an eye cap: the basal antennal segment is enlarged and concave beneath and fits over the eye when the antenna is depressed. The head is smooth-scaled in most moths in this group, but in some is rather bristly, especially on the vertex.

OPOSTEGID MOTHS   Family Opostegidae   **Not illus.**
Identification: Small to minute. Antennae with an eye cap. Maxillary palps small and folded. Venation greatly reduced, FW with only 3 or 4 unbranched veins.

Only 6 species of opostegids occur in N. America, and they are not common. Larvae are leaf miners.

NEPTICULID MOTHS   Family Nepticulidae   **Not illus.**
Identification: Minute moths. Antennae with an eye cap. Maxillary palps well developed, long, folded. FW with branched veins.

Some nepticulids have a wingspread of only a few mm. They are fairly common, but because of their small size are often overlooked. Larvae are mostly leaf miners, but a few form galls on the twigs or leaf petioles of various trees.

LYONETIID MOTHS   Family Lyonetiidae
Identification: Small to very small moths. Antennae with *an eye cap*. Vertex usually *rough and bristly*. Labial palps *very short* and drooping. Ocelli absent.

Some species in this fairly large group are quite common. Larvae are leaf miners, or live in webs between leaves. The largest genus is *Bucculatrix*, the larvae of which form whitish, longitudinally ribbed cocoons attached to twigs.

LEAF BLOTCH MINERS   Family Gracilariidae   **p. 255**
Identification: Small to minute moths. Antennae with or without an eye cap. Maxillary palps usually absent; if present, small and projecting forward. Scaling on vertex rough or smooth.

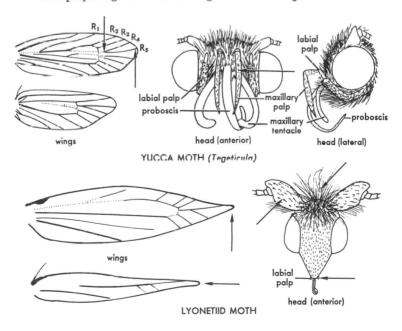

wings

R₁ R₂ R₃ R₄ R₅

labial palp
proboscis

labial palp

maxillary palp
maxillary tentacle

proboscis

head (anterior)

head (lateral)

YUCCA MOTH *(Tegeticula)*

wings

labial palp

head (anterior)

LYONETIID MOTH

HW usually *without a discal cell*, and costal margin *humped slightly near base*. FW generally without an accessory cell, but if one is present (*Parornix*) it is less than ⅓ as long as discal cell and there are 5 veins extending from discal cell to costal margin of wing. Venation of FW usually reduced, with 8 or fewer veins extending from discal cell to wing margin, *not more than 4 of which normally extend to costal margin*.

This is a large group, with over 200 N. American species. Larvae are leaf miners, and usually make blotch mines. Larvae of *Phyllocnistis* make narrow winding mines containing a dark central line of excrement; *P. populiella* Chambers is a common species attacking aspens and poplars.

OINOPHILID MOTHS   Family Oinophilidae   **Not illus.**
Identification: Very small, strongly flattened moths. Head smooth-scaled. R_5 in FW, when present, extends to costal margin of wing.

Our single oinophilid occurs in Mississippi and Louisiana. It is brownish gray, with a wingspread of about 9 mm. Larvae feed on fungi and decaying plant materials.

DIAMONDBACK MOTHS   Family Plutellidae
Identification: Small moths with wings *narrowly rounded at apex* and HW about as wide as FW. Rs and M_1 in HW stalked or *very close together in basal* ⅓ of their length. Head smooth-scaled. R_4 and R_5 in FW *not stalked*.

The front wings of these moths are often brightly patterned; light marks along costal margin of the wing in some species form diamond-shaped spots when the wings are folded over the abdomen. Larvae of most species are leaf miners or leaf tiers; a few are pests of garden plants.

DOUGLASIID MOTHS   Family Douglasiidae
Identification: Small moths. HW *without a discal cell*, R vein near middle of wing *with a branch to costal margin* at about ⅔ the wing length; R_5 in FW, when present, *free from R_4 but stalked with M_1*. Ocelli very large.

Only 4 species of douglasiids occur in N. America. Larvae are leaf miners.

CASEBEARERS   Family Coleophoridae
Identification: Small to minute moths. Discal cell of FW *oblique*, its apex *much closer to hind margin* of wing than to costal margin. R_2 in FW rises *about halfway between R_1 and R_3*, but not at apex of discal cell. Front tibiae slender, without a movable pad on inner surface.

About 100 species of casebearers occur in N. America, and some are fairly common. Young larvae are usually leaf miners, but older larvae construct portable cases of bits of leaves and excrement; older larvae feed by protruding the head from end of the case and eating holes in leaves of the host plant. The Pistol Casebearer, *Coleophora malivorella* Riley, and the Cigar Casebearer, *C. serratella* (Linn.), are pests of apple and other fruit trees; the common names refer to the shape of the cases.

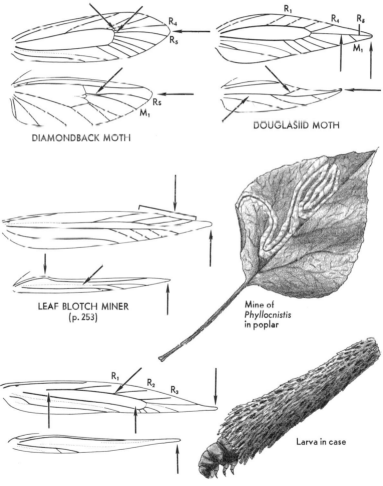

DIAMONDBACK MOTH

DOUGLASIID MOTH

LEAF BLOTCH MINER
(p. 253)

Mine of
Phyllocnistis
in poplar

CASEBEARER

Larva in case

COSMOPTERYGID MOTHS   Family Cosmopterygidae
Identification: Head smooth-scaled. R_1 in FW usually rises *about middle of discal cell* or slightly beyond, and R_2 rises *before apex of cell*, usually a little more than halfway between R_1 and R_3. R_4 and R_5 in FW *stalked* or fused. Discal cell in FW variable in position, but if oblique (as in Coleophoridae) front tibiae have a well-developed pad on inner surface at about half their length.

Cosmopterygids are small moths that vary somewhat in wing shape and venation and in color; a few have apex of the front wing elongated. Larvae of most species are leaf miners; 1 species feeds in the bolls of cotton.

BLASTOBASID MOTHS   Family Blastobasidae
Identification: Small moths. FW with *a stigmalike thickening* between C and R_1. R_2 in FW rises at or *near apex of discal cell*. R_4 and R_5 in FW *long-stalked*.

About 100 species of blastobasids occur in N. America, but they are not common. Some larvae are scavengers; the larva of 1 species feeds inside acorns that have been hollowed out by acorn weevils.

SHIELD BEARERS   Family Heliozelidae   Not illus.
Identification: Small moths. Head smooth-scaled. Palps short and drooping. Discal cell absent in HW, sometimes (*Coptodisca*) also absent in FW.

Larvae of most shield bearers are leaf miners and when ready to pupate cut out an oval section of the leaf and make it into a pupal case that is attached to another part of the host plant. Most species attack trees.

HELIODINID MOTHS   Family Heliodinidae
Identification: Very small moths. Head smooth-scaled. Venation variable, FW sometimes (*Cycloplasis*) with only 3 or 4 veins, usually with a well-developed discal cell and 8–10 veins extending from cell to wing margin and R_1 rising *about middle of cell. At least 3 veins rise from apex of discal cell in FW. HW without a forked vein* at apex.

These moths rest with the hind or middle legs outstretched or elevated above the wings. Larvae vary in habits: some are leaf miners, some are external feeders on foliage, and 1 species is an internal parasite of oak scales (Kermidae).

EPERMENIID MOTHS   Family Epermeniidae   Not illus.
Identification: Small moths. FW with 10 veins from discal cell to wing margin, 4 of them to costal margin. Head smooth-scaled. R_1 in FW rises before middle of discal cell. Rs in HW ends at or before apex of wing. Ocelli absent.

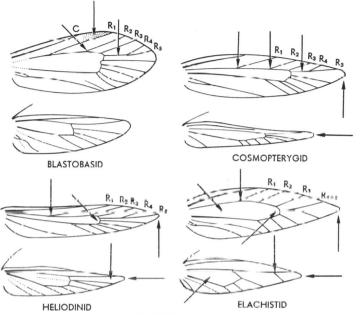

BLASTOBASID

COSMOPTERYGID

HELIODINID

ELACHISTID

MOTHS

This is a small and uncommon group. The moths are yellowish, grayish, or brownish.

SCYTHRID MOTHS Family Scythridae   **Not illus.**
Identification: Small moths. FW with 9 veins extending from discal cell to wing margin, 4 of them to costal margin. R_1 in FW rises about ⅔ the length of discal cell. Forked vein at tip of HW.

This is a small group, and not common. Larvae are leaf folders or leaf tiers.

ELACHISTID MOTHS Family Elachistidae
Identification: Mostly dark brown or gray, with silvery spots or bands. Head smooth-scaled. Discal cell *present* in both FW and HW. FW with only 8 or 9 veins extending from discal cell to wing margin, 4 or fewer of them to costal margin, and *only 1 or 2 veins* rising from apex of discal cell. R_1 in FW rises *about middle of discal cell*. HW with *a forked vein at apex*.

Larvae are leaf miners in grasses and sedges. Elachistids are not common.

TISCHERIID MOTHS   Family Tischeriidae
Identification: Face smooth-scaled, but vertex with rough bristly hair. Labial palps small. Costal margin of FW somewhat arched and apex *prolonged* to a sharp point. FW with *5 veins* extending from discal cell to costal wing margin, and with an accessory cell at least half as long as the discal cell.

Larvae of most tischeriids make blotch mines in oak leaves. The Apple Leaf Trumpet Miner, *Tischeria malifoliella* Clemens, is a common pest in the East; its larva makes a trumpet-shaped mine. Other species mine in leaves of blackberry or raspberry.

Jugate Moths: Suborder Jugatae

Venation in FW and HW similar, Rs in HW with *as many branches as in FW*. No frenulum, the wings on each side united by a jugum (a fingerlike lobe at base of FW). This is a small group, with about 30 N. American species. Most are quite rare.

ERIOCRANIID MOTHS   Family Eriocraniidae   **Not illus.**
Identification: Wingspread 12 mm. or less. Maxillary palps well developed, 5-segmented. Mandibles vestigial. Middle tibiae with 1 spur. Sc in FW forked near its tip.

Eriocraniids are somewhat similar to clothes moths. Larvae are leaf miners, usually attacking trees. Larva of *Mnemonica auricyanea* Walsingham makes blotch mines in oak and chestnut, and pupates in the soil. None of our 5 N. American species is common.

MANDIBULATE MOTHS   Family Micropterygidae
Identification: Similar to Eriocraniidae, but with functional mandibles, middle tibiae without spurs, and Sc in FW *forked near its middle*.

These moths differ from all other Lepidoptera in having functional mandibles. They feed chiefly on pollen. Larvae whose habits are known feed on mosses and liverworts. Only 3 species of mandibulate moths occur in N. America, and they are quite rare.

GHOST MOTHS or SWIFTS   Family Hepialidae
Identification: Wingspread 1–3 in. Maxillary palps well developed. *No tibial spurs.*

Ghost moths are relatively uncommon, but are the jugates most likely to be encountered. The name "swift" refers to the very fast flight of most of these moths. Some of the larger species are similar to sphinx moths. Larvae bore in the roots of various trees; *Sthenopis argenteomaculatus* Harris attacks alder and *S. thule* Strecker bores in willow.

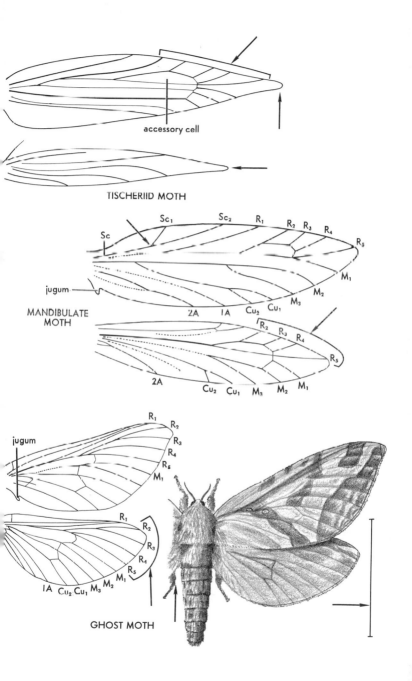

accessory cell

TISCHERIID MOTH

Sc_1 Sc_2 R_1 R_2 R_3 R_4

Sc

R_5

jugum

MANDIBULATE
MOTH

M_1

M_2

M_3

2A   1A Cu_2 Cu_1

R_2 R_3 R_4

R_5

2A Cu_2 Cu_1 M_3 M_2 M_1

jugum

R_1

R_2

R_3

R_4

R_5

M_1

R_1

R_2

R_3

R_4

R_5

M_1

1A Cu_2 Cu_1 M_3 M_2

GHOST MOTH

Flies: Order Diptera

Identification: *One pair* of membranous wings, borne by meso-thorax (wings rarely reduced or lacking). HW reduced to small knobbed structures (halteres). Antennae *variable, often short, inconspicuous, and 3-segmented.* Compound eyes large, sometimes meeting on dorsal side of head. Mouth parts *sucking* (rarely vestigial), maxillary palps well developed, labial palps lacking. Tarsi nearly always *5-segmented*. Relatively soft-bodied. Metamorphosis complete.

Similar orders: Most insects in other orders likely to be confused with Diptera have 2 pairs of wings. The few with 1 pair (certain grasshoppers, beetles, mayflies, and others) generally do not resemble flies. Male scale insects resemble midges, but have 1-segmented tarsi and 1 or 2 long stylelike processes at end of the abdomen.

Immature stages: Larvae are usually legless and wormlike, and often lack a well-developed head; they are commonly called maggots. They live in water, soil, decaying materials, or in plant or animal tissues. Many are aquatic and occur in a variety of aquatic habitats. The plant-feeding species generally live in the roots, fruit, leaves, or other parts of the plant. Many are parasitic, living in the bodies of other animals.

Habits: Flies occur in many different habitats; each species is usually found near the habitat of its larvae. Adults often occur on flowers. Many are bloodsucking, and are to be found on or near the animals on which they feed.

Importance: Flies constitute one of the larger orders of insects and are abundant in individuals as well as species; they occur almost everywhere. They are an important food of many larger animals. Many species are parasitic or predaceous on other insects and are of value in keeping noxious species under control; others are of value as scavengers. Large numbers are a nuisance because they bite; some are important as vectors of disease. Many attack and damage cultivated plants; a few of these serve as vectors of plant diseases.

Classification: Three suborders — Nematocera, Brachycera, and Cyclorrhapha — differing principally in wing venation and antennal structure. Wing venation provides useful characters for separating families throughout the order; the venational terminology usually used is that of Comstock, but many terms of an older terminology, particularly those referring to cells, are frequently used; these 2 terminologies are illustrated opposite. Other characters used in separating families of Diptera are discussed in the accounts of the groups in which they are used.

No. of species: World, 87,000; N. America, 16,144.

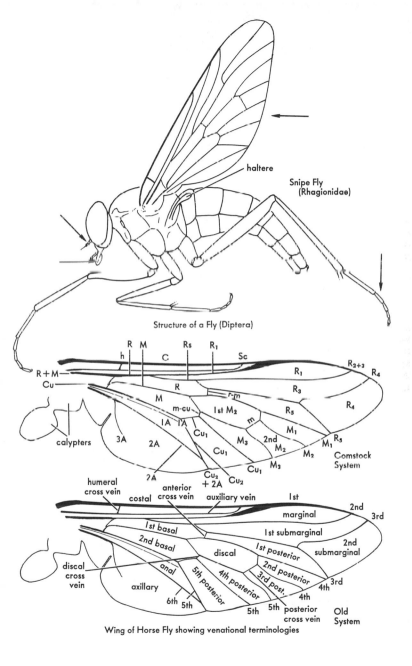

haltere

Snipe Fly
(Rhagionidae)

Structure of a Fly (Diptera)

R M   Rs R₁
h   C   Sc   R₂₊₃   R₄
R + M   R₁   R₄
Cu   R   R₃
            M   r-m   R₃
                    1st M₂   R₅   R₄
calypters   m-cu   m   M₁
        3A   2A   1A 1A Cu₁   M₃   2nd   M₁   R₅
                    Cu₁   M₂   M₂
                ?A   Cu₁   M₃   M₂   Comstock
                    Cu₂ Cu₂   Cu₁   M₃   System
                    + 2A

humeral   anterior auxiliary vein
cross vein   costal   cross vein   1st
                                            marginal   2nd
                                                            3rd
        1st basal   1st submarginal
        2nd basal   discal   2nd
discal   1st posterior   submarginal
cross   anal   4th posterior 2nd posterior   3rd
vein   axillary   5th posterior   3rd post.   4th
        6th 5th   5th   4th
                            5th posterior   Old
                            cross vein   System

Wing of Horse Fly showing venational terminologies

Suborder Nematocera

Antennae apparently with 6 or more segments (3rd subdivided), plumose in some males. Wing venation varies from complete (with R 5-branched) to greatly reduced. R_{2+3} often forked (never forked in other suborders). Mostly slender, soft-bodied, midgelike, with relatively long legs and antennae. Larvae usually aquatic or living in moist soil, the nonaquatic larvae generally being gall makers.

CRANE FLIES   Family Tipulidae
Identification: Mosquitolike, with *very long legs*. Mesonotum with a *V-shaped suture*. Ocelli *absent*. R with 4 or fewer branches. *2 anal veins reach wing margin.*

This is a large group, with nearly 1500 N. American species. Many of its members are very common flies. Most species are 10–25 mm. and brownish or gray; a few have dark markings on the wings. Larvae live in water or in moist soil, and generally feed on decaying plant material. Adults are most common near water or where there is abundant vegetation. Crane flies do not bite.

WINTER CRANE FLIES   Family Trichoceridae
Identification: Similar to Tipulidae, but *with ocelli.*

These crane flies are most likely to be seen in early spring or on mild days in winter. They are not common. Larvae live in decaying plant materials.

PRIMITIVE CRANE FLIES   Family Tanyderidae
Identification: Similar to Tipulidae, but R *5-branched*. M_3 cell with *a cross vein*. Anal angle of wing *well developed*.

This group contains 4 N. American species, 1 in the East and 3 in the West, none of them common. The eastern species occurs from Quebec to Florida; it is 7–10 mm., and grayish brown with brown crossbands on the wings; it generally occurs in dense vegetation near streams, and the larvae live in wet sand along stream shores.

PHANTOM CRANE FLIES   Family Ptychopteridae
Identification: Similar to Tipulidae, but wings with *only 1 anal vein reaching margin* and *without a closed discal cell.*

Our most common ptychopterid is *Bittacomorpha clavipes* (Fabricius), which has legs banded with black and white and the basal tarsal segment swollen; it often flies with the legs extended. Other species lack leg bands and do not have a swollen basal tarsal segment. Larvae live in decaying plant materials and adults occur in swampy areas.

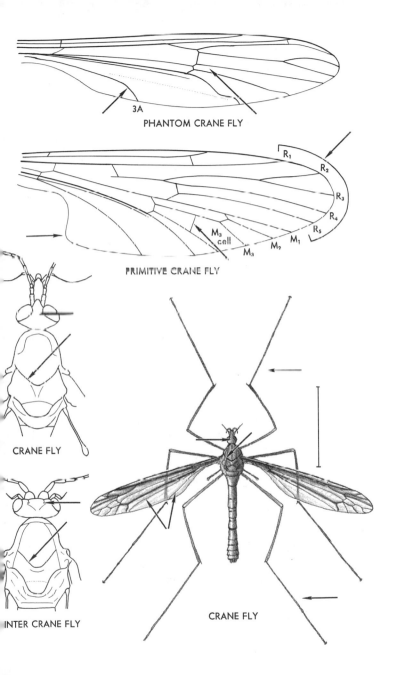

PHANTOM CRANE FLY

3A

PRIMITIVE CRANE FLY

R_1
R_2
R_3
R_4
R_5
M_3 cell
M_3
M_2
M_1

CRANE FLY

INTER CRANE FLY

CRANE FLY

MOTH and SAND FLIES   Family Psychodidae
Identification: Small, *very hairy* flies, mostly 5 mm. or less.
Wings usually broad, *pointed apically*, and at rest are held roof-
like over body (moth flies) or together above body (sand flies).
R *5-branched*.

Moth flies occur near drains and sewers, often in considerable
numbers; larvae live in places where there are decaying materials.
Sand flies (subfamily Phlebotominae) occur near water; larvae
live in moist soil. Moth flies are common and widely distributed,
and do not bite; sand flies occur in the South and the tropics,
and they bite. Sand flies serve as vectors of several diseases —
pappataci fever, kala-azar, oriental sore, espundia, and oroya
fever or verruga peruana — in S. America and other tropical
areas of the world.

NET-WINGED MIDGES   Family Blephariceridae
Identification: Mosquitolike in size, with long legs and resembling crane flies but without a V-shaped suture on mesonotum. Anal angle of wing *projects*. Wings sometimes with a network of fine lines between the veins. Ocelli present.

These flies occur along swift-moving streams, in which the
larvae live. They are relatively rare.

NYMPHOMYIID FLIES   Family Nymphomyiidae   **Not illus.**
Identification: Wings vestigial. Antennae 5-segmented, 3rd
segment large and club-shaped. Head elongate. Mouth parts
vestigial. Legs long and slender and widely separated.

One species in this group has been reported from rapid streams
in New Brunswick. Larvae are assumed to be aquatic.

MOUNTAIN MIDGES   Family Deuterophlebiidae   **Not illus.**
Identification: Wings broad (broadest in basal ¼), pubescent,
almost veinless but with a fanlike development of folds. Antennae very long, at least 3 times as long as body. Ocelli and
mouth parts lacking.

Four species of mountain midges have been reported from the
West (Colorado to California), where the larvae occur in swift-
flowing mountain streams.

DIXID MIDGES   Family Dixidae
Identification: Similar to mosquitoes (p. 266) but wings lacking
scales and body bare (not scaly).

Dixid midges are common and widely distributed insects.
Larvae occur in pools and ponds, and adults are usually found
near these habitats. Larvae are slender and wormlike, and feed
on surface of the water; the body is generally bent into a U, and
the larvae move by alternately straightening and bending the
body. Adults are blackish and 5–6 mm. They do not bite.

PHANTOM MIDGES Family Chaoboridae
  Identification: Similar to mosquitoes (p. 266) but wing scales mostly confined to margin and *proboscis short*.

  These insects are quite common, and generally occur near pools and ponds in which the larvae live. The common name of the group is derived from the appearance of some larvae, which

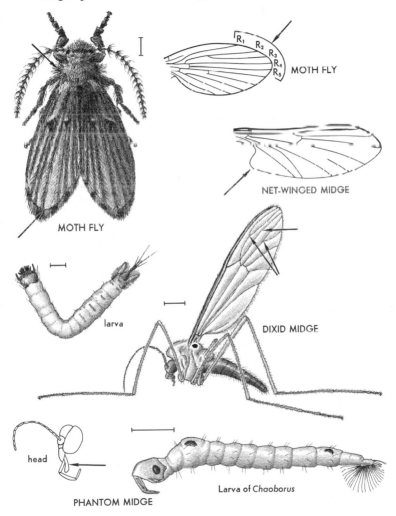

MOTH FLY

MOTH FLY

NET-WINGED MIDGE

larva

DIXID MIDGE

head

PHANTOM MIDGE

Larva of *Chaoborus*

are nearly transparent; these larvae are predaceous, capturing prey with their antennae. Adults do not bite.

MOSQUITOES   Family Culicidae
Identification: Wings long, narrow, with *scales along veins and wing margin.* Distal part of wing with *an unforked vein between 2 forked veins.* Proboscis *long.* Ocelli absent.

Mosquitoes are common, widely distributed, and well-known insects. Males have very plumose antennae and do not bite; females, which have only a few short hairs on the antennae, do bite and are often serious pests. Mosquitoes serve as vectors of several important diseases: malaria, yellow fever, dengue, filariasis, and encephalitis; these diseases are chiefly tropical but some may occur in this country.

This family is divided into 3 subfamilies, Anophelinae, Culicinae, and Toxorhynchitinae. Anophelinae have palps long in both sexes (clubbed in ♂) and the scutellum rounded; they rest with the body and proboscis in an almost straight line, at an angle to the substrate (some appear almost to "stand on their heads"). Culicinae have palps of the female short and those of the male usually long, and the scutellum 3-lobed; they rest with the proboscis bent down and the body more or less parallel to the substrate. Toxorhynchitinae are large mosquitoes, with the scutellum rounded and palps of the female short; the basal part of the proboscis is stout, the apical part is slender and decurved. Anophelinae are represented in N. America by 1 genus, *Anopheles,* most species of which have patches of light and dark scales on the wings. Most of our mosquitoes (*Culex, Aedes, Psorophora,* and others) belong to the Culicinae. Toxorhynchitinae are represented by 1 genus, *Toxorhynchites,* and are not very common.

Mosquito larvae are aquatic and occur in ponds, pools, various containers of water, and in tree holes containing water. Larvae breathe at the surface; larvae of Toxorhynchitinae and Culicinae have *a breathing tube* at the posterior end of the body; larvae of *Anopheles lack a breathing tube,* and spend most of their time at the surface. Most larvae feed on organic debris; a few are predaceous. Pupae are aquatic and generally very active. Eggs are usually laid on surface of the water, either singly or in rafts. A few lay eggs near water, and the eggs hatch when flooded.

SOLITARY MIDGES   Family Thaumaleidae   **Not illus.**
Identification: Bare, reddish-yellow or brownish flies, about 8 mm. Venation reduced, and only 7 veins reach wing margin (R 2-branched, M unbranched). Ocelli absent. Antennae short, about as long as head, the 2 basal segments enlarged.

This group includes 5 rare N. American species, 2 in the East (Quebec to N. Carolina) and 3 in the West (British Columbia

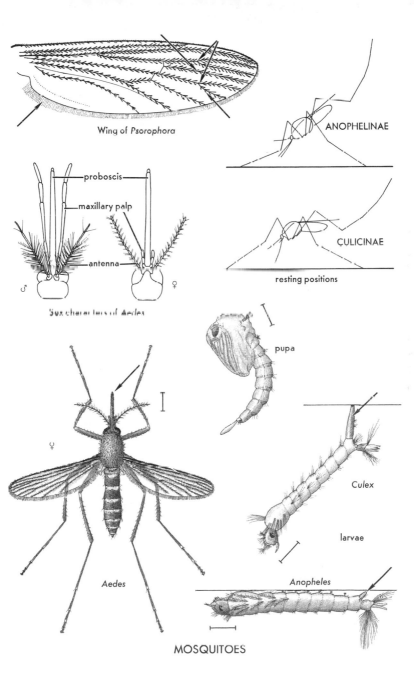

Wing of *Psorophora*

ANOPHELINAE

CULICINAE

resting positions

proboscis

maxillary palp

antenna

♂   ♀

Sex characters of *Aedes*

♀

pupa

Aedes

Culex

larvae

Anopheles

MOSQUITOES

and Idaho). Larvae occur on rocks in streams, and adults are usually found on vegetation near streams.

PUNKIES or BITING MIDGES   Family Ceratopogonidae
Identification: Minute flies, generally *less than 3 mm*. Ocelli absent. Radial branches prominent. Thickened part of C usually ends ½ or ¾ way to wing tip; M with 2 branches. Front tarsi *not lengthened*.

Many punkies bite man and may be very annoying. A few live as ectoparasites on the bodies of other insects and a few are predaceous. Larvae are mostly slender and snakelike, aquatic or semiaquatic.

MIDGES   Family Chironomidae
Identification: Ocelli absent. Thickened part of C ends near wing tip. M unbranched. Front tarsi usually *lengthened*. Wings long and narrow. ♂ antennae generally *plumose*.

Midges are very common insects, found almost everywhere and often in considerable numbers. Larvae of most species are aquatic, and many live in tubes constructed of debris; some larvae are red. Adults are generally soft-bodied, with long legs and antennae and a short proboscis; they do not bite. Midges frequently occur in large swarms, particularly near ponds and lakes.

BLACK FLIES   Family Simuliidae
Identification: Generally 4 mm. or less, stocky in build, and somewhat *humpbacked*. Antennae *short*. Ocelli absent. Wings *broad at base*, narrowing distally, anterior veins *heavy* and remaining veins weak. Usually gray.

Female black flies are vicious biters and are serious pests in many parts of the country. Larvae live in streams, often in large numbers, where they attach to objects in the water. Black flies are widely distributed; adults appear chiefly in late spring and early summer. In parts of the tropics these insects serve as the vector of onchocerciasis, a disease caused by a roundworm that sometimes also causes blindness.

WOOD GNATS   Family Anisopodidae
Identification: Mosquitolike in appearance, 4–6 mm. Ocelli present. Thickened section of C ends *near wing tip*. A discal cell and 5 posterior cells usually *present* (Anisopodinae); if a discal cell is lacking (Mycetobiinae) the base of M is *lacking*, the 2 basal cells *coalesce*, and Rs forks *opposite the r-m cross vein*.

The most common wood gnats (Anisopodinae) have faint spots on the wings, and usually occur in moist places where there is abundant vegetation; larvae occur in decaying materials. Wood gnats are often attracted to sap, and the larvae of some species live in fermenting sap.

PACHYNEURID GNATS Family Pachyneuridae

Identification: Similar to Anisopodidae, but Rs 3-branched; R_2 resembles *a cross vein* and extends from R_{2+3} to about the end of R_1.

There are 2 rather rare species of pachyneurids in the U.S., 1 in the East and 1 in the West. The eastern species, *Axymyia furcata* McAtee, occurs from Massachusetts to Virginia. It is

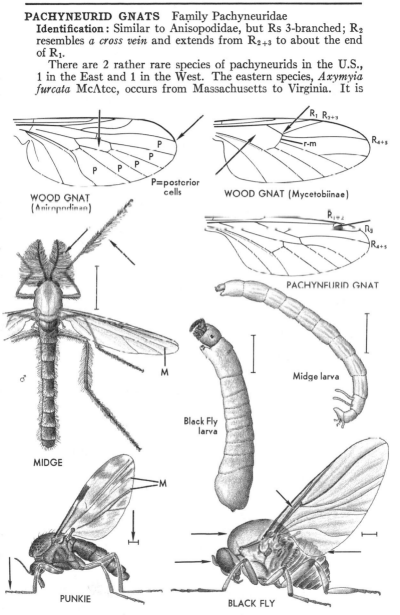

WOOD GNAT (Anisopodinae)

P=posterior cells

WOOD GNAT (Mycetobiinae)

R_1 R_{2+3}

r-m

R_{4+5}

R_{1+2}

R_3

R_{4+5}

PACHYNEURID GNAT

MIDGE

Midge larva

Black Fly larva

PUNKIE

BLACK FLY

dark brown, about 5–6 mm., and has been taken in low vegetation bordering mountain streams.

MARCH FLIES   Family Bibionidae
Identification: Small to medium-sized, stout-bodied, and usually black. Ocelli present. Antennae short, rising low on face. Thickened section of C ends near wing tip. Anal angle of wing usually *well developed*. Pulvilli (pads beneath tarsal claws) present. Tibiae with *apical spurs*.

March flies are common insects usually found on flowers. Larvae feed on roots and in decaying vegetation. Adults are most common in spring and early summer; some species are abundant in March (hence the common name).

FUNGUS GNATS   Family Mycetophilidae
Identification: Slender, long-legged, mosquitolike, with *elongated coxae*. Ocelli present. Thickened section of C ends near wing tip. Rs *simple* or 2-branched; if 2-branched the fork is beyond r-m, or r-m is obliterated by the fusion of Rs and M. Eyes do not meet above antennae. Pulvilli (see above) absent or minute.

This is a large group whose members are common in areas where there is an abundance of decaying vegetation or fungi. Most species are between 5 and 10 mm. and have relatively long antennae. Larvae live in fungi, decaying vegetation, or moist soil.

DARK-WINGED FUNGUS GNATS   Family Sciaridae
Identification: Similar to Mycetophilidae, but eyes meet *above bases of antennae*. R-m cross vein *appears as a basal continuation of Rs*, the base of Rs resembling a cross vein.

Sciarids are common insects usually found in moist shady places. Most species are 5 mm. or less, and dark-colored. Larvae feed in fungi, decaying vegetation, or on plant roots; a few species are pests in mushroom cellars.

MINUTE BLACK SCAVENGER FLIES   Family Scatopsidae
Identification: Black or dark brown, 3 mm. or less, rather stocky in build, with short antennae. Veins in anterior part of wing heavy, remaining veins weak. Rs *not forked*. Palps 1-segmented.

Scatopsid larvae occur in decaying material and excrement, and adults are usually found around such materials; adults often enter houses, where they are generally seen on windows. These flies are fairly common.

HYPEROSCELIDID GNATS   **Not illus.**
Family Hyperoscelididae
Identification: Similar to Scatopsidae, but Rs forked and the palps 3- or 4-segmented.

A single rare species in this group has been recorded from Quebec, Alaska, and Washington. Its larva is unknown.

GALL GNATS. Family Cecidomyiidae

Identification: Minute flies, rarely over 3 mm., usually slender, with long legs and antennae. Venation reduced, with 7 or fewer veins reaching wing margin. Ocelli present or absent.

This is a large and widely distributed group whose members occur in a variety of situations. Larvae of most species are gall makers; some are plant feeders but do not form a gall, some occur in decaying materials, and several are parasitic or predaceous. A few species are important crop pests. The Hessian

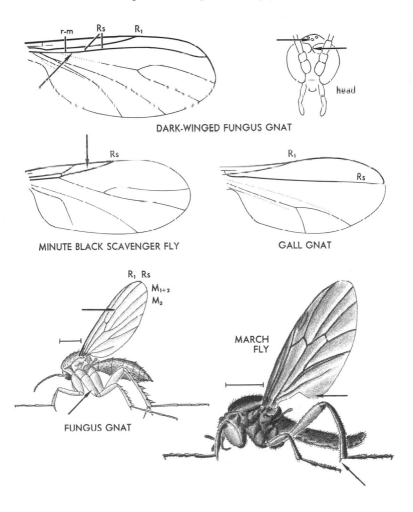

DARK-WINGED FUNGUS GNAT

MINUTE BLACK SCAVENGER FLY

GALL GNAT

FUNGUS GNAT

MARCH FLY

Fly, *Mayetiola destructor* (Say), is a serious pest of wheat, and other species attack clovers.

Suborder Brachycera

Five or fewer (usually 3) antennal segments, the 3rd *sometimes annulated* (subdivided by rings), sometimes with a terminal style (illus., p. 275), only rarely with an arista (see opp.). Rs usually 3-branched (R_{4+5} forked). No frontal suture. Anal cell generally *longer than 2nd basal cell*, and in most cases closed near wing margin. Mostly medium-sized to large, relatively robust flies.

The families in this suborder may be divided into 3 groups on the basis of form of 3rd antennal segment and number of pads on last tarsal segment:

1. 3rd antennal segment annulated or subdivided; tarsi with 3 pads: Xylophagidae, Xylomyidae, Stratiomyidae, Tabanidae, and Pelecorhynchidae.
2. 3rd antennal segment not annulated; tarsi with 3 pads: Rhagionidae, Nemestrinidae, and Acroceridae.
3. 3rd antennal segment not annulated; tarsi with 2 pads or none: the remaining families (Hilarimorphidae through Dolichopodidae).

XYLOPHAGID FLIES Family Xylophagidae
Identification: Third antennal segment *elongate*. M_3 cell *open* (except in *Rachicerus*, in which the antennae appear many-segmented). Calypters (illus., p. 275) small or vestigial. At least middle tibiae with apical spurs.

Xylophagids are not common; the ones most often seen are in the genera *Xylophagus* and *Coenomyia*. *Xylophagus* flies are slender and ichneumonlike, generally 10–13 mm., with eyes bare, the scutellum lacking spinelike protuberances, and the posterior cells usually longer than wide. *Coenomyia* flies are robust, 14–25 mm., with eyes pubescent, the scutellum bearing 2 spinelike protuberances and the posterior cells about as long as wide. Other xylophagids are 2–9 mm. and rare. Larvae occur in decaying wood or in the soil.

XYLOMYID FLIES Family Xylomyidae
Identification: Similar to Xylophagidae (most resemble *Xylophagus*), but with the M_3 cell *closed*.

These flies are not very common, but those most likely to be encountered are slender, ichneumonlike, and 10 mm. or less. Larvae occur in decaying wood or under bark.

SOLDIER FLIES Family Stratiomyidae
Identification: Third antennal segment rounded or *elongate*. Branches of R crowded toward anterior part of wing, with R_5 *ending in front of wing tip*.

Members of this group vary in appearance. The most common species are 10–15 mm. and wasplike, with the 3rd antennal segment elongate. Others are brownish or metallic blue-black, often less than 10 mm., and some have the 3rd antennal segment rather rounded, with a long style or an arista. Adults are usually found on flowers. Larvae occur in a variety of situations: some are aquatic, some live in decaying materials, and some are found in other situations.

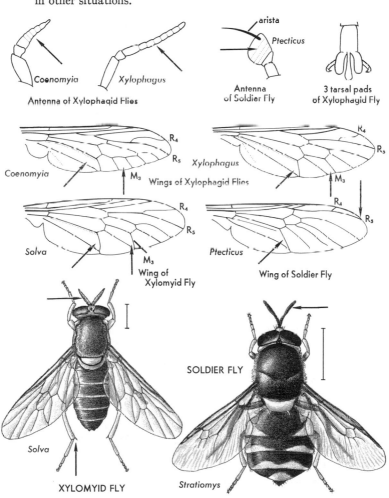

Coenomyia Xylophagus

Antenna of Xylophagid Flies

arista
Ptecticus
Antenna of Soldier Fly

3 tarsal pads of Xylophagid Fly

Coenomyia R_4 R_5 M_3

Wings of Xylophagid Flies

Xylophagus R_4 R_5 M_3 R_4 R_5

Solva R_4 R_5 M_3

Wing of Xylomyid Fly

Ptecticus

Wing of Soldier Fly

Solva

XYLOMYID FLY

SOLDIER FLY

Stratiomys

HORSE and DEER FLIES Family Tabanidae **See also Pl. 13**
Identification: Third antennal segment *elongate.* Calypters large. R₄ and R₅ *divergent, enclosing wing tip.*

Tabanids are relatively stout-bodied, medium-sized to large (mostly 10–25 mm.), and many are very common. Females are bloodsucking, and are often serious pests of man and animals. Males, much less often seen, feed on flowers. The eyes meet dorsally in the male and are separated in the female; the eyes are brightly colored or iridescent in many species. Larvae of most species are aquatic. Adults are often abundant near swamps or ponds where the larvae occur, but are strong fliers and may range many miles from their breeding places. Deer flies (*Chrysops*) are smaller than most other tabanids (House Fly size or slightly larger), black or brownish, and usually have dark spots on the wings; the 3rd antennal segment lacks a basal toothlike process, and there are apical spurs on the hind tibiae. The larger horse flies, most of which are in the genus *Tabanus*, are usually gray or blackish, generally without dark spots on the wings (some species have entirely dark wings); the 3rd antennal segment has *a basal toothlike process* and there are no apical spurs on the hind tibiae. Some species of *Tabanus*, commonly called greenheads, have bright green eyes in life.

PELECORHYNCHID FLIES   **Not illus.**
Family Pelecorhynchidae
Identification: Similar to Tabanidae, but 2A somewhat S-shaped (straight or gently curved in Tabanidae), anal cell open (closed in Tabanidae), and eyes densely hairy (usually bare in Tabanidae).

The only U.S. member of this family is *Bequaertomyia jonesi* (Cresson), which occurs in the Pacific Coast states. This fly is 13–15 mm., black, with wings smoky and antennae and palps orange.

SNIPE FLIES Family Rhagionidae   **See also p. 261**
Identification: Third antennal segment *more or less rounded* and bearing a long slender terminal style. Calypters small or vestigial. Venation *normal.*

Snipe flies are mostly 8–15 mm., with the abdomen somewhat tapering posteriorly, the legs long, and the head more or less rounded; some species have spots on the wings. Most are black or gray; some are black with a yellow mesonotum. Snipe flies are common, and generally occur in wooded areas or areas of fairly dense vegetation. Eastern species in this group do not bite, but some western species do. Larvae generally occur in decaying vegetation; some are aquatic.

TANGLE-VEINED FLIES Family Nemestrinidae
Identification: Venation peculiar, with most branches of M

ending in front of wing tip. 3rd antennal segment short, rounded, and with a long slender terminal style. Tibiae without apical spurs.

Nemestrinids are medium-sized, stout-bodied flies that do quite a bit of hovering and are fast fliers. They are quite rare but are most likely to be found in weedy fields where the vegetation is high.

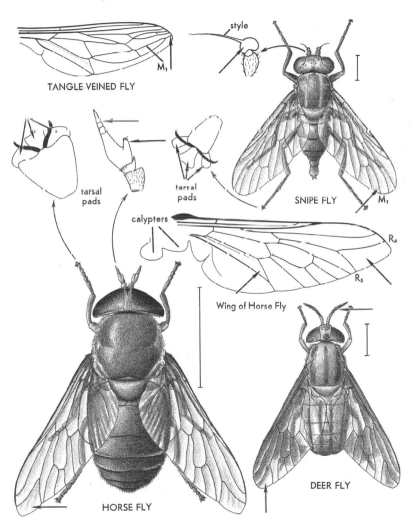

TANGLE-VEINED FLY

style

SNIPE FLY

M₁

tarsal pads

tarsal pads

calypters

Wing of Horse Fly

R₄

R₅

HORSE FLY

DEER FLY

SMALL-HEADED FLIES Family Acroceridae
Identification: Appearance distinctive: head *very small* and attached low on thorax, the body appearing *humpbacked;* calypters very large. Venation often *reduced.*

Acrocerids are medium-sized to small and are rather rare. Some species have a long slender proboscis. Larvae are parasites of spiders.

HILARIMORPHID FLIES Family Hilarimorphidae **Not illus.**
Identification: Third antennal segment oval, with a short 2-segmented style. Tibiae without apical spurs. 4 posterior cells and no closed discal cell.

Three very rare species are known from the U.S. They have been taken from New England to the West Coast.

STILETTO FLIES Family Therevidae
Identification: Medium-sized, usually grayish flies, with abdomen *somewhat tapering.* 3rd antennal segment *slightly elongated,* often with a short terminal style. *5 posterior cells,* the 4th *sometimes closed.*

Therevids resemble some robber flies (Asilidae), but differ in having the vertex flat or slightly convex (hollowed out in robber flies). They are relatively uncommon, and are usually found in open areas. Larvae occur in soil or decaying materials and are predaceous.

FLOWER-LOVING FLIES Family Apioceridae **Not illus.**
Identification: Similar to Therevidae but larger, and M_1 curves forward and ends in front of wing tip.

This is a small group of rare flies occurring in arid regions of the West and usually found on flowers.

WINDOW FLIES Family Scenopinidae
Identification: Moderately robust flies somewhat smaller than a House Fly, and usually grayish or blackish. 3 posterior cells. M_{1+2} ends *in front of wing tip.*

Window flies are so named because some are occasionally found on windows. They are relatively rare. Larvae usually occur in decaying wood or fungi; most are predaceous, and feed on a variety of insects.

ROBBER FLIES Family Asilidae **See also Pl. 13**
Identification: Top of head *hollowed out* between eyes; *3 ocelli.* 3rd antennal segment usually *elongate,* often bearing a short terminal style. Body varies from very hairy to nearly bare, but face usually *bearded.*

Robber flies are common insects, 5–30 mm., with legs and thorax relatively large. Most are relatively bare, with a long abdomen that tapers posteriorly; some are robust and hairy, and resemble bumble bees; some have a *very long and slender* abdomen, and resemble damselflies. They occur in a variety of habitats and are predaceous, often attacking insects larger than themselves. Larger species can inflict a painful bite if handled carelessly. Larvae occur chiefly in soil or decaying wood, and some are predaceous on larvae of other insects.

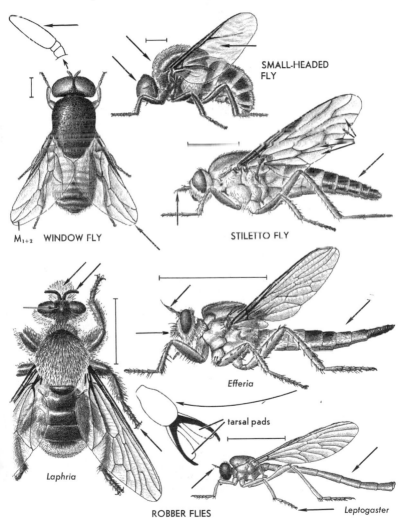

M_{1+2} WINDOW FLY

SMALL-HEADED FLY

STILETTO FLY

Efferia

tarsal pads

Laphria

ROBBER FLIES

Leptogaster

MYDAS FLIES Family Mydidae   **See also Pl. 13**
Identification: Large flies, about 1 in. Elongate, bare, blackish, some with 2nd abdominal segment yellowish or orange. Antennae long, 4-segmented, last segment *somewhat swollen. 1 ocellus* or none. Top of head somewhat hollowed out between eyes. M₁ ends at or *in front of wing tip.*

Members of this small group are not very common. Adults are predaceous. Larvae occur in decaying wood or in soil.

BEE FLIES Family Bombyliidae   **See also Pl. 13**
Identification: Usually stout-bodied and hairy. 3rd antennal segment *variable in shape.* M₁ ends *behind wing tip.* 3 or 4 posterior cells. Discal cell present. Anal cell *open* or closed near wing margin.

Bee flies are mostly medium to large, and are usually found on flowers or resting on the ground in open areas. The wings at rest are held outstretched. Some have patterned wings, and some have a very long proboscis (but do not bite). Larvae are parasites of other insects.

DANCE FLIES Family Empididae
Identification: Third antennal segment usually rounded, with *a long terminal style.* Rs 2- or 3-branched. Anal cell often shorter than 2nd basal cell, rarely absent. R-m cross vein located *beyond basal ¼ of wing.* ♂ genitalia not folded forward under abdomen.

Members of this large group vary somewhat in appearance and wing venation. Most are small (some minute) and have *a stout thorax* and a tapering abdomen. Many resemble small muscoids but lack a frontal suture. Empidids are common flies occurring in many different situations; some occur in swarms, flying with an up-and-down or circular movement (hence the common name). Most are predaceous but many occur on flowers. Larvae live in the soil, decaying vegetation, under bark, in decaying wood, and in water.

LONG-LEGGED FLIES   **See also Pl. 13**
Family Dolichopodidae
Identification: Small to medium-sized, and usually metallic green or coppery. Rs 2-branched, *slightly swollen at the fork.* R-m cross vein in basal ¼ of wing or absent. Anal cell *small,* sometimes absent. ♂ genitalia often *large and folded forward* under abdomen. Antennae usually *aristate.*

Dolichopodids are very common, occurring in many different situations but most frequently in marshy places and meadows. Males often have very large genitalia, and sometimes have the legs peculiarly ornamented. Most adults are predaceous on

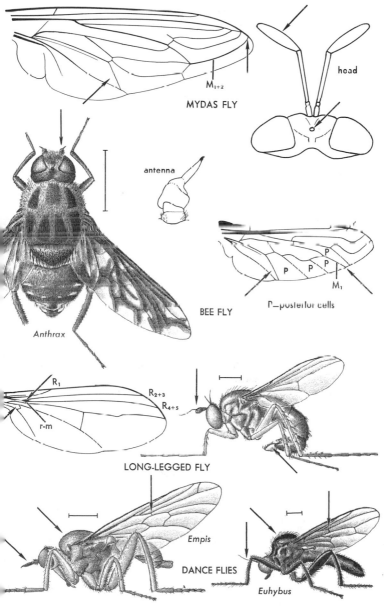

M_{1+2}

MYDAS FLY

head

antenna

BEE FLY

P—posterior cells

M_1

Anthrax

R_1

R_{2+3}

R_{4+5}

r-m

LONG-LEGGED FLY

Empis

DANCE FLIES

Euhybus

279

smaller insects. Larvae occur in wet soil, under bark, in decaying vegetation, and in water.

Suborder Cyclorrhapha

Antennae 3-segmented, aristate. Rs 2-branched. Frontal suture absent (Aschiza) or present (Schizophora).

Division Aschiza

Without a frontal suture.

SPEAR-WINGED FLIES Family Lonchopteridae
Identification: Slender, yellowish to brownish, 2–5 mm. Wings *pointed at apex* and without cross veins except at base.

These flies are generally found in moist shady or grassy places, and are fairly common. Larvae occur in decaying vegetation. This family is a small one. Our 4 species belong to the genus *Lonchoptera.*

HUMPBACKED FLIES Family Phoridae
Identification: Small to minute, usually dark-colored, with distinctive appearance and wing venation: appear *humpbacked* and wings have *strong veins in costal area*, the remaining veins being *weaker and oblique.* Hind femora *flattened.* Antennae very short, the 2 basal segments very small and 3rd segment globular and bearing a long arista (bristle).

Humpbacked flies are quite common. They occur in a variety of habitats but probably most often where there is decaying vegetation. Larvae vary in habits; some live in decaying materials or in fungi, some are parasites of other insects, and some live in the nests of ants and termites.

FLAT-FOOTED FLIES Family Platypezidae
Identification: Small black or brown flies with the hind tibiae and tarsi *dilated.* Anal cell *longer than 2nd basal cell*, and *pointed apically.*

Platypezids are rather uncommon flies usually found in wooded areas. Adults of some species are attracted to smoke. Larvae live in fungi.

BIG-HEADED FLIES Family Pipunculidae
Identification: Head *very large*, hemispherical, and consisting almost entirely of eyes. Wings somewhat narrowed basally. Anal cell *long*, closed near wing margin.

Members of this group are small and not very common. They are usually found in meadows or along the edges of woods. Larvae are parasites of various Homoptera, chiefly leafhoppers and planthoppers.

SYRPHID FLIES   Family Syrphidae   **See also Pl. 13**
   Identification : A spurious vein usually *present between R and M*.
R_5 cell (and usually also M_2 cell) *closed*. Anal cell *long*, closed
near wing margin. Proboscis short and fleshy.

   Many members of this large group are very common flies.
Syrphids occur in many habitats, usually on flowers. All are
good fliers, and often do a great deal of hovering. Adults vary
greatly in size, color, and appearance; many are brightly colored
with yellow, brown, and black; others are uniformly black or

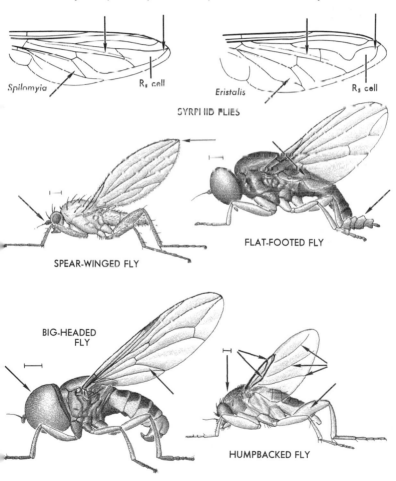

Spilomyia R_5 cell

Eristalis R_5 cell

SYRPHID FLIES

SPEAR-WINGED FLY

FLAT-FOOTED FLY

BIG-HEADED
FLY

HUMPBACKED FLY

brown. Some species are relatively bare and strongly resemble
wasps; others are quite hairy, and resemble bees. The resemblance of many syrphids to various Hymenoptera is often very
striking, and it may require a second look to determine that they
are flies and not Hymenoptera. Syrphids do not bite or sting.
Larvae also vary in appearance and habits; many are predaceous
on aphids, many are scavengers (living in dung, carrion, decaying
vegetation, or in highly polluted aquatic habitats), and some
live in ant nests. Aphid-feeding larvae are maggotlike and
usually greenish; some larvae living in polluted aquatic habitats
have the posterior end extended as a long tail-like structure and
are called rat-tailed maggots; larvae living in ant nests are oval
and very flat.

THICK-HEADED FLIES   Family Conopidae

Identification: Venation as in Syrphidae, but lacking a spurious
vein. Proboscis *long, slender,* and often folding. Abdomen
usually *narrowed at base.*

Thick-headed flies are usually medium-sized and brownish;
many resemble small thread-waisted wasps. The head is slightly
wider than the thorax and the antennae are generally *long.*
Adults are commonly found on flowers. Larvae are parasites of
adult bees and wasps.

Muscoid Flies: Division Schizophora

A frontal suture present.

Adult flies in the suborder Cyclorrhapha emerge from the
puparium through a circular opening at the anterior end. This
opening is made with a saclike structure called the *ptilinum,* which
is everted from the head of the fly. After emergence the ptilinum
is withdrawn into the head, and in the Schizophora the break in
the head wall through which the ptilinum was everted is marked
by a suture called the *frontal suture.* This suture is in the form of
an inverted U or V, with its apex just above the base of the antennae, and the 2 arms extending downward toward the cheeks.

Muscoid flies have *aristate antennae,* numerous bristles on the
head and body, and most of them are rather stout-bodied. Muscoids are usually small; many are very small. This group makes
up about ⅓ of the order, and its members occur almost everywhere, often in considerable numbers. Because of their small size
and the large number of species and families, their identification
can be difficult.

The principal characters used in separating families of muscoid
flies are those of the bristles and wing venation. The characteristic
bristles and areas of the head are shown opposite and those of the
thorax are shown on p. 287. The chief venational characters used
are the size and shape of various cells, the development of Sc,

and the presence of breaks in the costa. Sc may be complete (extending to costa) or incomplete (not extending to costa, fading out distally or fusing with R_1). Costal breaks are points on the costa where the sclerotization is weak or the vein appears to be broken; they are best seen with transmitted light.

The muscoid flies are divided into 2 groups, the acalyptrates (Section Acalyptratae) and calyptrates (Section Calyptratae, p. 302), chiefly on the basis of the development of the calypters and the *structure of the 2nd antennal segment*.

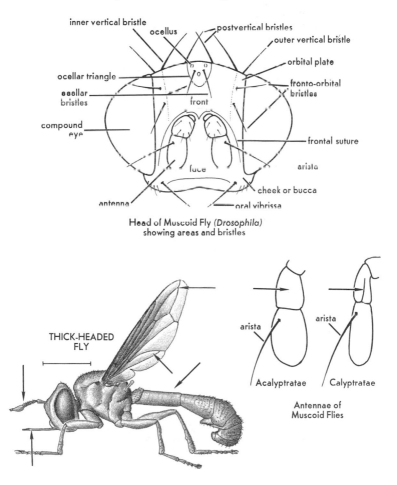

Head of Muscoid Fly *(Drosophila)*
showing areas and bristles

THICK-HEADED
FLY

Acalyptratae / Calyptratae

Antennae of
Muscoid Flies

Acalyptrate Muscoids: Section Acalyptratae

Calypters absent or vestigial. Second antennal segment *without a longitudinal suture* (except in some Psilidae); see p. 283.

This Section includes a large number of families, many of which are difficult to separate. The key below should be of value in separating these families. See illustrations on pp. 283 and 287 for technical characters.

1.	Coxae widely separated; wingless ectoparasites of the Honey Bee	**Braulidae (p. 295)**
1'.	Coxae close together; not ectoparasitic in habit; wings nearly alway present	2
2(1').	Sc complete or nearly so	3
2'.	Sc incomplete	17
3(2).	Posterior spiracle of thorax with at least 1 bristle; palps vestigial; head spherical; abdomen elongate and usually narrowed at base	**Sepsidae (p. 290)**
3'.	Without this combination of characters	4
4(3').	Dorsum of thorax flattened; legs and abdomen very bristly; seashore species	**Coelopidae (p. 290)**
4'.	Without this combination of characters	5
5(4').	Oral vibrissae present	6
5'.	Oral vibrissae lacking	8
6(5).	2nd basal and discal cells confluent	**Aulacigastridae (p. 294); Curtonotidae (p. 302)**
6'.	2nd basal and discal cells separated	7
7(6').	Costa broken only near humeral cross vein; eyes with microscopic pubescence	**Acartophthalmidae (p. 294)**
7'.	Without this combination of characters	**Piophilidae, Neottiophilidae (p. 292); Thyreophoridae, Lonchaeidae (p. 294); Clusiidae (p. 295); Trixoscelididae, Heleomyzidae, Chyromyidae (p. 296)**
8(5').	Costa broken near humeral cross vein	**Acartophthalmidae (p. 294)**
8'.	Costa entire, broken near end of R_1, or broken near humeral cross vein and near end of R_1	9
9(8').	R_5 cell narrowed distally; legs long and slender	**Micropezidae, Tanypezidae, Neriidae (p. 286)**
9'.	R_5 cell not narrowed distally, or legs not particularly long and slender	10
10(9').	Eyes prominently bulging, vertex sunken; femora enlarged	**Ropalomeridae, Rhinotoridae (p. 286)**
10'.	Without this combination of characters	11
11(10').	Tibiae with preapical bristles	12
11'.	Tibiae without preapical bristles	13
12(11).	Postvertical bristles converging; 2A short, not reaching wing margin	**Lauxaniidae (p. 292)**
12'.	Postvertical bristles parallel, diverging, or absent; 2A reaching wing margin, at least as a fold	**Sciomyzidae, Dryomyzidae (p. 290); Helcomyzidae (p. 291)**
13(11').	Ocelli absent; medium-sized to large flies, wings usually patterned	**Pyrgotidae (p. 288)**

13'.	Ocelli present; size and wing color variable	14
14(13').	Anal cell with an acute distal projection posteriorly; wings usually patterned **Otitidae, Tephritidae (p. 288)**	
14'.	Anal cell truncate or rounded apically, without an acute distal projection posteriorly	15
15(14').	Head extended on each side into a lateral process bearing eye, with antennae widely separated on eye stalks; scutellum with 2 tubercles **Diopsidae (p. 286)**	
15'.	Without this combination of characters	16
16(15').	Costa broken near end of Sc or R_1 **Tephritidae, Richardiidae (p. 288); Lonchaeidae, Pallopteridae (p. 294); Canaceidae (p. 298)**	
16'.	Costa not broken near end of Sc or R_1 **Platystomatidae, Otitidae (p. 288); Chamaemyiidae (p. 292)**	
17(2').	Basal segment of hind tarsi short and swollen, shorter than 2nd segment **Sphaeroceridae (p. 296)**	
17'.	Basal segment of hind tarsi not swollen, and usually longer than 2nd segment	18
18(17').	R_{2+3} ending in C just beyond R_1 **Asteiidae (p. 296)**	
18'.	R_{2+3} longer, ending beyond middle of wing	19
19(18').	3rd antennal segment large, reaching lower edge of head, with arista absent or represented by only a small apical tubercle **Cryptochetidae (p. 297)**	
19'.	Without this combination of characters	20
20(19').	Costa broken once or twice, at least near end of R_1	21
20'.	Costa not broken **Neriidae (p. 286); Periscelididae, Chamaemyiidae (p. 292)**	
21(20).	Costa broken once, near end of R_1	22
21'.	Costa broken twice, near end of R_1 and near humeral cross vein	26
22(21).	Anal cell lacking; ocellar triangle usually large and distinct **Chloropidae (p. 298)**	
22'.	Anal cell present or ocellar triangle small	23
23(22')	Sternopleural bristles present	24
23'.	No sternopleural bristles **Psilidae (p. 290); Canaceidae (p. 298)**	
24(23).	Eyes prominently bulging, vertex sunken; front femora thickened **Rhinotoridae (p. 286)**	
24'.	Without this combination of characters	25
25(24').	Postvertical bristles converging **Trixoscelididae (p. 296); Tethinidae, Anthomyzidae (p. 298)**	
25'.	Postvertical bristles diverging or absent **Opomyzidae (p. 298); Agromyzidae, Odiniidae (p. 300)**	
26(21').	Antennae retractile into deep grooves, face receding; eyes small and round **Thyreophoridae (p. 294)**	
26'.	Without this combination of characters	27
27(26').	Anal cell absent; oral vibrissae absent; postvertical bristles diverging **Ephydridae (p. 300)**	
27'.	Anal cell and oral vibrissae usually present; postvertical bristles parallel or converging (rarely absent) **Milichiidae, Drosophilidae (p. 300); Curtonotidae, Diastatidae, Camillidae (p. 302)**	

STILT-LEGGED FLIES Family Micropezidae
Identification: Medium-sized, slender, usually black. Legs *long and stiltlike*. Sc complete. Oral vibrissae absent. R$_5$ cell *narrowed* or closed apically. Anal cell usually *long and pointed*. Arista dorsal.

These flies are relatively uncommon, and are generally found in moist places. Larvae occur in dung.

TANYPEZID FLIES Family Tanypezidae   **Not illus.**
Identification: Similar to Micropezidae, but head in profile higher than long and anal cell rounded apically.

Two rare species of tanypezids occur in the East, and are usually found in moist woods. Larvae are unknown.

CACTUS FLIES Family Neriidae
Identification: Similar to Micropezidae, but with the antennae *long and projecting forward*, and the arista *apical*. Grayish flies with brown markings.

Two species of cactus flies occur on or near decaying cacti from Texas to California. Larvae live in decaying cacti.

ROPALOMERID FLIES Family Ropalomeridae   **Not illus.**
Identification: Medium-sized and usually brownish or grayish. Sc complete. Oral vibrissae lacking. Eyes prominently bulging, the vertex sunken. Femora enlarged. R$_1$ ends far beyond Sc. R$_5$ cell narrowed distally. Posterior thoracic spiracle with a group of bristles. Palps broad.

Our only species, *Rhytidops floridensis* (Aldrich), occurs about fresh exudates of palm trees in Florida.

RHINOTORID FLIES Family Rhinotoridae   **Not illus.**
Identification: Similar to Ropalomeridae, but R$_1$ ends close to Sc, the R$_5$ cell is not narrowed distally, the posterior thoracic spiracle is without a group of bristles, and the palps are narrow.

The single U.S. species in this family has been taken at banana-baited traps in New Mexico and Arizona. Its larva is unknown.

STALK-EYED FLIES Family Diopsidae
Identification: Small blackish flies. Sc complete. Oral vibrissae lacking. Head *slightly extended on each side* into a short stalklike process bearing the eye, the antennae *widely separated*. Scutellum with 2 tubercles. Front femora swollen.

This group is represented in N. America by a single rare species, *Sphyracephala brevicornis* (Say), which occurs from Quebec to Colorado and N. Carolina. Larva breeds in decaying vegetation. Adults are sometimes found on skunk cabbage. Some tropical species in this group have long and slender eye stalks.

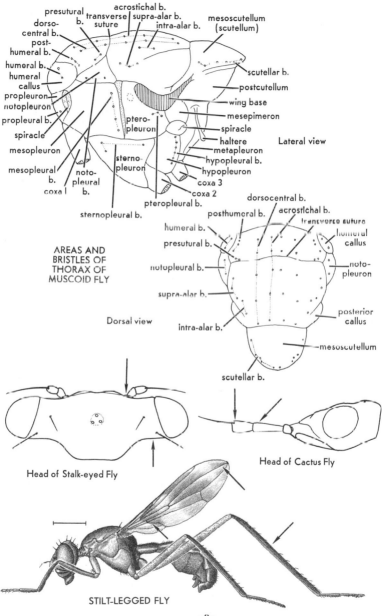

presutural b.
dorso-central b.
acrostichal b.
transverse suture
supra-alar b.
intra-alar b.
mesoscutellum (scutellum)

post-humeral b.
humeral b.
humeral callus
propleuron
notopleuron
propleural b.
spiracle
mesopleuron
mesopleural b.
coxa I
noto-pleural b.
sterno pleuron
pteropleuron
pteropleural b.
sternopleural b.

scutellar b.
postcutellum
wing base
mesepimeron
spiracle
haltere
metapleuron
hypopleural b.
hypopleuron
coxa 3
coxa 2

Lateral view

AREAS AND BRISTLES OF THORAX OF MUSCOID FLY

Dorsal view

dorsocentral b.
posthumeral b.
acrostichal b.
transverse suture
humeral b.
presutural b.
humeral callus
notopleural b.
notopleuron
supra-alar b.
posterior callus
intra-alar b.
mesoscutellum
scutellar b.

Head of Stalk-eyed Fly

Head of Cactus Fly

STILT-LEGGED FLY

287

PICTURE-WINGED FLIES   See also Pl. 14
Family Otitidae

Identification: Small to medium-sized, usually blackish and shiny. Wings often *banded or patterned*. Sc *complete, only slightly curved at tip*. Anal cell generally with *an acute distal projection posteriorly* (if anal cell is rounded apically, then costa is not broken, postvertical bristles are parallel or convergent, sternopleural bristles are usually present, and anterior side of anal cell is usually less than ¼ as long as posterior side of discal cell). Oral vibrissae lacking. Without well-developed bristles on anterior half of front. Tibiae usually without preapical bristles.

These flies are common in moist places, and are sometimes quite abundant. Habits of the larvae vary — some are plant feeders (a few of these are pests of cultivated plants) and some occur in decaying materials.

PLATYSTOMATID FLIES   Family Platystomatidae

Identification: Similar to Otitidae, but anal cell always *rounded apically*, its anterior side more than ¼ as long as posterior side of discal cell. R_1 with bristles. Sternopleural bristles lacking; propleural bristles weak or lacking. Costa not broken near end of Sc.

This is a much smaller group than the Otitidae but some species are fairly common. The wings are usually marked with narrow dark bands, and the costa is not broken.

RICHARDIID FLIES   Family Richardiidae   Not illus.

Identification: Similar to Otitidae, but anal cell rounded or truncate apically, never with a pointed distal projection posteriorly. Costa broken near end of Sc.

These flies are uncommon in our area but are common in the tropics. Larvae occur in decaying vegetation.

PYRGOTID FLIES   Family Pyrgotidae

Identification: Medium-sized to large flies. Head *large and rounded*. Ocelli *lacking*. Wings usually spotted or patterned. Sc complete. Oral vibrissae lacking.

Members of this group are mostly nocturnal, and often are attracted to lights; they are not common. Larvae are parasites of June beetles.

FRUIT FLIES   Family Tephritidae   See also Pl. 14

Identification: Small to medium-sized, often brightly colored. Wings usually *spotted or banded*. Apex of Sc *bent abruptly forward* at almost a right angle, and usually not quite reaching costa. Anal cell often with *an acute distal projection posteriorly*. Anterior half of front with 1 or more erect bristles.

Many species in this large group are quite common, and are

usually found on flowers or vegetation. Most species have attractively patterned wings; the wings may be banded or spotted, and the spotting sometimes forms intricate patterns. Some species move their wings slowly up and down while resting. Larvae are plant feeders, and a few are pests of fruits. Larva of *Rhagoletis pomonella* (Walsh) tunnels in the fruit of apple, and is called an Apple Maggot (Pl. 14). The Mediterranean Fruit Fly, *Ceratitis capitata* (Wiedemann), is a serious pest of citrus.

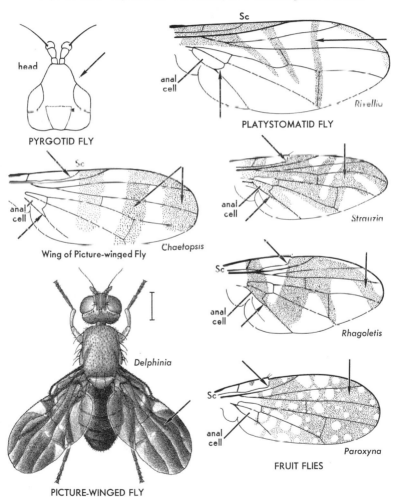

PYRGOTID FLY

PLATYSTOMATID FLY
Rivellia

Wing of Picture-winged Fly
Chaetopsis

Strauzia

Delphinia

Rhagoletis

Paroxyna
FRUIT FLIES

PICTURE-WINGED FLY

Species of *Eurosta* form round, thick-walled stem galls on golden-rod. Many species breed in the flower heads of Compositae.

RUST FLIES   Family Psilidae
Identification: Small to medium-sized, usually slender and brownish, with relatively long antennae. Species with a longitudinal suture on 2nd antennal segment have *3rd antennal segment very long*, longer than arista. Species without this suture on antennae have Sc incomplete, costa broken near end of R_1, no sternopleural bristles, and no oral vibrissae.

Rust flies are fairly common insects. Larvae are plant feeders, and some species are pests of garden crops.

SEAWEED FLIES   Family Coelopidae
Identification: Small to medium-sized, brown to blackish. Dorsum of thorax *somewhat flattened*, and body and legs *very bristly*. Sc complete.

Seaweed flies occur along the seashore, often in large numbers; larvae live in seaweed washed up on the shore. Adults occur around this seaweed, or on flowers and vegetation near the shore.

BLACK SCAVENGER FLIES   Family Sepsidae
Identification: Small, usually shining black or purple flies, with head *rounded* and abdomen usually *narrowed at base*. Palps vestigial. Posterior spiracle of thorax with at least 1 bristle. Sc complete. Oral vibrissae lacking.

Larvae of these flies live in manure and similar materials, and adults are usually found around such materials. Sepsids are common flies and are often abundant around manure piles.

MARSH FLIES   Family Sciomyzidae   **See also Pl. 14**
Identification: Small to medium-sized, usually yellowish or brownish, often with *spotted* or patterned wings. Sc *complete*. R_1 usually ends *at middle of wing*. Oral vibrissae *absent*. Postverticals *slightly divergent*. Preapical tibial bristles present. Antennae generally *project forward*, often long. Femora with bristles, and middle femur usually with a characteristic bristle near middle of anterior surface.

Marsh flies are common insects that occur in marshy areas near ponds and streams. Larvae feed on aquatic snails, usually as predators.

DRYOMYZID FLIES   Family Dryomyzidae
Identification: Yellowish or brownish, similar to Sciomyzidae but antennae usually not projecting, R_1 ending *beyond middle of wing*, and femoral bristles not developed. 3rd antennal segment longer than wide, more or less flattened laterally.

Dryomyzids are small to medium-sized and often have brownish spots on the cross veins. They occur in moist woods and along the seashore but are not common. Larval stages are unknown.

HELCOMYZID FLIES Family Helcomyzidac **Not illus.**
Identification: Similar to Dryomyzidae but blackish and 3rd antennal segment spherical.

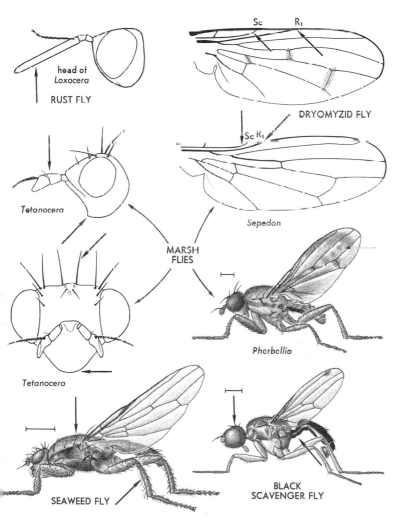

head of *Loxocera*

RUST FLY

Sc   R₁

DRYOMYZID FLY

Tetanocera

Sc R₁

Sepedon

MARSH FLIES

Phorbellia

Tetanocera

SEAWEED FLY

BLACK SCAVENGER FLY

Helcomyzids are uncommon flies occurring along the Pacific Coast from Oregon to Alaska. Larvae develop in rotting seaweed.

LAUXANIID FLIES Family Lauxaniidae   **See also Pl. 14**
Identification: Small, usually rather stout-bodied flies, often with spots on the wings. Sc *complete.* 2A *short, not reaching wing margin.* Oral vibrissae *absent.* Postvertical bristles *converge.* Tibiae with *preapical bristles.*

Lauxaniids are common flies that usually occur in wooded areas or in places where the vegetation is fairly dense. Larvae breed in decaying vegetable matter and occur in leaf litter, bird nests, and similar situations.

CHAMAEMYIID FLIES Family Chamaemyiidae
Identification: Very small flies, usually grayish with *black spots on abdomen.* Sc variable, complete or incomplete. Costa not broken. Postvertical bristles *converge.* Tibiae without preapical bristles. Oral vibrissae *absent.* Arista bare or pubescent.

These flies are usually *less than 4 mm.* and are relatively common. Larvae of some species are predaceous on aphids and mealybugs.

PERISCELIDID FLIES Family Periscelididae   **Not illus.**
Identification: Small flies, similar to Chamaemyiidae but postvertical bristles diverging, Sc incomplete, and arista plumose.

The 3 U.S. species are widely distributed but rare. Adults may occur at fresh sap flows on trees.

SKIPPER FLIES   Family Piophilidae
Identification: Small flies, usually *less than 5 mm.,* black or bluish, and rather metallic. Sc complete. 2A does not reach wing margin. Oral vibrissae *present.* Postvertical bristles *diverge.* 2nd basal and discal cells separated. *2 or fewer pairs* of fronto-orbital bristles; 2 sternopleural bristles. Arista rises near base of 3rd antennal segment.

Larvae of skipper flies live in decaying animal materials, and some occasionally are pests in meats and cheese. Piophilids are called skipper flies because the larvae can jump. Adults are fairly common.

NEOTTIOPHILID FLIES Family Neottiophilidae   **Not illus.**
Identification: Similar to Piophilidae, but with 4 or 5 sternopleural bristles, vein 2A reaching wing margin, and costa spiny.

This family contains only 2 species, both European, but 1 has been recorded in n. Quebec. Nothing is known of the habits of this species; the larva of the other (European) is a bloodsucking ectoparasite of nestling birds.

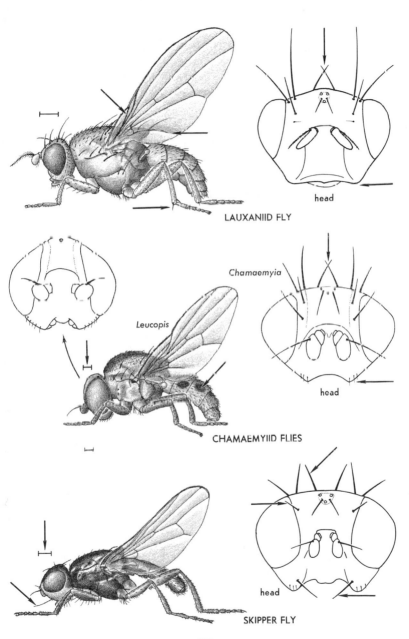

LAUXANIID FLY

head

Chamaemyia

Leucopis

CHAMAEMYIID FLIES

head

SKIPPER FLY

head

293

THYREOPHORID FLIES Family Thyreophoridae **Not illus.**
Identification : Sc incomplete. Costa broken near tip of R_1. Antennae retractile into deep grooves below bases of antennae, the face receding. Postvertical bristles diverge. Costa spinose (as in the heleomyzid fly, p. 297).

This group is represented in the U.S. by 2 rare species occurring in California and Arizona.

PALLOPTERID FLIES Family Pallopteridae **Not illus.**
Identification: Small, usually pale-colored flies with patterned wings. Sc complete. Costa broken near end of Sc. Anal cell rounded apically. Oral vibrissae absent. Tibiae lack preapical bristles. Head in profile rounded, eyes round. 3rd antennal segment oval. Postvertical bristles parallel.

Pallopterids are rare flies usually found in moist shady places. Our 9 species are widely distributed.

LONCHAEID FLIES Family Lonchaeidae
Identification: Small, shiny black flies. Abdomen in dorsal view somewhat rounded but pointed apically. Oral vibrissae absent. Sc complete. Costa broken near end of Sc. 2A usually *wavy*. Tibiae without preapical bristles. Head in profile hemispherical, the eyes *large and oval*. 3rd antennal segment *elongate*. Postvertical bristles diverge.

Lonchaeids are fairly common flies that occur in shady places. Larvae are found in live, injured, or dead plant tissue, often along with other insects attacking the plant.

ACARTOPHTHALMID FLIES **Not illus.**
Family Acartophthalmidae
Identification: Sc complete. Costa broken near humeral cross vein. Oral vibrissae weakly developed. Eyes slightly pubescent. Postvertical bristles widely separated and diverging.

Adults of this small group are quite rare but have been taken on rotting fungi and carrion from Massachusetts to Oregon and Alaska. *Acartophthalmus nigrinus* (Zetterstedt), a widely distributed species, is about 2 mm., blackish, with the front coxae and halteres yellowish.

AULACIGASTRID FLIES **Not illus.**
Family Aulacigastridae
Identification: Sc complete or nearly so. Costa broken near humeral cross vein and near end of R_1. 2nd basal and discal cells confluent. Oral vibrissae present. Postvertical bristles lacking. Arista pubescent.

This family is represented in the U.S. by a single species, *Aulacigaster leucopeza* Meigen, a small black fly that occurs from the east coast to Kansas and Texas. It is fairly common and usually found in sap flows.

BEE LICE   Family Braulidae   **Not illus.**
  Identification: Wingless, 1.5 mm. Coxae widely separated.
Abdominal segmentation somewhat obscure. Mesonotum short,
resembling abdominal segments. No scutellum.

  Braulids are represented in this country by a single very rare
species, *Braula coeca* Nitzsch, which is an ectoparasite of the
Honey Bee.

CLUSIID FLIES   Family Clusiidae
  Identification: Sc complete. Oral vibrissae *present*. 2nd basal
and discal cells separated. Postvertical bristles diverging or
absent. *2–4 pairs* of fronto-orbital bristles. Arista rises near
apex of 3rd antennal segment. 2nd antennal segment often with
an angular projection on outer side.

  Clusiids are small flies, mostly 3–4 mm., that often have
brownish or smoky wings. The body color varies but many

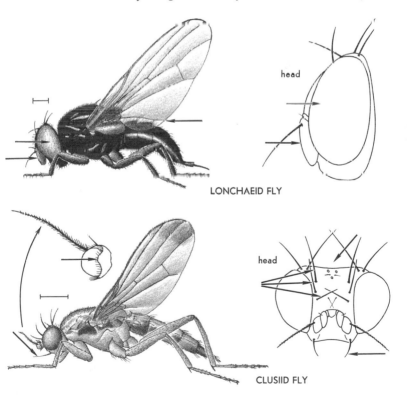

LONCHAEID FLY

CLUSIID FLY

species have the thorax black dorsally and yellowish laterally.
Larvae occur in decaying wood. These flies are not very
common.

TRIXOSCELIDID FLIES Family Trixoscelididae

Identification: Sc usually complete. Oral vibrissae *present*.
2nd basal and discal cells separated. Postvertical bristles
converge. Tibiae with *preapical bristles*. Costa *spiny*. Orbital
plates *long*, extending nearly to antennae.

Trixoscelidids are very small flies, generally *2–3 mm.*, most
of which occur in the West; only 2 rather rare species occur in
the East, 1 in Florida and 1 in Maryland. Some species occur in
grassy areas and woodlands; others, including many western
species, live in desert areas. Larvae of 1 eastern species inhabit
bird nests.

HELEOMYZID FLIES Family Heleomyzidae

Identification: Similar in general appearance to Sciomyzidae
(p. 290), but antennae *smaller* and not projecting forward, oral
vibrissae *present*, and postvertical bristles *converging*. Costa
usually *spiny*. Orbital plates *short*, not reaching antennae.

Heleomyzids are fairly common flies usually found in areas of
abundant vegetation. Most are brownish and some have spots
on the wings. Larvae occur in fungi, bird nests, mammal burrows, and in other places where there are decaying materials.

CHYROMYID FLIES Family Chyromyidae   **Not illus.**

Identification: Small yellow flies with golden eyes. Sc complete.
Oral vibrissae weakly developed. 2nd basal and discal cells
separated. Tibiae without preapical bristles. Postvertical
bristles converge.

Adults of this small but widely distributed group usually
occur on vegetation. Larvae probably are scavengers.

SMALL DUNG FLIES Family Sphaeroceridae

Identification: Very small and blackish or brownish. Sc *incomplete*. Basal segment of hind tarsi *somewhat swollen* and
shorter than 2nd segment.

Sphaerocerids usually occur near manure or other refuse, and
are very common. Larvae live in dung and various decaying
materials.

ASTEIID FLIES Family Asteiidae

Identification: Mostly 2 mm. or less, and usually light-colored.
Sc incomplete. R_{2+3} ends in costa *just beyond end of R_1*. Postvertical bristles diverge.

The group is a small one, its members relatively rare. Immature stages are unknown.

CRYPTOCHETID FLIES

Not illus.

Family Cryptochetidae

Identification: Sc incomplete. 3rd antennal segment large, extends almost to lower edge of head. Arista absent, but a short spine or tubercle at apex of 3rd antennal segment. Eyes large, vertically elongate.

One species of cryptochetid, *Cryptochetum iceryae* (Williston),

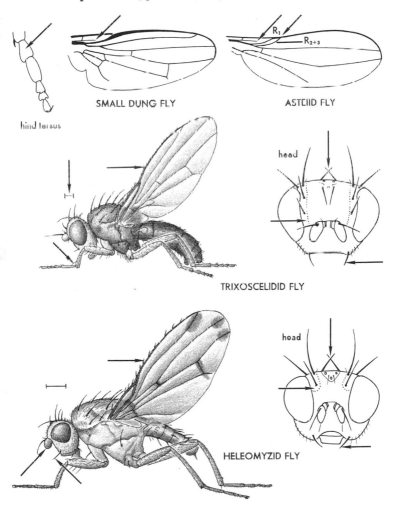

hind tarsus

SMALL DUNG FLY

ASTEIID FLY

R_1

R_{2+3}

TRIXOSCELIDID FLY

head

HELEOMYZID FLY

head

occurs in California, where it has been introduced from Australia; its larva is parasitic on scale insects in the family Margarodidae. This insect is about 1.5 mm., stout-bodied, with head and thorax dark metallic blue and abdomen metallic green.

FRIT FLIES Family Chloropidae

Identification: *Small to very small flies.* Color variable, some species blackish or grayish, some brightly colored with black and yellow. Sc incomplete. Costa *broken near end of* R_1. Anal cell *lacking.* Ocellar triangle usually *large*, shining. Postvertical bristles *converging*, parallel, or absent.

This is a large group, and many of its members are common flies. They occur in a variety of habitats but are most common in grassy areas. Larvae of many species live in grass stems; a few live in decaying materials. Adults of a few species (called eye gnats) are attracted to the eyes or to sores.

BEACH FLIES Family Canaceidae   **Not illus.**

Identification: Sc incomplete. Costa broken near end of R_1. Anal cell present. Ocellar triangle large and shining, reaching to near base of antennae. Oral vibrissae present.

Canaceids are small flies, 3.5 mm. or less, occurring along the seashore, chiefly in the intertidal zone, and they are not common. Larvae feed on algae.

TETHINID FLIES Family Tethinidae

Identification: Sc incomplete. Costa broken near end of R_1. Anal cell present. Sternopleural bristles present. Postvertical bristles *converge*. Oral vibrissae *present*. All fronto-orbital bristles *directed outward*. At least some dorsocentral bristles in anterior part of mesonotum.

This family is represented in the U.S. by 22 relatively uncommon species, most of which occur along the seashore; the inland species usually inhabit alkaline areas. The majority of the seashore species are found along the Pacific Coast. Larvae are unknown.

ANTHOMYZID FLIES Family Anthomyzidae

Identification: Similar to Tethinidae, but with *at least 1 pair of fronto-orbital bristles bent upward* and with no dorsocentral bristles in anterior part of mesonotum.

Anthomyzids are *small*, somewhat elongate flies that occur in grass and low vegetation, especially in marshy areas. They are widely distributed and fairly common.

OPOMYZID FLIES Family Opomyzidae

Identification: Similar to Anthomyzidae and Tethinidae, but postvertical bristles *absent* or diverging. Oral vibrissae present or *absent*. More stout-bodied than Anthomyzidae.

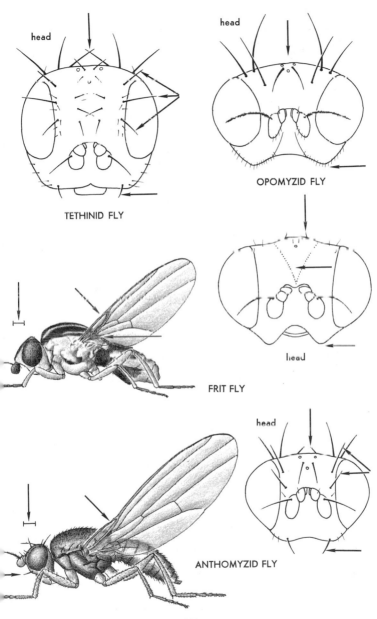

head

TETHINID FLY

head

OPOMYZID FLY

FRIT FLY

head

head

ANTHOMYZID FLY

299

These are rare flies, small to minute, usually dark-colored, and generally found in grassy places. Most of the 13 N. American species occur in Canada and w. U.S.

LEAF MINER FLIES   Family Agromyzidae

Identification: *Small* to very small flies, usually blackish or yellowish. Sc incomplete or fused with R_1. Costa *broken near end of R_1.* Anal cell *present.* Sternopleural bristles present. Postvertical bristles *diverge.* Oral vibrissae *present.* No preapical tibial bristles. 6 abdominal segments.

Leaf miner flies are common insects usually occurring on vegetation. Larvae are mostly leaf miners and generally make a narrow winding mine; some feed in stems and seeds.

ODINIID FLIES   Family Odiniidae   Not illus.

Identification: Similar to Agromyzidae, but with preapical tibial bristles and 5 abdominal segments.

Adults of this small group occur around rotting logs and sap flows but are uncommon. Larvae are scavengers.

SHORE FLIES   Family Ephydridae

Identification: Most species blackish and relatively small; some very small. Sc incomplete. Costa *broken near end of R_1* and *near humeral cross vein.* Anal cell *absent.* Face usually *somewhat bulging.* Oral vibrissae absent. Postvertical bristles diverging (sometimes small and difficult to see).

Shore flies are common insects often occurring in large numbers along the shores of ponds and streams and along the seashore. Larvae are usually aquatic. Larvae of the seashore species live in brackish water and adults often cluster in large numbers on the surface of pools just above the high tidemark.

MILICHIID FLIES   Family Milichiidae

Identification: Very small flies, usually blackish. Sc *incomplete.* Costa *broken near end of R_1* and *near humeral cross vein.* Anal cell *present.* Postvertical bristles *converging* or parallel. Oral vibrissae *weakly developed.* At least 1 pair of fronto-orbital bristles *bent inward.*

Milichiids are fairly common in grassy areas. Larvae are scavengers.

POMACE FLIES   Family Drosophilidae   See also p. 283

Identification: Usually yellowish or brownish, *3–4 mm.* Sc incomplete. Costa *broken near end of R_1* and *near humeral cross vein;* not spiny. Anal cell *present.* Postvertical bristles *converge.* Oral vibrissae *well developed.* Arista *plumose.* Sternopleural bristle present; no mesopleural bristles.

Pomace flies are very common, and are usually found near

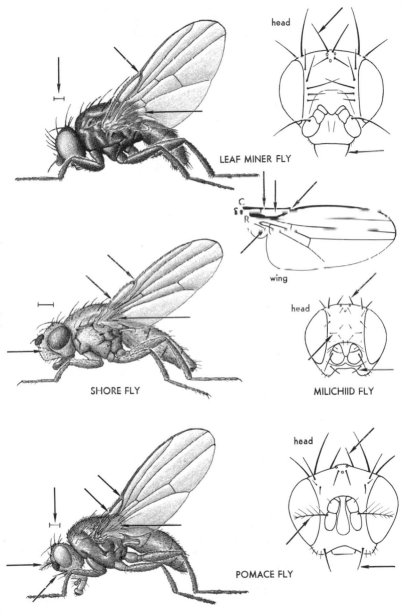

LEAF MINER FLY

head

wing

C
R

SHORE FLY

MILICHIID FLY

head

POMACE FLY

head

decaying vegetation or fruit. Larvae of most species breed in decaying fruit or fungi. Species of *Drosophila* have been used extensively in studies of heredity.

CAMILLID FLIES   Family Camillidae   **Not illus.**
Identification: Similar to Drosophilidae, but body metallic, no sternopleural bristles, and mesopleura bristly.

A single species in this group has been recorded in Ontario. Its immature stages are unknown.

DIASTATID FLIES   Family Diastatidae
Identification: Similar to Drosophilidae, but costa *spiny* and mesopleura with bristles. Arista *short-plumose*. Fronto-orbital bristles close to eyes, the downward-bent pair positioned *above the upward-bent pair*.

These flies are similar to the Drosophilidae, but little is known of their habits. The U.S. has 6 species.

CURTONOTID FLIES   Family Curtonotidae
Identification: Similar to Diastatidae, but arista *long-plumose*, fronto-orbital bristles remote from eyes, the downward-bent pair positioned *below upward-bent pair* and *no cross vein* separating 2nd basal and discal cells.

Curtonotids are represented in the U.S. by 1 species, *Curtonotum helvum* Loew, occurring in the East. It is found in high grass in moist places and its larva is unknown.

Calyptrate Muscoids: Section Calyptratae

Calypters usually well developed. Second antennal segment *with a longitudinal suture* (except in Gasterophilidae); see p. 283.

This is a large group that contains many common and well-known flies. The 11 families of calyptrates may be divided into 4 groups to aid identification:

1. Body somewhat leathery and flattened dorsoventrally; coxae separated; abdominal segmentation usually indistinct; often wingless; ectoparasites of birds and mammals: Hippoboscidae, Streblidae, and Nycteribiidae (louse and bat flies).
2. Mouth opening small, mouth parts vestigial or lacking; robust, hairy, beelike: Gasterophilidae, Cuterebridae, and Oestridae (bot and warble flies).
3. R_5 cell usually parallel-sided, only rarely narrowed distally; hypopleura usually without bristles (if hypopleural bristles are present, there are no pteropleural bristles or the proboscis is slender, rigid, and piercing): Anthomyiidae and Muscidae.
4. R_5 cell narrowed or closed distally; hypopleura and pteropleura with bristles; proboscis not slender and piercing: Tachinidae, Calliphoridae, and Sarcophagidae.

LOUSE FLIES Family Hippoboscidae

Identification: *Winged or wingless;* wings when present, with *strong veins anteriorly* and weak veins posteriorly. Palps slender, elongate, and forming a sheath for the proboscis. Eyes well developed.

Winged louse flies are usually found on birds. A common wingless species in this group is the Sheep Ked, *Melophagus*

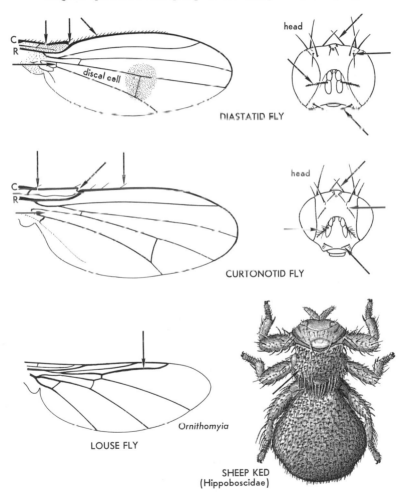

DIASTATID FLY

CURTONOTID FLY

LOUSE FLY

Ornithomyia

SHEEP KED
(Hippoboscidae)

ovinus (Linn.), about 6 mm. and reddish brown; it is an ecto-parasite of sheep.

BAT FLIES   Not illus.
Families Streblidae and Nycteribiidae
Identification: Streblidae: winged or wingless, wings when present uniformly veined; head with a fleshy movable neck, and not folding back into a groove on mesonotum; palps broad and extending leaflike in front of head; eyes small or vestigial. Nycteribiidae: wingless, somewhat spiderlike in appearance; head small and narrow and at rest folding back into a groove on mesonotum.

Members of these 2 families are ectoparasites of bats, and are seldom encountered.

HORSE BOT FLIES   Family Gasterophilidae
Identification: Beelike flies resembling the Honey Bee. 2nd antennal segment without a longitudinal suture. R_5 cell *widened distally.*

Larvae of these flies develop in the stomach or intestine of horses, and when full grown pass out with the feces and pupate in the ground. Eggs are usually laid on the lips or jaws of the horse, and either burrow through the skin into the mouth, or get into the mouth when licked by the horse. These insects are rather serious pests of horses. Adults are not common, but are generally found in the vicinity of horse barns and pastures.

ROBUST BOT FLIES   Family Cuterebridae
Identification: Scutellum extends *beyond metanotum.* Postscutellum *not developed.* R_5 cell narrowed or closed distally. Large robust flies resembling bumble bees.

Larvae mostly are parasites of rabbits and rodents, usually developing just under the skin. Adults are rare.

BOT and WARBLE FLIES   Family Oestridae
Identification: Scutellum *very short.* Postscutellum usually *well developed.* R_5 cell narrowed or closed distally.

Oestrid larvae are parasites of various animals, including some domestic animals. Ox warbles (*Hypoderma*) parasitize cattle; larvae develop in boil-like swellings just under the skin, usually on the back of the animal; adults are very annoying to cattle though they do not bite. The Sheep Bot Fly, *Oestrus ovis* Linn., is a parasite of sheep; its larva develops in the nostrils of the sheep. Adult bot and warble flies are not very common; they are most likely to be found in the vicinity of their hosts.

ANTHOMYIID FLIES   Family Anthomyiidae
Identification: R_5 cell *parallel-sided.* Hypopleura *without bristles.* 2A *reaches wing margin,* at least as a fold. Often *only 1*

sternopleural bristle, or fine erect hairs on undersurface of scutellum.

This is a large group that includes many common flies. Most are similar to a House Fly in general appearance and vary from being smaller to larger than a House Fly. Larval habits vary: many are plant feeders, and some of these are serious pests of

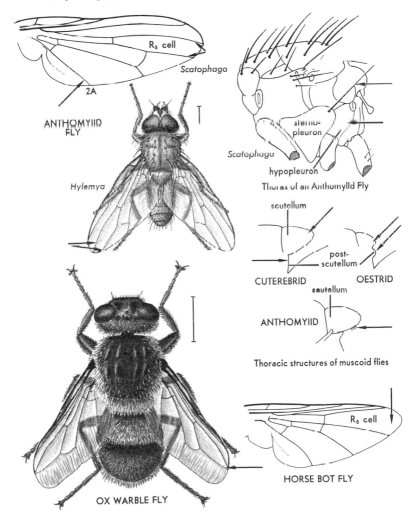

R₅ cell

Scatophaga

2A

ANTHOMYIID FLY

Hylemya

sterno-pleuron

Scatophaga

hypopleuron

Thorax of an Anthomyiid Fly

scutellum

post-scutellum

CUTEREBRID   OESTRID

scutellum

ANTHOMYIID

Thoracic structures of muscoid flies

R₅ cell

HORSE BOT FLY

OX WARBLE FLY

cultivated plants; many are scavengers, living in excrement or decaying materials; some are aquatic.

MUSCID FLIES   Family Muscidae   **See also Pl. 14**
Identification: Similar to Anthomyiidae, but 2A *short* and *not reaching wing margin*, undersurface of scutellum usually without fine hairs, and generally more than 1 sternopleural bristle. R_5 cell *parallel-sided or narrowed distally.*

This group includes many common flies, some of which are important pests. Most of them resemble the House Fly. The House Fly (Pl. 14), *Musca domestica* Linn., is a common household pest that breeds in various sorts of filth; it may serve as a vector of several diseases (typhoid fever, dysentery, cholera, and others); it has a short fleshy proboscis (and does not bite), and the R_5 cell is narrowed distally. The Little House Fly, *Fannia canicularis* (Linn.), is often a nuisance in houses and around poultry yards; adults can be recognized by the form of 3A (bends toward wing tip distally); its larva breeds in filth. The Stable Fly, *Stomoxys calcitrans* (Linn.), is very similar to the House Fly, but has a slender proboscis and it bites; it breeds in decaying vegetation.

TACHINID FLIES   Family Tachinidae   **See also Pl. 14**
Identification: R_5 cell *narrowed* or closed *distally.* Postscutellum *developed*. Hypopleura with *bristles*. Arista usually bare.

This is one of the largest families of Diptera, and its members are to be found almost everywhere. Most tachinids resemble the House Fly but many are larger; some are hairy and beelike. Tachinids can usually be distinguished from other calyptrates by the bare arista; some have the arista plumose, and can be recognized as tachinids by the well-developed postscutellum. Larvae are parasites of other insects, and many are of value in keeping noxious species under control.

BLOW FLIES   Family Calliphoridae   **See also Pl. 14**
Identification: Similar to Tachinidae, but postscutellum not developed. Arista plumose. Body often metallic. Usually *2 notopleural bristles*, and hindmost posthumeral bristle located *lateral to presutural bristle*.

This is a large group of flies, and its members are often common and abundant. Most species are as large as the House Fly and some are larger. Many are metallic bluish or green. Larvae are generally scavengers, living in carrion, dung, and similar materials; most maggots one finds in the body of a dead animal are blow fly larvae. Larvae usually feed on dead tissue, but a few, such as the Screw-worm, *Cochliomyia hominivorax* (Coquerel), may attack living tissue (in an animal's nostrils or in wounds).

FLESH FLIES Family Sarcophagidae   See also Pl. 14
Identification: Similar to Calliphoridae, but arista plumose only in basal half, body generally blackish with gray thoracic stripes (never metallic), usually *4 notopleural bristles,* and hindmost posthumeral bristle located even with or *toward midline from presutural bristle.*

Flesh flies are fairly common and usually resemble the House Fly. Larvae of most species are scavengers, feeding in the same sorts of materials as blow fly larvae; a few are parasites of other insects, a few develop in skin sores of vertebrates, and some feed on the insects stored in the nests of various wasps.

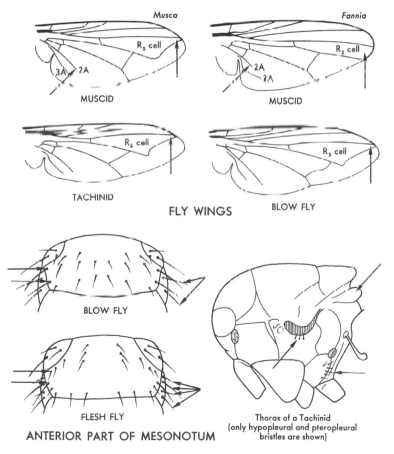

Musca

R₅ cell

3A   2A

MUSCID

Fannia

R₅ cell

2A
?A

MUSCID

R₅ cell

TACHINID

R₅ cell

BLOW FLY

FLY WINGS

BLOW FLY

FLESH FLY

ANTERIOR PART OF MESONOTUM

Thorax of a Tachinid
(only hypopleural and pteropleural
bristles are shown)

Fleas: Order Siphonaptera

Identification: Small wingless insects, generally less than 5 mm. and usually living as ectoparasites on birds or mammals. Body laterally flattened, rather bristly, and heavily sclerotized. Antennae short, 3-segmented, usually fitting into grooves on head. Ocelli lacking. Compound eyes present or absent. Legs relatively long, coxae *large*, tarsi *5-segmented*. Mouth parts sucking, the palps well developed. Metamorphosis complete. Usually jumping insects.

Similar orders: (1, 2) Mallophaga and Anoplura (pp. 106, 108): dorsoventrally flattened; legs short, tarsi 1- or 2-segmented; not jumping insects.

Immature stages: Larvae are slender, whitish, and legless, with a well-developed head and 2 small hooks on posterior end of body; they are usually found in dirt or debris, often in nest of the host. Larvae feed on various organic materials and pupate in silken cocoons.

Habits: Fleas are active insects that generally move freely over body of the host and from one host to another; some may spend considerable time off the host. Eggs are laid on the host or in dirt of the host nest; if laid on the host they eventually fall off, and larvae develop off the host. Adults feed on blood of the host. Many species (including those attacking man) are not very specific in their choice of a host and may feed on various animals.

Importance: Many fleas are annoying pests because of their bloodsucking habits. A few act as vectors of disease (bubonic plague and endemic typhus) and a few burrow into the skin of man or animals.

Classification: There are differences of opinion regarding the number and arrangement of the families of fleas; we follow an arrangement in which the fleas are grouped in 7 families. Families are separated principally by head and abdominal characters and on the character of various bristles. Some of these characters are difficult to see unless the specimen is mounted on a microscope slide.

No. of species: World, 1100; N. America, 238.

COMMON FLEAS Family Pulicidae

Identification: Abdominal *terga 2–6 with a single row of bristles*. Compound eyes *well developed*. Genal comb (row of strong bristles on lower front border of head) present or absent.

This is a large group, and many of its members are fairly common; the fleas most often attacking man and domestic animals, and those most important as disease vectors, belong to this family. Many are not very specific in their selection of a host,

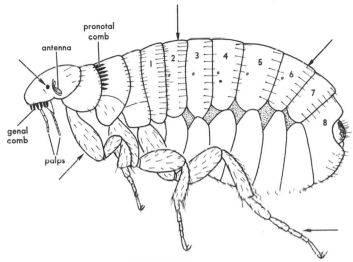

pronotal comb

antenna

genal comb

palps

DOG FLEA (Pulicidae)

and though they may be named from their principal host they often attack man and other animals indiscriminately. Many are worldwide in distribution. The Cat and Dog Fleas, *Ctenocephalides felis* (Bouché) and *C. canis* (Curtis), often occur in houses where these animals are kept and may attack man; the Dog Flea serves as the intermediate host of a dog tapeworm. Cat and Dog Fleas possess both genal and pronotal combs (pronotal comb is a row of strong spines on posterior margin of pronotum). The Human Flea, *Pulex irritans* Linn., and the Rat Flea, *Xenopsylla cheopis* (Rothschild), lack genal and pronotal combs; both attack man and other animals.

The most important flea-borne disease is bubonic plague, transmitted chiefly by the Rat Flea; this is primarily a rodent disease but occasionally occurs in man, sometimes in epidemic form. Plague infection usually occurs at the time of biting, but may also result from the ingestion of an infected flea or by scratching its feces into the skin.

BAT FLEAS Family Ischnopsyllidae p. 311
Identification: Parasites of bats. Genal comb *present and consisting of 2* or *3 broad lobes* on each side. Head elongated. Eyes absent or vestigial. Some or all of abdominal terga 2–6 with 2 rows of bristles.
This is a small group whose members are not often seen.

STICKTIGHT and CHIGOE FLEAS Family Tungidae
Identification: The 3 thoracic terga together *shorter* than the 1st abdominal tergum.

The most common species in this group is the Sticktight Flea, *Echidnophaga gallinacea* (Westwood), an important pest of poultry; it also attacks other birds and mammals. Adults usually occur on head of the host, often in dense masses, and remain attached for long periods. This insect is more common in the southern states. Another member of this family occasionally occurring in the southern states is the Chigoe, *Tunga penetrans* (Linn.); females burrow into the skin of man and other animals, usually on the feet. Males and newly emerged females of the Chigoe live much like other fleas, and feed on various hosts; the females burrow into the skin of man and other animals after mating. Once under the skin, the female's abdomen becomes greatly distended, and the surrounding tissues of the host swell to form a boil-like sore.

RODENT FLEAS Family Dolichopsyllidae
Identification: Some or all of abdominal terga 2–6 *with 2 rows of bristles* (if these terga bear only 1 row of bristles, eyes are absent or vestigial and there is no genal comb). Genal comb usually *absent*, but if present consists of 3 or more narrow lobes on each side. Pronotal comb *present*. *No suture* on dorsal surface of head between antennae.

This is a large group whose members are chiefly parasites of rodents; a few attack birds, and one of these is sometimes a pest of poultry. Some fleas in this family act as vectors of bubonic plague but they generally transmit the plague from one rodent to another and are not important as a vector of plague to man.

RAT AND MOUSE FLEAS Family Hystrichopsyllidae
Identification: Similar to Dolichopsyllidae, but usually with *a suture* on dorsal surface of head between antennae, *a genal comb*, and *2 or 3 rows of bristles* on anterior part of head.

This is a small family whose members are parasites of rats, mice, and shrews.

MALACOPSYLLID FLEAS Family Malacopsyllidae
Identification: Some or all of abdominal terga 2–6 with 2 rows of bristles. Pronotal and genal combs absent. Clypeal tubercle (in middle of front part of head) *well developed* and *somewhat pointed*. 1 long bristle on each side of next to last abdominal segment.

This group is represented in the U.S. by 2 species of *Rhopalopsyllus*, which are parasites of opossums and rats. They occur from Georgia and Florida to Texas.

CARNIVORE FLEAS   Family Vermipsyllidae

Identification: Similar to Malacopsyllidae, but with clypeal tubercle small and rounded or *lacking*, and without long bristles near end of abdomen. Parasites of carnivores.

This is a small group whose members are parasites of bears, wolves, and other large carnivores. They resemble some Pulicidae in lacking genal and pronotal combs, but have 2 rows of bristles on most abdominal terga. They occur in the West.

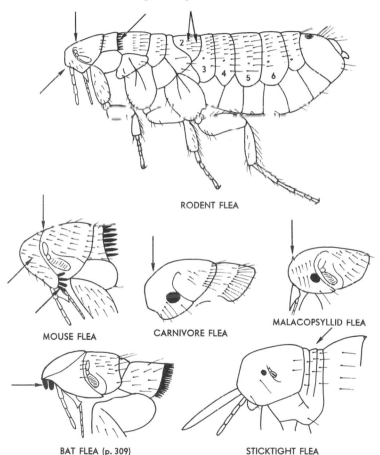

RODENT FLEA

MOUSE FLEA

CARNIVORE FLEA

MALACOPSYLLID FLEA

BAT FLEA (p. 309)

STICKTIGHT FLEA

Sawflies, Ichneumons, Chalcids, Ants, Wasps, and Bees: Order Hymenoptera

Identification: Wings, when present, 4 in number, membranous; FW a little larger than HW; wings with relatively few veins, venation in some minute forms nearly lacking. Antennae usually fairly long, generally with 10 or more segments. ♀ with well-developed ovipositor, which is sometimes longer than the body, and is sometimes modified into a sting. Tarsi 5-segmented (except in a few minute forms). Mouth parts chewing, sometimes with the maxillae and labium modified into a tonguelike sucking structure. Metamorphosis complete.

Similar orders: (1) Diptera (p. 260): only 2 wings; antennae often 3-segmented; mouth parts sucking; softer-bodied. (2) Lepidoptera (p. 218): wings covered with scales, at least in part; usually with a coiled proboscis. (3, 4) Ephemeroptera and Odonata (pp. 65, 68): antennae very short; wings with a more extensive venation.

Immature stages: Larvae caterpillarlike (Symphyta) or maggotlike with a well-developed head (Apocrita). Larvae of Symphyta resemble those of Lepidoptera, but usually have 6 or more pairs of prolegs that lack crochets, and 1 pair of large ocelli. Many larvae are plant feeders, feeding on or in foliage, stems, fruits, and other parts of the plant; many live as parasites in or on bodies of other insects; others live in nests constructed by the adults and feed on material put in nest by the adults. Pupation occurs in a cocoon, in special cells, or (in many parasitic species) in the host.

Habits: Adults are found in many habitats; most occur on flowers or vegetation, but some live on the ground or in debris, and many nest in the ground. Species whose larvae are plant feeders usually lay their eggs in or on the food plant; parasitic species generally lay their eggs on or in the host. Many Hymenoptera construct a nest of some sort and lay their eggs in this nest. Ovipositor of some species is modified into a sting, which is often used to paralyze prey and is an effective means of defense.

Importance: Some Hymenoptera, especially bees, are important plant pollinators. The Honey Bee provides us with useful products (honey and wax). Many parasitic and predaceous species aid in keeping noxious insects under control. Some plant-feeding species are serious pests of cultivated plants.

Classification: Two suborders, Symphyta and Apocrita, which differ in body shape and wing venation. Each suborder is divided into superfamilies.

No. of species: World, 105,000; N. America, 16,300.

Characters used in identification: Principal characters used in separating families of Hymenoptera are those of wing venation, legs, antennae, pronotum, certain sutures on the thorax, and the

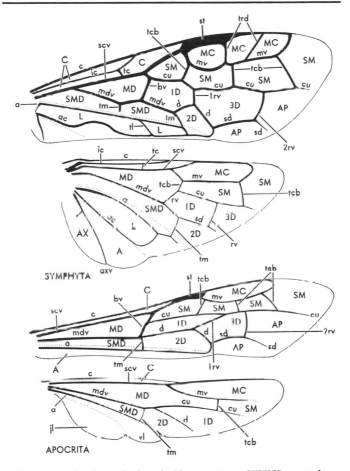

Wing venational terminology in Hymenoptera. VEINS: *a*, anal;
ac, accessory; *axv*, axillary; *bv*, basal; *c*, costal; *cu*, cubital; *d*,
discoidal; *ic*, intercostal; *mdv*, median; *mv*, marginal; *rv*, recurrent;
scv, subcostal; *sd*, subdiscoidal; *st*, stigma; *tc*, transverse costal;
tcb, transverse cubitals; *tl*, transverse lanceolate; *tm*, transverse
median; *trd*, transverse marginal. CELLS: *A*, anal; *AP*, apical;
AX, axillary; *C*, costal; *D*, discoidal; *L*, lanceolate; *MC*, marginal;
MD, median; *SM*, submarginal; *SMD*, submedian. The basal
cells in the hind wing of Symphyta are *MD*, *SMD*, and *L*, and
those in Apocrita are *MD* and *SMD*. Lobes in the hind wing:
jl, jugal; *vl*, vannal.

ovipositor. Not all entomologists agree on how the standard venational terminology (p. 35) should be applied in this order, so we use an older terminology, which is shown on p. 313. Numbers of cells mentioned in descriptions of Hymenoptera refer to the number of *closed* cells, unless otherwise indicated. Leg characters involve the number of trochanter segments (1 or 2), spurs at apex of the tibia, and occasionally other features. Antennae may vary in shape, number of segments, or in location on the face. Shape of the pronotum is useful in distinguishing superfamilies of Apocrita and some families of Symphyta, and the character of various thoracic sutures is used to separate families in a few superfamilies. Ovipositor in many cases rises anterior to apex of the abdomen and cannot be withdrawn into body (Hymenoptera with such an ovipositor generally do not sting); the ovipositor in others issues from the apex of the abdomen, and is withdrawn when not in use (most Hymenoptera with this type of ovipositor can sting).

Sawflies, Horntails, and Wood Wasps: Suborder Symphyta

Base of abdomen broadly joined to thorax. FW with 1–3 marginal cells and nearly always with an accessory vein. HW with 3 basal cells. Trochanters 2-segmented. Larvae plant feeders (except Orussidae). ♀ nonstinging.

The sawflies in the first 7 families below (Tenthredinidae through Diprionidae), and the Orussidae, have *2 apical spurs* on the front tibiae; other families have only 1. Sawflies commonly occur on flowers or are found in association with their food plants.

COMMON SAWFLIES   See also Pl. 15
Family Tenthredinidae
  Identification: Antennae *threadlike,* usually *9-segmented,* the segments similar. FW with 1 or *2 marginal cells,* and without an intercostal vein.
    This is the largest family of sawflies, with about 800 N. American species; it contains most of the species the general collector will encounter. These sawflies are 5–20 mm.; some are black, some brownish, and some are brightly patterned. They are usually found on flowers or vegetation. Larvae of most species are external feeders on foliage; a few are leaf miners and a few are gall makers. Some members of this group cause considerable damage to cultivated plants and forest trees.

XYELID SAWFLIES   Family Xyelidae
  Identification: Usually brownish and less than 10 mm. 3rd antennal segment *very long,* often longer than the remaining segments combined. FW with *an intercostal vein* and *3* (rarely *2*) *marginal cells.*

Xyelids are uncommon. Larvae are external feeders on elm, hickory, and staminate flowers of pine or bore in new pine shoots.

WEB-SPINNING and LEAF-ROLLING SAWFLIES
Family Pamphiliidae

Identification: Relatively stout-bodied, 15 mm. or less, and usually brightly colored. Antennae long and slender, with *13 or more similar segments*. FW with *an intercostal vein, 2 marginal cells*, and 2 transverse median veins.

Pamphiliids are not common. Larvae roll up leaves, tie them with silk, and feed inside the shelter so formed; they feed on various trees and shrubs.

PERGID SAWFLIES   Family Pergidae   Not illus.
Identification: Small sawflies, usually less than 10 mm. Antennae 6-segmented and threadlike.

Pergids occur from the eastern states west to Arizona, but are uncommon. Larvae feed on foliage of oak and hickory.

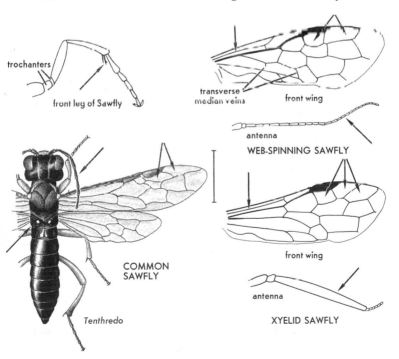

trochanters

front leg of Sawfly

transverse median veins   front wing

antenna

WEB-SPINNING SAWFLY

front wing

COMMON SAWFLY

Tenthredo

antenna

XYELID SAWFLY

ARGID SAWFLIES Family Argidae
Identification: Relatively stout-bodied, and 15 mm. or less. Antennae 3-segmented, the *3rd segment very long* (U-shaped in many males).
Larvae of most species feed on various trees and shrubs.

CIMBICID SAWFLIES Family Cimbicidae **See also Pl. 15**
Identification: Our largest sawflies, 18–25 mm. Look somewhat like bumble bees (but are not hairy) or hornets. Antennae with 7 or fewer segments and *slightly clubbed.*
Cimbicids do not sting but can inflict a strong pinch with the mandibles. Most larvae feed on willow and elm; these are greenish yellow, with black spiracles and a black stripe down the back; the body is often in a spiral position; when disturbed they may eject a fluid from glands above the spiracles.

CONIFER SAWFLIES Family Diprionidae
Identification: *Stout-bodied.* Antennae with *13 or more segments,* usually serrate in ♀ and *pectinate in* ♂.
Most diprionids are 12 mm. or less, and their larvae feed on conifers. They are scarce in the Midwest but common in most areas where conifers are abundant.

HORNTAILS Family Siricidae **See also Pl. 15**
Identification: Large insects (mostly 25–35 mm.), usually brownish or black, and some species have dark wings. Pronotum in dorsal view *wider than long* and *shorter along midline than laterally.* Apex of abdomen with a dorsally located spear or spine (♀ with 2 long slender structures at apex of abdomen, lower one the ovipositor).
This and the remaining families of Symphyta (except Orussidae) have a *single apical spur* on the front tibiae. Larvae are wood-boring, and attack both deciduous trees and conifers.

CEDAR WOOD WASPS Family Syntexidae
Identification: Pronotum in dorsal view *trapezoidal, about twice as wide as long.* FW with a costal cell.
Our only syntexid, *Syntexis libocedrii* Rohwer, occurs in n. California and s. Oregon. It is black and about 8 mm., and its larva bores in the wood of the incense cedar.

WOOD WASPS Family Xiphydriidae
Identification: Blackish and 20–25 mm. Pronotum in dorsal view *U-shaped, much longer laterally than along midline.* FW with *a costal cell* and *a transverse costal vein.*
Larvae bore in the dead and decaying wood of deciduous trees.

PARASITIC WOOD WASPS Family Orussidae
Identification: FW with *only 1 submarginal cell.* Antennae rise below eyes, just above mouth. 8–14 mm.

This is a small and relatively rare group. Adults resemble horntails but are much smaller; larvae are parasites of metallic wood-boring beetles (Buprestidae).

STEM SAWFLIES Family Cephidae

Identification: Pronotum in dorsal view *trapezoidal, about as long as wide.* Costal cell in FW lacking or *very narrow.* Abdomen somewhat flattened laterally.

Cephids are usually black and 9–13 mm. Larvae bore in the stems of various grasses and shrubs.

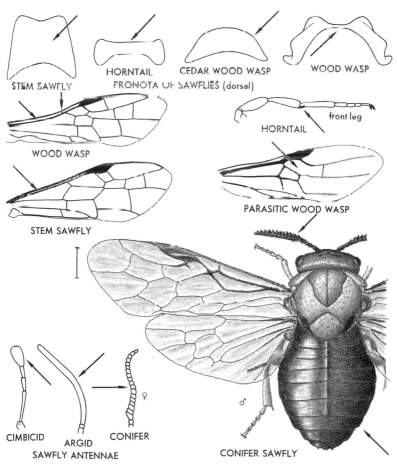

STEM SAWFLY   HORNTAIL   CEDAR WOOD WASP   WOOD WASP

PRONOTA OF SAWFLIES (dorsal)

WOOD WASP

HORNTAIL   front leg

STEM SAWFLY

PARASITIC WOOD WASP

CIMBICID   ARGID   CONIFER
SAWFLY ANTENNAE

♀

♂

CONIFER SAWFLY

Parasitic Hymenoptera, Ants, Wasps, and Bees: Suborder Apocrita

Base of abdomen constricted, sometimes distinctly stalked. Thorax contains a 4th segment, the *propodeum* (actually the basal abdominal segment, fused to thorax). Wings without an accessory vein and HW with not more than 2 basal cells; venation in some minute forms greatly reduced, almost lacking.

Many Apocrita larvae live as parasites in or on bodies of other insects or spiders, and many are plant feeders. Parasitic forms generally lay their eggs on, in, or near body of the host; many have a long ovipositor with which they reach hosts in cocoons or other seemingly protected situations; a few lay their eggs on vegetation, and the newly hatched larvae remain on the vegetation until they can attach to a host passing by. From 1 to many parasites may develop in a single host individual (depending on species of parasite); the host is nearly always killed by the parasite, but usually not until the parasite has completed its larval development. Many parasitic Apocrita are very valuable insects, since they aid in the control of noxious species.

This suborder is divided into 11 superfamilies, adults of which can usually be separated by the character of the pronotum, antennae, and ovipositor, the number of trochanter segments, and the wing venation. Pronotum in profile may appear (a) more or less triangular and extending nearly to the *tegulae* (small scalelike structures overlying bases of front wings), (b) more or less squarish and not extending to the tegulae, or (c) short and collarlike, with a small rounded lobe on each side that does not reach the tegulae. Antennae vary in number of segments they contain and in whether or not they are elbowed; the 1st segment of an elbowed antenna is much longer than any other segment, and in nonelbowed antennae is comparatively short. The distinction between "elbowed" and "not elbowed" is occasionally not very sharp. The terminal antennal segments are usually slender, but are sometimes swollen to form a club and are pectinate in a few Chalcidoidea. The ovipositor in some groups (a) rises anterior to apex of the abdomen (on the ventral side) and is more or less permanently extruded, and usually cannot be withdrawn into abdomen. In other groups (b) the ovipositor issues from apex of abdomen and is withdrawn when not in use. Apocrita with an ovipositor of the 1st type (a) usually do not sting; those with an ovipositor of the 2nd type (b) usually do. Trochanters may be 1- or 2-segmented. Venation varies from almost no veins to a maximum of 10 closed cells in front wing (see p. 313). If there are at least 6 closed cells in the front wing the venation may be described as "normal"; if there are fewer, it may be described as "reduced."

Larvae of Ichneumonoidea, Evanioidea, Pelecinoidea, Bethy-

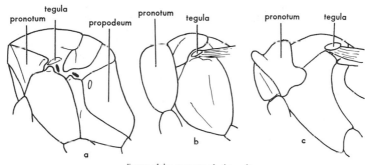

Forms of the pronotum in Apocrita

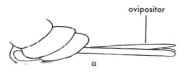

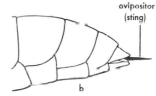

Forms of the ovipositor in Apocrita

Characters of the Superfamilies of Apocrita

Character	Ichneumonoidea	Chalcidoidea	Cynipoidea	Evanioidea	Pelecinoidea	Proctotrupoidea	Bethyloidea	Scolioidea	Vespoidea	Sphecoidea	Apoidea
Pronotum[1]	a	b	a	a	a	a	a,b	a,b	a	c	c
Ovipositor[1]	a	a[4]	a	a	b	b	b	b	b	b	b
Sting	no[4]	no	no	no	no	no	yes[4]	yes[4]	yes	yes	yes[4]
No. of hind trochanter segments	2	2	1[4]	2	1	1-2	1[4]	1	1	1	1
Antennae[2]	T	E	T	T	T	E,T	T[4]	T[4]	T	T	T
No. of antennal segments	16+[4]	5-13	13-16	13-14	14	7-15	10+	12-13	12-13	12-13	12-13
Venation[3]	N[4]	R	R	N	R	N,R	N,R	N[4]	N	N	N

[1] Letters refer to the types illustrated above.
[2] E, elbowed; T, not elbowed (usually threadlike).
[3] N, normal; R, reduced.
[4] Exceptions occur.

loidea, most Scolioidea, most Chalcidoidea, and a few Cynipoidea are parasites of other insects or spiders; adults of these generally seek out and oviposit in or on a host, then go on seeking other hosts; they do not make a nest. Adults of Vespoidea and Sphecoidea generally build a nest, or cells of some sort, then go out and capture prey (other insects or spiders) with which they provision the nest or cells; after doing this they go their way, leaving the young to fend for themselves (except in the social wasps). There is no sharp line of distinction between these 2 methods of operation, and in the family Pompilidae one may find a range in habits from one extreme to the other. Bees (Apoidea) resemble Vespoidea and Sphecoidea in building and provisioning nests or cells, but provision them with nectar and pollen. Social wasps and bees do not provision the nest, but feed the young as they grow. Ants, which are also social, generally construct a nest and feed the young as they grow. The food of young ants may be plant or animal material, depending on the species. Some wasps (certain Sphecidae and Chrysididae) and cuckoo bees are *inquilines:* they build no nest and collect no food for their young, but lay their eggs in nests of other wasps or bees, where their young feed on the food with which these nests are provisioned. The term "cuckoo" is derived from the European Cuckoo (and our cowbirds), which lays its eggs in the nests of other birds.

Apocrita that are plant-feeding in the larval stage include bees, gall wasps (Cynipinae), and some Chalcidoidea. Gall wasp larvae live in and feed on plant galls; the plant-feeding chalcid larvae feed in various ways on plants (a few feed in galls).

Superfamily Ichneumonoidea

Pronotum in profile more or less triangular, and extending to tegulae or nearly so. Antennae *threadlike*, usually with *16 or more segments*. Hind trochanters *2-segmented*. Ovipositor rises in front of apex of abdomen, not capable of being withdrawn, often long, sometimes longer than body. Venation usually normal. FW *without a costal cell* (except in Stephanidae).

This is one of the largest superfamilies in the order, and its members are to be found almost everywhere. The known larvae are parasites of other insects or spiders.

BRACONIDS   Family Braconidae   **See also Pl. 15**
   Identification: Most are brownish or black, not brightly colored. 1 recurrent vein (see p. 313) or none, the 2nd recurrent vein *absent*. 2–15 mm. 1st submarginal and 1st discoidal cells either coalesce or are separated by base of cubital vein.
   This is a large and widely distributed group found almost everywhere. Larvae are parasites of a great variety of insects, and many are important agents in the control of noxious insects.

Some species pupate in silken cocoons on the outside of the body
of the host.

STEPHANIDS Family Stephanidae   **Not illus.**
Identification: Costal cell present. Head spherical on long neck
and bearing a crown of teeth. 5–19 mm. Gasteruptiids and
aulacids (p. 334) are similar but do not have a crown of teeth,
and abdomen is attached to propodeum high above hind coxae.
    This is a small group of very rare insects. Larvae are parasites
of wood-boring beetles.

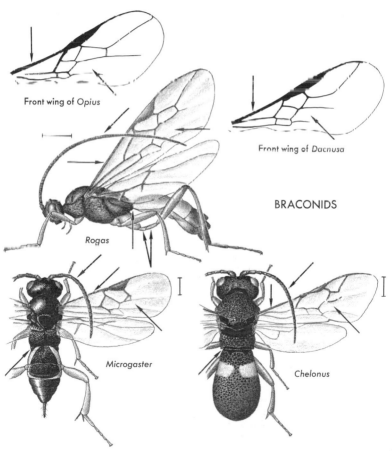

Front wing of *Opius*

Front wing of *Dacnusa*

BRACONIDS

Rogas

Microgaster

Chelonus

ICHNEUMONS Family Ichneumonidae **See also Pl. 15**
Identification: Slender, wasplike, 3–40 mm. *2 recurrent veins.*
2nd submarginal cell *small* or lacking. Base of cubital vein
lacking, 1st submarginal and 1st discoidal cells fused. Antennae
with 16 or more segments and usually *at least half as long as body.*
Braconids have only 1 recurrent vein or none; gasteruptiids and
aulacids (p. 334) have head set out on a slender neck, the
abdomen attached high above hind coxae, and FW with a costal
cell; wasps (Scolioidea, Vespoidea, and Sphecoidea, pp. 340, 346,
348) are usually stouter-bodied, with shorter antennae (that
nearly always have 13 or fewer segments), 1 trochanter segment,
a costal cell, and the cubital vein complete; trigonalids (p. 340),
which have long many-segmented antennae, have a costal cell
and a complete cubital vein.

This is the largest family of insects, with over 3000 N. American species. They are common insects, found almost everywhere.
They vary greatly in size and color: many are uniformly colored,
from yellowish to black, and many are brightly patterned with
black and brown or black and yellow; many have middle
segments of antennae yellowish or whitish. Most species have
a long ovipositor.

Most ichneumons do not sting, though they generally try to
do so when handled. The few that will sting are usually large,
yellowish brown, and have a laterally flattened abdomen. These
ichneumons have a short, sharp ovipositor capable of piercing
the skin. They must be able to move the abdomen in order to
sting, and if grasped by the abdomen are quite harmless.

The family Ichneumonidae is divided into a number of subfamilies, many of which are further divided into tribes. Each
subfamily or tribe is often parasitic on a particular group of
insects. Many ichneumons are of value in the control of noxious
insects.

Chalcids: Superfamily Chalcidoidea

Pronotum in profile *somewhat squarish* and not quite reaching
tegulae. Antennae *elbowed* and usually short, with 5–13 segments.
At least 1 pair of trochanters 2-segmented. Ovipositor generally
short, occasionally as long as body, and usually rising in front of
apex of abdomen. Wing venation *greatly reduced.* Mostly 5 mm.
or less. Small Proctotrupoidea (p. 335) with elbowed antennae
and a similarly reduced venation have the pronotum more triangular and the ovipositor apical.

This is a large group occurring almost everywhere. Most
chalcids are black, blue-black, or greenish, and many are metallic.
Wings are usually held flat over abdomen at rest; a few are wingless or have wings greatly reduced. Most larvae are parasites of
other insects and some are hyperparasites; many are plant feeders.

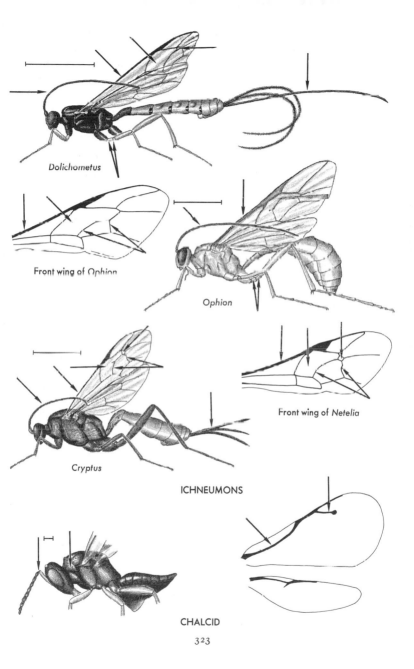

Dolichometus

Front wing of *Ophion*

Ophion

Cryptus

Front wing of *Netelia*

ICHNEUMONS

CHALCID

323

The small size of most chalcids is likely to make their identification difficult; the key below is designed to aid the beginner in separating the families of chalcids.

1. HW very narrow, almost linear **Mymaridae (p. 324)**
1'. HW elongate-oval, not linear 2
2(1'). Tarsi 3-segmented **Trichogrammatidae (p. 324)**
2'. Tarsi 4- or 5-segmented 3
3(2'). Tarsi 4-segmented; apical spur of front tibiae small and straight **Eulophidae (p. 324)**
3'. Tarsi 5-segmented; apical spur of front tibiae large and curved 4
4(3'). Head of ♀ oblong, with a deep longitudinal groove above; ♂ wingless; front and hind legs stout, middle legs slender **Agaonidae (p. 328)**
4'. Without this combination of characters 5
5(4'). Mesopleura convex; apical spur of middle tibiae large and stout **Tanaostigmatidae, Thysanidae, Eupelmidae, Encyrtidae, Eutrichosomatidae (p. 326)**
5'. Mesopleura with a broad shallow groove; apical spur of middle tibiae small 6
6(5'). Mandibles sickle-shaped; thorax strongly arched, the pronotum usually not visible from above; scutellum large; abdomen laterally flattened **Eucharitidae (p. 326)**
6'. Without this combination of characters 7
7(6'). Hind femora greatly enlarged and usually toothed beneath **Chalcedectidae, some Torymidae, Leucospididae, Chalcididae (p. 328)**
7'. Hind femora slender, not greatly enlarged 8
8(7'). Hind coxae considerably larger than front coxae **Torymidae, Ormyridae (p. 328)**
8'. Hind coxae little if any larger than front coxae **Pteromalidae, Perilampidae, Eurytomidae (p. 330)**

FAIRYFLIES Family Mymaridae
Identification: HW *very narrow, almost linear.*
Fairyflies are minute insects, mostly less than 1 mm. (smallest is only 0.21 mm.), usually blackish, with relatively long legs and antennae. Larvae are egg parasites.

TRICHOGRAMMATIDS Family Trichogrammatidae
Identification: Minute insects, 1 mm. or less, and rather stocky in build. Tarsi *3-segmented.* Tiny hairs of wings usually *in rows.*
Larvae are parasites of insect eggs, and some are of value in controlling insect pests.

EULOPHIDS Family Eulophidae
Identification: 1–3 mm. and varying in shape and color; most are black, a few brilliantly metallic. Tarsi *4-segmented.* Apical spur of front tibiae *small,* straight. Axillae *extend forward beyond tegulae.* ♂ antennae often pectinate.

This is a large, fairly common group of about 600 N. American species. Larvae are parasites of various other insects: Eulophinae are parasites of leaf miners; Tetrastichinae are egg parasites or attack larvae and pupae of Colcoptera, Lepidoptera, and Diptera; Entedontinae are chiefly parasites of small larvae in cases or leaf mines; Aphelininae are parasites of aphids, scale insects, and whiteflies; Elachertinae and Elasminae are parasites of caterpillars, the latter as hyperparasites attacking braconids and ichneumons in the caterpillars.

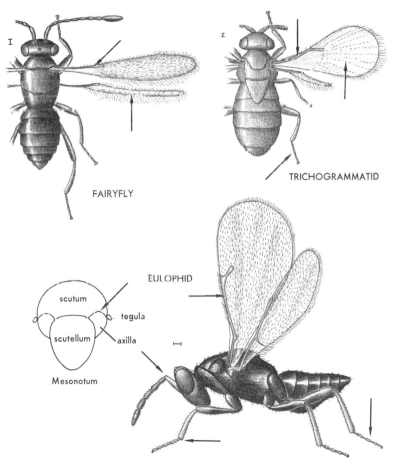

FAIRYFLY

TRICHOGRAMMATID

EULOPHID

scutum

tegula

scutellum

axilla

Mesonotum

ENCYRTIDS Family Encyrtidae
Identification: 1–2 mm. and usually black or brown; some are wingless. *Mesopleura* and *mesonotum convex*. Parapsidal sutures indistinct or *lacking*. Marginal vein *very short*. Scutellum *not much wider than long*. Apical spur of front tibiae *large and curved*, of middle tibiae large and stout.

The encyrtids are a large and widely distributed group. Larvae of most species are parasites of scale insects and white-flies. Many species are polyembryonic, with 10 to over 1000 young developing from a single egg.

THYSANIDS Family Thysanidae **Not illus.**
Identification: Similar to Encyrtidae but marginal vein as long as submarginal, and scutellum much wider than long.

Thysanids are small and stout-bodied. Larvae are mostly hyperparasites, attacking parasites of scale insects and other Homoptera. The group is small and rare.

EUPELMIDS Family Eupelmidae
Identification: Similar to Encyrtidae but marginal vein *long*, mesonotum flattened or *concave*, and parapsidal sutures usually distinct and nearly straight.

Eupelmids are black or brown and usually over 2 mm. Some species have the wings greatly reduced. Larvae are parasites of various insects and spiders. Eupelmids are fairly common, but generally not as common as encyrtids.

TANAOSTIGMATIDS Family Tanaostigmatidae **Not illus.**
Identification: Similar to Eupelmidae but mesonotum slightly convex, parapsidal sutures complete and curved laterally, scapulae short. 1st antennal segment of ♀ somewhat dilated or flattened laterally; terminal antennal segments of ♂ with 4 branches.

Four rare species have been reported from Florida, Arizona, and California. Most larvae are gall makers.

EUTRICHOSOMATIDS **Not illus.**
Family Eutrichosomatidae
Identification: Similar to Tanaostigmatidae but 1st antennal segment of ♀ slender and terminal antennal segments of ♂ without branches.

This group is represented in the U.S. by 2 rare species of *Eutrichosoma*, 1 occurring in Georgia and Texas and 1 from Maryland west to New Mexico, Idaho, and Montana. Larvae are parasites of snout beetles.

EUCHARITIDS Family Eucharitidae
Identification: Shape characteristic: head *short*, thorax *hump-*

backed, pronotum usually not visible from above, scutellum often *extending backward* over base of abdomen, and abdomen *stalked and attached low on thorax*.

This is a small group of rather uncommon insects, usually black, with interesting habits. Larvae are parasites of ant pupae; eggs are laid in large numbers on vegetation and the larvae on hatching attach themselves to a passing ant that carries them

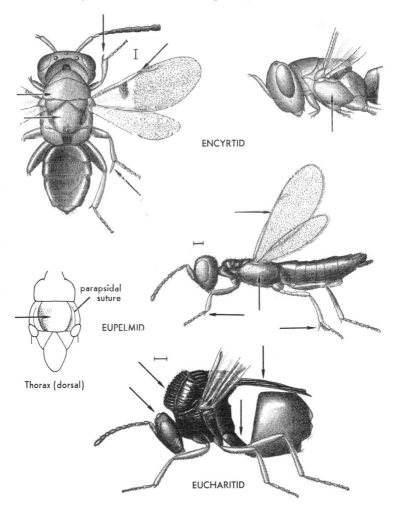

ENCYRTID

parapsidal suture

EUPELMID

Thorax (dorsal)

EUCHARITID

to its nest, where the eucharitid larvae leave the ant and attack the ant pupae.

CHALCIDIDS   Family Chalcididae
Identification: Generally uniformly dark-colored. Hind femora *greatly swollen*, toothed beneath. Hind coxae *considerably larger* than front coxae, long and cylindrical. Ovipositor usually short. Wings not folded longitudinally at rest.

Chalcidids are mostly 2–7 mm., and are relatively common. Larvae are parasites of a variety of insects.

CHALCEDECTIDS   Family Chalcedectidae   **Not illus.**
Identification: Hind femora greatly swollen. Hind coxae little if any larger than front coxae.

Four rare species have been reported from the s. U.S. Larvae are parasites of Buprestidae.

LEUCOSPIDIDS   Family Leucospididae
Identification: Similar to Chalcididae but larger (8–12 mm.). Brightly colored with black and yellow. Somewhat hump-backed in appearance. Wings folded longitudinally at rest. Ovipositor *curves upward and forward* along dorsal side of abdomen.

These are a small group of uncommon insects resembling small yellowjackets. They are usually found on flowers. Larvae are parasites of various wasps and bees.

TORYMIDS   Family Torymidae
Identification: Elongate, usually metallic green, and mostly 2–4 mm. Hind coxae *much larger* than front coxae. Ovipositor *as long as body or longer*. Hind femora *slender*, or if greatly swollen hind coxae are triangular in cross section. Parapsidal sutures *present*. Abdomen smooth, shiny.

Larvae of some species are parasites of various caterpillars, gall insects, or insect eggs; larvae of others feed on seeds.

ORMYRIDS   Family Ormyridae   **Not illus.**
Identification: Similar to Torymidae but parapsidal sutures indistinct or absent, ovipositor short and often hidden, and abdomen usually with rows of deep punctures.

This is a small group of rather rare chalcids. Larvae are parasites of gall insects.

FIG WASPS   Family Agaonidae
Identification: Black. Head of ♀ somewhat oblong, with *a deep dorsal longitudinal groove*. ♀ winged, ♂ wingless. Front and hind legs *stout*, middle legs slender.

This group is represented in the U.S. by 2 species, 1 occurring

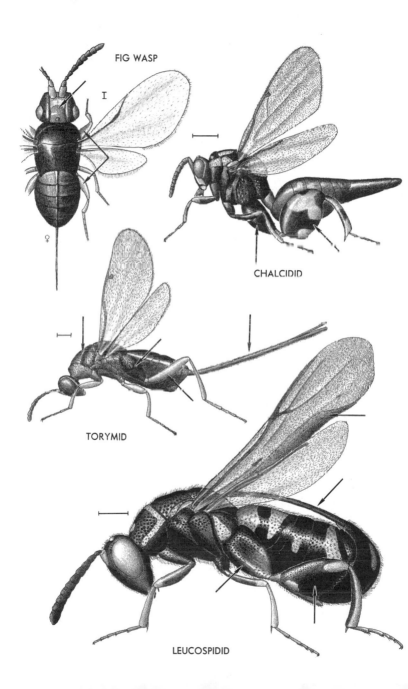

FIG WASP

♀

CHALCIDID

TORYMID

LEUCOSPIDID

in Florida and 1 in California and Arizona. Fig wasps pollinate certain varieties of figs.

PTEROMALIDS   Family Pteromalidae
Identification: Tarsi *5-segmented.* Apical spur of front tibiae *large and curved.* Mesopleura slightly concave, or with *a broad shallow groove.* Front and hind coxae about same size. Hind femora not enlarged. Pronotum in dorsal view *somewhat conical, narrowed toward front.*

Pteromalids are common insects, and most of the chalcids encountered by the general collector will probably belong to this family. They are mostly 2–4 mm., and are usually black or metallic green. Larvae are parasites of a variety of insects, and many species are important agents in the control of crop pests.

EURYTOMIDS or SEED CHALCIDS   Family Eurytomidae
Identification: Similar to Pteromalidae but pronotum in dorsal view *squarish,* not particularly narrowed anteriorly. Usually dull black, not shiny. Thorax rough or pitted. Abdomen of ♀ rounded or oval. Often somewhat hairy.

Eurytomids are fairly common chalcids. Some larvae are parasites of other insects and some feed on plants; a few are parasites as young and plant feeders when older. Plant-feeding species attack seeds and stems or are gall makers.

PERILAMPIDS   Family Perilampidae
Identification: Relatively large chalcids; most are 6–8 mm. and metallic blue or blue-black. Thorax *stout, pitted* with punctures. Pronotum in dorsal view *transversely linear.* Abdomen *small,* in profile *triangular, shining.* Perilampids strongly resemble cuckoo wasps (Chrysididae; p. 338), but have a differently shaped abdomen and a more reduced wing venation.

Larvae are mostly hyperparasites, attacking the Diptera and Hymenoptera that are parasitic in caterpillars.

Gall Wasps and Others: Superfamily Cynipoidea

Pronotum in lateral view (see p. 333) *more or less triangular,* extending to tegulae or nearly so. Antennae *threadlike, 13- to 16-segmented.* Trochanters usually 1-segmented. Ovipositor rises anterior to apex of abdomen. Wing venation *reduced.* (For additional characters see p. 319.) The most distinctive features of the Cynipoidea are the wing venation and the threadlike antennae.

Most members of this group are black and 2–8 mm.; many have the abdomen shiny and often laterally flattened. More than 800 species of Cynipoidea occur in the U.S. The vast majority are gall wasps (Cynipinae) and the rest, as far as known, are parasites of other insects.

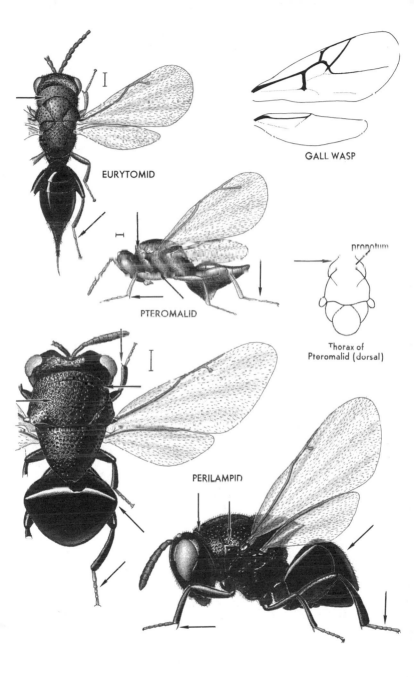

EURYTOMID

GALL WASP

PTEROMALID

pronotum

Thorax of
Pteromalid (dorsal)

PERILAMPID

IBALIIDS Family Ibaliidae

Identification: Usually black and yellowish brown, the abdomen banded. Length 7–16 mm.; abdomen elongate and laterally flattened. 1st segment of hind tarsi *twice as long* as other segments combined; 2nd segment with *a long apical process* extending to tip of 4th segment. FW usually with brownish spots in middle and at apex.

This is a small and rare group. Larvae are parasites of horntails (Siricidae).

LIOPTERIDS Family Liopteridae   **Not illus.**

Identification: Abdomen stalked and attached far above base of hind coxae. Propodeum with a median furrow.

Three rare species occur in Texas, Idaho, and California. Their larval stages are unknown.

FIGITIDS Family Figitidae

Identification: A small group of shining black insects, mostly 3–6 mm. Anacharitinae: abdomen *distinctly stalked*, attached near base of hind coxae. Aspiceratinae: 2nd abdominal tergum (1st tergum forms stalk, and 2nd is 1st tergum in swollen part of abdomen) *somewhat shovel-shaped*, in lateral view *narrow dorsoventrally*, shorter than 3rd along middorsal line. Figitinae: 2nd abdominal tergum *not particularly shovel-shaped*, *wider dorsoventrally* in side view, and about as long as 3rd along middorsal line.

Figitids are not common. Larvae are parasites of pupae of various Diptera (Aspiceratinae and Figitinae) or lacewings (Anacharitinae).

CYNIPIDS Family Cynipidae

Identification: Eucoilinae: 4–6 mm., shining black, scutellum with *a dorsal O-shaped elevation*. Charipinae (not illus.): 2 mm. or less; thorax smooth. Cynipinae: usually over 2 mm. (mostly 6–8 mm.), the thorax rather rough, the abdomen *oval and shining*.

Eucoilinae and Charipinae are parasitic in the larval stage, Eucoilinae on pupae of Diptera and Charipinae on the braconids that parasitize aphids. These cynipids are more common than the figitids.

The vast majority of Cynipoidea the general collector will encounter will be Cynipinae (gall wasps), many of which are common insects. They either are gall makers or live in galls formed by another organism; each gall maker forms a characteristic gall on a particular part of a particular plant. The galls are much more often seen, and are usually more distinctive, than the gall wasps themselves. Many gall wasps form galls on oak leaves.

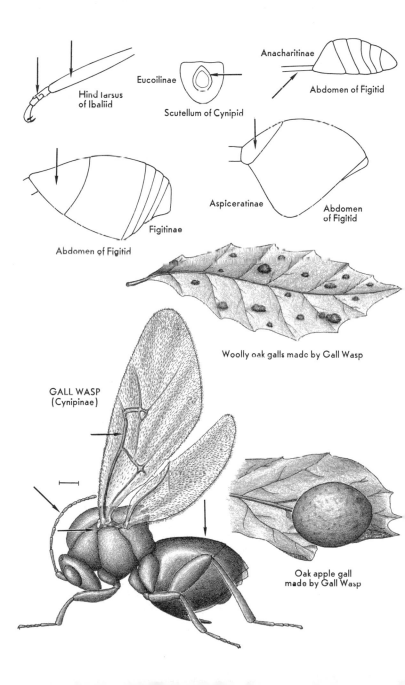

Hind tarsus of Ibaliid

Eucoilinae

Scutellum of Cynipid

Anacharitinae

Abdomen of Figitid

Figitinae

Abdomen of Figitid

Aspiceratinae

Abdomen of Figitid

Abdomen of Figitid

Woolly oak galls made by Gall Wasp

GALL WASP (Cynipinae)

Oak apple gall made by Gall Wasp

Superfamily Evanioidea

Abdomen attached high above hind coxae. Antennae threadlike, 13- or 14-segmented. Trochanters *2-segmented.* Venation fairly complete. FW with *a costal cell.* (For additional characters see p. 319.)

ENSIGN WASPS   Family Evaniidae

Identification: Black, somewhat spiderlike in appearance, and 6–10 mm.; the small abdomen is carried like a flag (hence the common name). Characteristic appearance: thorax stout, abdomen *small* and attached by *a slender stalk high above hind coxae.* Ovipositor very short, not protruding.

Larvae are parasitic in the egg capsules of cockroaches, and adults are likely to be found in places where cockroaches occur.

GASTERUPTIIDS   Family Gasteruptiidae

Identification: Similar to Ichneumonidae (p. 322) but head set out *on a slender neck,* abdomen attached *high above hind coxae,* antennae *short,* and FW with *a costal cell,* only *1 recurrent vein,* and cubital vein complete. Ovipositor long, often as long as body. Black, 13–20 mm.

Adults are fairly common, and are usually found on flowers. Larvae are parasites of solitary wasps and bees.

AULACIDS   Family Aulacidae

Identification: Similar to Gasteruptiidae but antennae longer and FW with *2 recurrent veins.* Black, or black with a reddish abdomen, and wings often banded or spotted.

Aulacid larvae parasitize various wood-boring insects, and adults are generally found around logs in which the hosts occur. These insects are moderately common.

Superfamily Pelecinoidea

Female about 2 in., shining black, with wings short and abdomen *very long and slender.* ♂ about 1 in., abdomen club-shaped. (For additional characters, see p. 319.)

PELECINIDS   Family Pelecinidae

Identification: By the characters of the superfamily.

This group is represented in the U.S. by a single species, *Pelecinus polyturator* (Drury), which occurs in the eastern part of the country. Females, which are very distinctive in appearance, are fairly common but males are very rare. Larvae are parasites of white grubs.

Superfamily Proctotrupoidea

Trochanters 1- or 2-segmented; if 2-segmented 2nd segment is often poorly defined. Small, often minute, usually shining black. (For additional characters see p. 319.) Differ from Chalcidoidea and Bethyloidea (pp. 322, 338) in form of the pronotum, and from most Bethyloidea in lacking a jugal lobe in hind wing. Differ from Cynipoidea (p. 330) in location of the ovipositor. All known larvae are parasites of other insects.

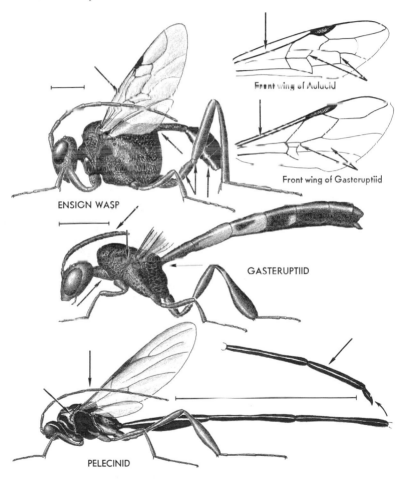

Front wing of Aulacid

Front wing of Gasteruptiid

ENSIGN WASP

GASTERUPTIID

PELECINID

HELORIDS   Family Heloridae   **Not illus.**
Identification: Antennae 15-segmented, and rising at middle of
face. Venation normal. 1st discoidal cell in FW triangular.
4 mm.
   One very rare but widely distributed species of helorid occurs
in the U.S. Its larva is a parasite of lacewings.

VANHORNIIDS   Family Vanhorniidae
Identification: Mandibles with *tips directed outward,* and not
meeting when closed. Abdomen somewhat elongate, the 1st
segment occupying most of abdomen. Antennae 13-segmented.
Trochanters 1-segmented.
   Our only species is very rare and occurs in the Northeast.
It is 6–8 mm., black, with a somewhat reduced wing venation.
Larvae parasitize the larvae of Eucnemidae.

ROPRONIIDS   Family Roproniidae
Identification: Appearance: about 10 mm., thorax moderately
robust and black, abdomen small, somewhat triangular, laterally
flattened, with a long petiole, and brownish. Venation normal,
1st discoidal cell in FW *irregularly 6-sided.* Antennae 14-seg-
mented, rising at middle of face. Trochanters 1-segmented or
indistinctly 2-segmented.
   Three very rare species occur in the U.S., 2 in the East and
1 in California and Oregon. Larvae parasitize sawflies.

PROCTOTRUPIDS   Family Proctotrupidae
Identification: Characteristic venation: FW with a costal cell,
a *large stigma,* and a *very small marginal cell.* Antennae *13-
segmented, threadlike,* and rising *at middle of face.* Abdomen
spindle-shaped, with a short cylindrical petiole. Trochanters
1-segmented.
   Proctotrupids are moderately common insects that vary from
about 3 to 10 mm.; the abdomen is often brownish. Little is
known of the larval stages, but some are known to parasitize
beetles.

CERAPHRONIDS   Family Ceraphronidae
Identification: Chalcidlike, but pronotum in lateral view triangular and extending to tegulae. Antennae *elbowed,* 9- to
11-segmented, and rise low on face. Venation reduced. HW
without a distinct jugal lobe. Trochanters 2-segmented. Abdomen
rounded laterally.
   This is a large group of small black insects, mostly less than
4 mm.; a few are wingless. Larvae are parasites of a variety of
insects.

DIAPRIIDS   Family Diapriidae
Identification: Small to minute, black and shining, fairly common. Antennae *11- to 15-segmented, threadlike,* and rising on

a shelflike protuberance in middle of face. Trochanters 2-segmented. Venation reduced, sometimes (Diapriinae) nearly absent. HW without a jugal lobe.

Larvae of most species are parasites of fungus gnats and other Diptera, and adults are usually found in wooded areas where there is decaying vegetation and fungi. Two of the 4 sub-

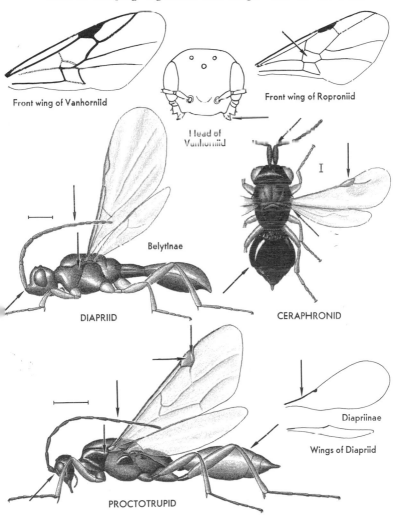

Front wing of Vanhorniid

Head of Vanhorniid

Front wing of Roproniid

Belytinae

DIAPRIID

CERAPHRONID

PROCTOTRUPID

Diapriinae

Wings of Diapriid

families are common; the Diapriinae are minute, with the venation *greatly reduced* (no closed cell in hind wing); the Belytinae are larger (5–7 mm.), with slightly more venation (a closed cell in hind wing).

SCELIONIDS Family Scelionidae
Identification: Minute insects, usually 2 mm. or less; black (rarely brown). Antennae *elbowed, rising low on face,* generally 11- or 12-segmented (occasionally 7- or 8-segmented and clubbed, the club unsegmented). Abdomen *flattened,* the lateral margins *sharp.* Venation *greatly reduced,* chalcidlike. HW without a jugal lobe.

This is a large, fairly common group. Larvae are chiefly egg parasites. Some females, which parasitize the eggs of grasshoppers or mantids, attach to the abdomen of a grasshopper or mantid and ride around on it until it lays its eggs; then the scelionid gets off and oviposits in the eggs of the host.

PLATYGASTERIDS Family Platygasteridae
Identification: Similar to Scelionidae but antennae 9- or 10-segmented, if clubbed the club segmented. Abdomen more or less flattened, less so than in Scelionidae.

Platygasterids are minute insects, shining black, with *almost no wing venation.* Some species have a peculiarly shaped abdomen. Females of the genus *Inostemma* have *a long handlelike process* extending from the base of the abdomen forward over the thorax (this process serves as a receptacle for the long ovipositor when it is withdrawn into the body). Most larvae are parasites of gall gnats (Cecidomyiidae); some parasitize mealybugs or whiteflies.

Superfamily Bethyloidea

Pronotum in lateral view variable, sometimes triangular, sometimes quadrate. Antennae generally 10- to 13-segmented and *threadlike.* Trochanters *1-segmented* (except Trigonalidae). Ovipositor issues from apex of abdomen. Venation usually *reduced.* HW of forms with reduced venation has *a jugal lobe.* Larvae, as far as known, parasites of other insects.

CUCKOO WASPS Family Chrysididae   See also Pl. 15
Identification: Body metallic blue or green, usually with coarse sculpturing. Abdomen with 4 or fewer segments, concave beneath, the last tergum often toothed apically. Hind wing with *a distinct lobe* at base and *without closed cells.*

Cuckoo wasps are about 6–12 mm. and have a brilliant metallic coloring; they are common insects. Some of them resemble perilampids (p. 330) and certain halictid bees (p. 356); they have more venation than a perilampid and not as much as a halictid. When disturbed they commonly curl up into a ball. They do not sting. Larvae are parasites or inquilines in nests of other wasps or bees.

BETHYLIDS Family Bethylidae

Identification: Small to medium-sized, usually black. Head *somewhat elongated.* Antennae *12- or 13-segmented, slightly*

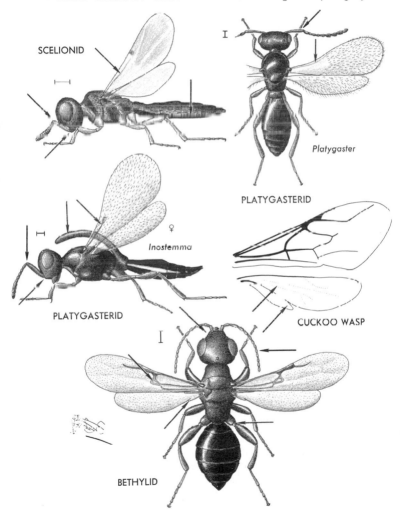

SCELIONID

Platygaster

PLATYGASTERID

♀

Inostemma

PLATYGASTERID

CUCKOO WASP

BETHYLID

elbowed. HW with a *jugal lobe.* Venation *reduced.* Abdomen 6- or 7-segmented.

Bethylids are generally 8 mm. or less and are seldom encountered. Larvae are parasites of other insect larvae, especially Coleoptera and Lepidoptera. A few bethylids will sting. Females of many species are wingless and antlike.

DRYINIDS   Family Dryinidae
Identification: Small black insects with a *reduced* wing venation. Antennae *10-segmented, rising low on face,* and *not distinctly elbowed.* HW *with a jugal lobe.* Front tarsi of ♀ often *pincerlike.*

Dryinids are relatively uncommon insects whose larvae are parasites of planthoppers, leafhoppers, or treehoppers. Most are 5–8 mm. Some species are polyembryonic, with 40 to 60 young developing from a single egg.

TRIGONALIDS   Family Trigonalidae
Identification: Stout-bodied, wasplike; usually brightly colored and 10–12 mm. Antennae *long, with 16 or more segments.* Trochanters 2-segmented, 2nd segment sometimes indistinct. Venation complete. FW *with a costal cell.* HW *without a jugal lobe.* Differ from wasps (Scolioidea, Vespoidea, Sphecoidea) in the long many-segmented antennae and from ichneumons by the costal cell in FW.

Trigonalids are relatively rare. Larvae parasitize vespid larvae or caterpillar parasites. Eggs are laid on foliage and hatch when eaten (with the foliage) by a caterpillar. The trigonalid larvae then attack a parasite in the caterpillar, or, if the caterpillar is eaten by a vespid and later regurgitated and fed to its larva, attack the vespid larva.

SCLEROGIBBIDS   Family Sclerogibbidae   **Not illus.**
Identification: Similar to Bethylidae (p. 339) but antennae of ♂ with 23 segments and rising low on face. ♂ winged, ♀ wingless. FW with a small marginal cell and 1 submarginal cell. HW with a closed cell.

One very rare species, known only from the male, has been taken in Arizona. The larva of this species is unknown, but sclerogibbids occurring elsewhere are known to parasitize webspinners (Embioptera).

Ants and Parasitic Wasps:
Superfamily Scolioidea

Pronotum in lateral view variable, more or less squarish to triangular, extending nearly to tegulae. Antennae 12- or 13-segmented (with fewer segments in some ants), distinctly elbowed only in ants. Ovipositor issues from apex of abdomen. Venation usually fairly complete.

The term "wasp" is used in this book (with a few exceptions) for members of the superfamilies Scolioidea (other than ants), Vespoidea, and Sphecoidea. Wasps are Hymenoptera in which the females usually sting, antennae are generally 13-segmented in the male and 12-segmented in the female, trochanters are 1-segmented, and larvae usually feed on animal food. Vespoid and sphecoid wasps (pp. 346, 348) generally build a nest, capture prey and provision the nest with it, and thus might be described as predaceous. Scolioid wasps are parasitic, their behavior like that of the parasitic members of the preceding superfamilies; they oviposit in or on the body of a host, then oviposit in other hosts, and make no further provision for their young. Bees (Apoidea, p. 354) differ from wasps in that the young are fed plant rather than animal food.

Scolioid wasps differ from Sphecoidea in the form of the pronotum (see illus., p. 319); Vespidae differ from Scolioidea in having the 1st discoidal cell of the front wings very long and in folding their wings longitudinally at rest; Pompilidae (p. 346) differ from scolioid wasps in having long legs and a transverse suture on the mesopleura. Females of many scolioid wasps are wingless and antlike; females of vespoid and sphecoid wasps have well-developed wings.

RHOPALOSOMATID WASPS   Not illus.
Family Rhopalosomatidae

 Identification: Slender, brownish, 6–25 mm. Antennae long, each segment with 2 apical spines. FW (species with normal wings) with transverse median vein considerably beyond base of basal vein (wings very short in 1 species).

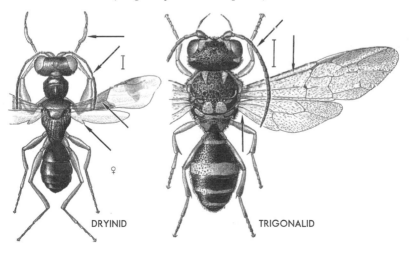

DRYINID   TRIGONALID

Two very rare species occur in the East. The larger has normal wings and resembles ichneumons in the genus *Ophion*, but the abdomen is not laterally flattened, the antennae contain only 12 or 13 segments, the trochanters are 1-segmented, and there is only 1 recurrent vein in the front wing. The smaller species (about 6 mm.) has very short and padlike wings. Larvae are parasites of crickets.

SIEROLOMORPHID WASPS   Not illus.
Family Sierolomorphidae
Identification: Shining black wasps, 4.5–6.0 mm. No jugal lobe in HW.

These wasps are widely distributed but are quite rare. Their immature stages are unknown.

TIPHIID WASPS   Family Tiphiidae   See also Pl. 15
Identification: Mesosternum with *2 posterior lobes*, or abdominal segments separated by rather strong constrictions.

Five of the 6 subfamilies of Tiphiidae occurring in the U.S. deserve special mention. The Tiphiinae, Myzininae, and Brachycistidinae have 2 posterior lobes on the mesosternum, the ♂ has an upcurved spine at end of the abdomen, and the ♀ is usually winged; the mesosternum of Methochinae and Myrmosinae lacks lobes (or has a pair of small toothlike projections posteriorly), the ♂ sometimes has an upcurved spine at the end of the abdomen, the abdominal segments are separated by fairly strong constrictions, and the ♀ is wingless.

Subfamily Tiphiinae. Black, *short-legged*, and generally *10–20 mm.* Middle tibiae with 1 apical spur. These wasps are fairly common and widely distributed; their larvae parasitize white grubs.

Subfamily Myzininae (Pl. 15). Black and yellow, longer-legged, and generally over 20 mm. Middle tibiae with 2 apical spurs. These wasps are also fairly common and widely distributed, and most species parasitize various beetle larvae.

Subfamily Brachycistidinae (not illus.). Brownish. Middle tibiae with 1 apical spur. ♀ wingless. These wasps are restricted to the western states and are mostly nocturnal.

Subfamily Methochinae. Thorax of ♀ divided into *3 parts.* ♂ with a spine at end of abdomen. ♂ 15 mm. or less, ♀ much smaller. Black or brownish. This is a small but widely distributed though not common group. Larvae parasitize tiger beetles.

Subfamily Myrmosinae. Similar to Methochinae but thorax of ♀ divided into 2 parts and ♂ without a spine at end of abdomen. This is a larger group than the Methochinae and somewhat more common. Larvae are parasites of various bees and wasps.

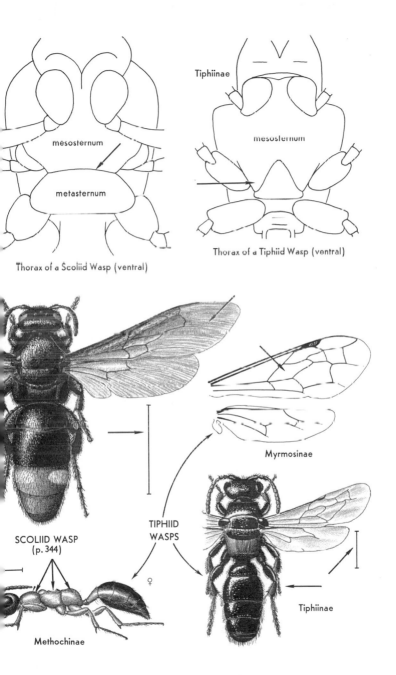

mesosternum

metasternum

Thorax of a Scoliid Wasp (ventral)

Tiphiinae

mesosternum

Thorax of a Tiphiid Wasp (ventral)

Myrmosinae

SCOLIID WASP
(p. 344)

TIPHIID
WASPS

♀

Methochinae

Tiphiinae

SCOLIID WASPS Family Scoliidae p. 343
Identification: *Large*, robust, hairy wasps, *mostly 20–30 mm.;* dark-colored, often with light markings. Mesosternum and metasternum form a large plate divided by *a transverse suture.* Hind coxae well separated. Wing membrane beyond closed cells with *numerous longitudinal wrinkles.*

Scoliid wasps are usually found on flowers. Larvae are parasites of white grubs.

VELVET ANTS Family Mutillidae See also Pl. 15
Identification: *Very hairy wasps*, often brightly colored. 6–20 mm. ♂ winged, ♀ wingless.

Mutillids are wasps that resemble ants but are quite hairy. Females inflict a very painful sting. Larvae are mostly parasites of larvae of ground-nesting bees and wasps; a few parasitize insects in other orders.

SAPYGID WASPS Family Sapygidae **Not illus.**
Identification: Slender, 15 mm. or less; black, marked with yellow. Eyes deeply notched. Differ from Myzininae (Tiphiidae) in being smaller and in lacking mesosternal lobes. Differ from mutillids in having the body bare.

This is a small but widely distributed and quite rare group. Larvae are parasites of leafcutting bees (Megachilidae).

ANTS Family Formicidae
Identification: First abdominal segment (or 1st 2 abdominal segments) *nodelike* or with a dorsal hump, differing from remaining segments. Antennae 6- to 13-segmented, and *strongly elbowed* (at least in ♀), the 1st segment quite long. Social insects, with different castes; queens and males usually winged, the workers wingless. Venation of winged forms normal or slightly reduced.

This is a large and widely distributed group occurring almost everywhere, often in considerable numbers. Ant colonies vary greatly in size, from a dozen or so up to many thousands of individuals. Most species nest in the ground but many nest in various natural cavities. Each colony usually consists of 1 or more queens (larger than other individuals and do all the egg laying), workers (larger colonies may contain 2 or more types of workers), and males; a few ants have no worker caste. Males and queens are produced at certain seasons, and mating usually occurs in a mating flight; males are generally much smaller than queens. After mating, the queen sheds its wings and either starts a new colony or enters an established colony.

Ants vary in habits: some are carnivorous, some are scavengers, and some are plant feeders. Most ants will bite when disturbed and many will sting; a few can eject a foul-smelling

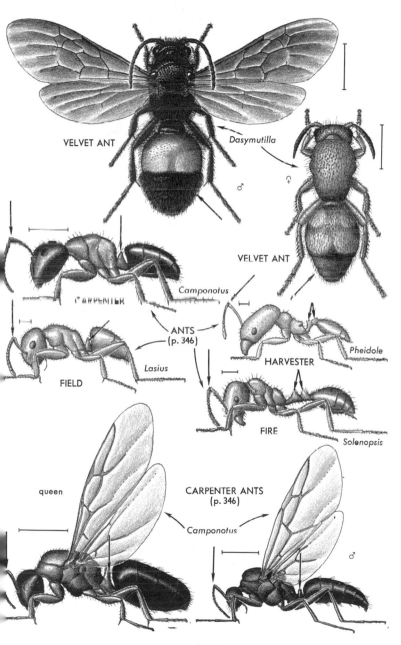

VELVET ANT

Dasymutilla ♂

♀

VELVET ANT

CARPENTER

Camponotus

ANTS
(p. 346)

Pheidole

HARVESTER

FIELD

Lasius

FIRE

Solenopsis

queen

CARPENTER ANTS
(p. 346)

Camponotus

♂

345

secretion from the anus. Seven subfamilies occur in the U.S., but about ⅚ of the species belong to 2 subfamilies — the Myrmicinae (abdominal stalk *2-segmented*) and Formicinae (abdominal stalk *1-segmented*). Myrmicinae include (see illus., p. 345) **fire ants, harvester ants,** and others; females sting. Formicinae include (see illus., p. 345) **carpenter ants, mound-building ants, field ants,** and others; females do not sting.

Vespoid Wasps: Superfamily Vespoidea

Pronotum in lateral view more or less triangular, extending to tegulae or nearly so. Antennae usually threadlike, 12-segmented in ♀ and 13-segmented in ♂. Trochanters 1-segmented. Ovipositor apical, and functioning as a sting. Venation normal.

SPIDER WASPS Family Pompilidae   **See also Pl. 16**
    Identification: Long-legged, hind femora usually extending to apex of abdomen. Mesopleura with *a transverse suture*. 1st discoidal cell in FW *not particularly long*. Wings not folded longitudinally at rest.
    Spider wasps are usually dark-colored, and many have dark wings; most are 10–25 mm., but some western species are 40 mm. or more. These wasps are quite common, and are usually found on flowers or on the ground in search of prey. Prey consists of spiders, which are generally placed in a specially constructed cell (in some species made of mud) or burrow in the ground; sometimes the spider is merely stung by the wasp and an egg laid on it and left in its own burrow. Spider wasps inflict a very painful sting.

VESPID WASPS Family Vespidae   **See also Pl. 16**
    Identification: First discoidal cell in FW *very long*, half as long as wing or nearly so. Wings folded longitudinally at rest.
    This is a large group that contains many very common wasps. Five of the 7 N. American subfamilies are widely distributed and the other 2 are restricted to the West. Members of the Vespinae, Polistinae, and Polybiinae are social wasps; the young are fed by the adults with chewed-up insects of various sorts. Because these wasps make nests of a papery material they are sometimes called paper wasps. Other subfamilies of Vespidae contain solitary wasps, which make nests of various types; their nests are usually provisioned with caterpillars.
    Subfamily Masarinae. Two submarginal cells (other subfamilies have 3 submarginal cells). Antennae *clubbed*. 10–20 mm. These wasps occur in the West. Their nests are made of mud or sand and attached to rocks or twigs, and are provisioned with pollen and nectar.
    Subfamily Euparagiinae. Second discoidal cell in FW *4-sided*.

Jugal lobe in HW *half as long* as submedian cell or longer. 6–7 mm. This is a small and rare group occurring in the Southwest. They nest in the ground and at least some species provision the nest with weevil larvae.

Potter Wasps, Subfamily Eumeninae. Most are 10–20 mm. and black with yellow or white markings. Middle tibiae with *1 apical spur.* Mandibles *elongate*, knifelike. This is a large and widely distributed group, and many species are very common.

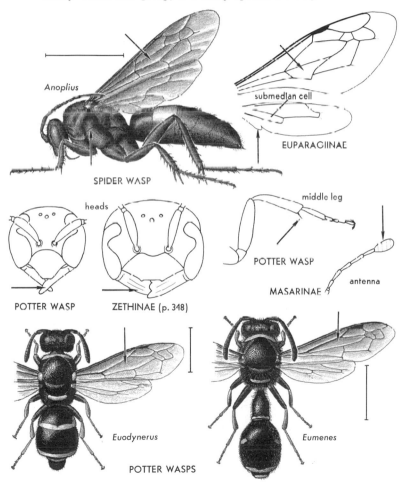

Anoplius

submedian cell

EUPARAGIINAE

SPIDER WASP

heads

middle leg

POTTER WASP   ZETHINAE (p. 348)

POTTER WASP

antenna

MASARINAE

Euodynerus   Eumenes

POTTER WASPS

Some species make nests of mud, others nest in burrows or in natural cavities. Nests are usually provisioned with caterpillars.

Subfamily Zethinae (see illus., p. 347). Similar to Eumeninae but mandibles *short and broad*. About 1 in., black, thorax narrowed in front of wings and narrower than head, and abdomen stalked. Zethinae occur in the South and are not common.

Yellowjackets and Hornets, Subfamily Vespinae. Middle tibiae with 2 apical spurs. HW *without a jugal lobe*. Clypeus *broadly truncate* and *slightly notched at apex*. Yellowjackets (see also Pl. 16) have the abdomen banded with black and yellow; hornets are largely black, with yellowish-white markings on the face, thorax, and end of abdomen. These wasps build nests of a papery material and the tiers of cells are surrounded by an outer covering; some species nest in the ground and others nest above ground in various protected situations. The nest is begun by the queen (the only individual to overwinter). Her nest is an inch or two in diameter with a single tier of cells; as workers are produced the nest is enlarged, and by the end of the summer the nest may contain several tiers of cells and be several inches to a foot in diameter. Females of these wasps inflict a very painful sting.

Paper Wasps, Subfamily Polistinae. Somewhat brownish and *long-legged*. Middle tibiae with 2 apical spurs. HW with *a small jugal lobe*. Clypeus usually *pointed at apex*. 1st abdominal segment *conical*, not stalklike. These wasps are common and widely distributed. The nest consists of a single more or less circular tier of cells, attached by a short stalk to the underside of some surface (eaves of a building, ceiling of a porch, or similar surface); there is no outer covering as in nests of Vespinae.

Subfamily Polybiinae. Similar to Polistinae but 1st abdominal segment *slender and stalklike*. These wasps occur in the Gulf states and in the West, and most make nests similar to those of the Polistinae.

Sphecoid Wasps: Superfamily Sphecoidea

Pronotum usually short and collarlike, with a small rounded lobe on each side that does not reach the tegulae. Venation complete or nearly so. Body hairs simple. 1st segment of hind tarsi slender. Bees (Apoidea, p. 354) have a similarly shaped pronotum, but have most body hairs branched, and usually have the 1st segment of the hind tarsi enlarged and flattened. Sphecoid wasps are solitary, and adults are usually found on flowers; they nest in the ground, in natural cavities, or make mud nests, and each provisions its nest with characteristic prey.

AMPULICID WASPS Family Ampulicidae
Identification: Black and 10–15 mm. Prothorax *narrowed*

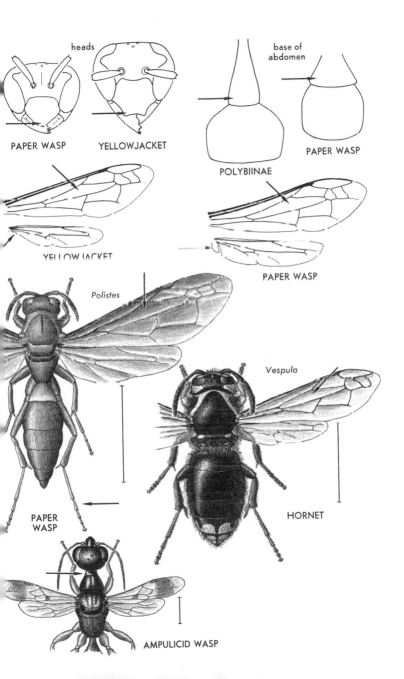

heads

PAPER WASP   YELLOWJACKET

base of abdomen

POLYBIINAE   PAPER WASP

YELLOWJACKET

PAPER WASP

Polistes

PAPER WASP

Vespula

HORNET

AMPULICID WASP

toward the front and somewhat *necklike*. Mesosternum with a forked process posteriorly. Parapsidal sutures distinct.

Ampulicids are rather rare. Nests are built in various protected situations, and are provisioned with immature cockroaches.

SPHECID WASPS Family Sphecidae **See also Pl. 16**
Identification: Pronotum short and collarlike. Mesosternum without a forked process posteriorly. Parapsidal sutures indistinct or absent.

This large family contains many common wasps. There are 9 subfamilies, differing in appearance and habits, in N. America.

Astatine Wasps, Subfamily Astatinae. Dark-colored, mostly about 15 mm. Eyes large, often meeting dorsally in male. Middle tibiae with 2 apical spurs. HW with vannal lobe large and jugal lobe small. Marginal cell in FW *truncate apically*, marginal vein continuing a short way beyond end of cell. 3 submarginal cells in FW. Astatinae nest in the ground and provision their nests with bugs; they are not common.

Sand-loving Wasps, Subfamily Larrinae. Most species are brownish and 10–20 mm. Middle tibiae with *1 apical spur*. FW with 3 submarginal cells, 3rd often *oblique*, and marginal vein usually continuing a way beyond tip of marginal cell. Mandibles generally notched on outer margin. Lateral ocelli often distorted, not round. These wasps nest in sandy areas and provision their nests with grasshoppers or crickets.

Organ-pipe Mud-daubers, Subfamily Trypoxyloninae. Slender, black. FW with *1 submarginal cell*, and marginal cell *pointed apically*. Inner margins of eyes *notched*. Some make tubular nests of mud, and others nest in natural cavities; all provision their nests with spiders.

Aphid Wasps, Subfamily Pemphredoninae. Usually black, 8–15 mm. Middle tibiae with 1 apical spur. Marginal cell in FW *pointed apically*. Abdomen sometimes with a slender basal stalk. Pemphredonini have 2 (rarely 1) submarginal cells in the front wing, and the antennae rise *very low on the face;* some nest in the ground, others in twigs, and they provision with aphids or thrips. Psenini have 3 submarginal cells in the front wing, and antennae rise *near middle of the face;* they nest in natural cavities or in the ground, and provision their nests with hoppers (Homoptera).

Thread-waisted Wasps, Subfamily Sphecinae. Abdomen *stalked at base.* HW with *a large vannal lobe.* Middle tibiae with 2 apical spurs. Mostly 20–30 mm. Common wasps, grouped in 3 tribes. (1) Ammobiini: blackish or brownish, wings clear or dark. 2 or more basal teeth on front tarsal claws, *only 1 recurrent vein* meets 2nd submarginal cell, and discoidal vein in HW rises at anterior end of transverse median vein. They nest in

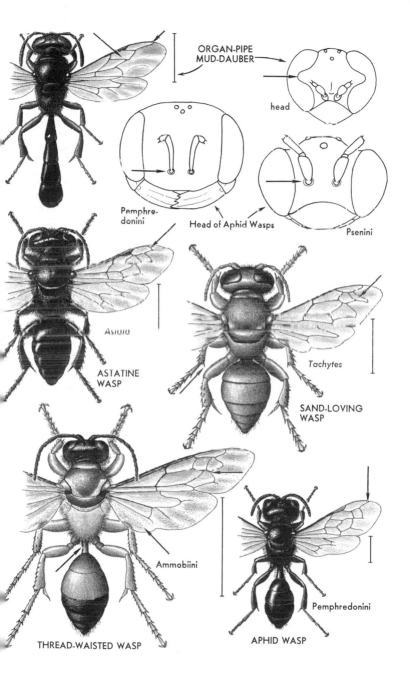

ORGAN-PIPE
MUD-DAUBER

head

Pemphre-
donini

Head of Aphid Wasps

Psenini

Astata

ASTATINE
WASP

Tachytes

SAND-LOVING
WASP

Ammobiini

THREAD-WAISTED WASP

Pemphredonini

APHID WASP

the ground and provision with grasshoppers. (2) Sphecini (see illus., p. 353): most are slender, and black with base of abdomen orange (*Sphex*, Pl. 16); *no teeth* on front tarsal claws, *both recurrent veins* meet 2nd submarginal cell, discoidal vein in HW rises *distinctly beyond anterior end* of transverse median vein, and abdominal petiole usually is 2-segmented; they nest in the ground and provision with caterpillars. (3) Sceliphronini: *1 tooth* on front tarsal claws, *both recurrent veins* meet 2nd submarginal cell (except *Chlorion*, blackish with blackish wings), and HW venation as in Ammobiini; most are mud-daubers, making nests of mud and provisioning them with spiders; some of these (*Sceliphron*, Pl. 16) are brownish with yellow markings and clear wings, others (*Chalybion*) are blue-black with blackish wings.

Subfamily Nyssoninae. FW with 3 submarginal cells. 1 or 2 (usually 2) apical spurs on middle tibiae. Rather diverse in appearance and habits. Three of the 5 tribes in this subfamily are fairly common. (1) Gorytini (2 apical spurs on middle tibiae, 2nd submarginal cell *squarish*, propodeum rounded, thorax smooth) are mostly 10–15 mm., and black with yellow markings. The largest species in this tribe is the Cicada Killer, *Sphecius speciosus* (Drury), which is about 30 mm.; it nests in the ground and provisions its nest with cicadas. Most Gorytini nest in the ground, and provision their nests with various Homoptera. (2) Bembicini, or sand wasps (see also Pl. 16), are mostly 20–25 mm., and often have pale greenish markings; labrum is *relatively long and triangular*. Sand wasps nest in sandy areas, usually in colonies; their nests are sometimes not completely provisioned, and the young are fed as they grow. (3) Nyssonini (2nd submarginal cell *triangular*, thorax coarsely punctate, and propodeum angled or spined) are inquilines, laying their eggs in nests of other wasps. The remaining tribes in this subfamily (not illus.), which are much less common, are the (4) Stizini (basal vein far toward wing base from stigma) and (5) Alyssonini (2nd submarginal cell triangular, and apex of hind femur produced into a process extending over base of tibia).

Subfamily Mellininae (not illus.). Similar to Nyssoninae, but no recurrent veins meet 2nd submarginal cell. Nest in the ground and provision with flies. A small group, widely distributed but uncommon.

Subfamily Philanthinae. Medium-sized, mostly 12–18 mm., black with yellow markings. 1 apical spur on middle tibiae. 3 submarginal cells. A constriction between 1st and 2nd abdominal segments. Nest in ground. Philanthini have 2nd submarginal cell *squarish* and clypeus *extended upward;* most provision their nest with bees. Cercerini have 2nd submarginal cell *triangular*, and provision their nests with various types of beetles (chiefly snout beetles).

Subfamily Crabroninae. This group includes 2 tribes of ground-nesting wasps, the Crabronini (square-headed wasps) and Oxybelini (spiny digger wasps); Crabronini are quite common but Oxybelini are rather rare. Most Crabronini are black with yellow markings, and 8–20 mm.; head and eyes are large, antennae rise *very low on face*, there is *only 1 submarginal cell* in the front wing, and apex of marginal cell is *truncate*. Different species use different types of prey in provisioning their nests.

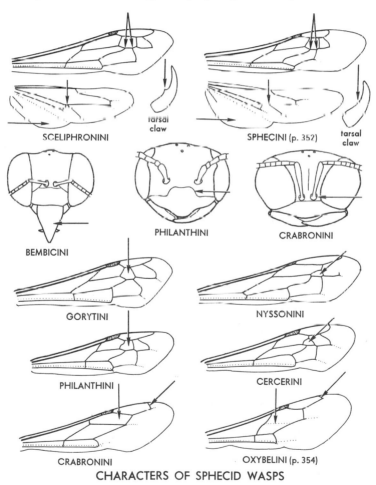

CHARACTERS OF SPHECID WASPS

Oxybelini are 10 mm. or less, stout-bodied, and dark-colored; the propodeum bears a long spine, or forked process, and the base of the cubital vein in the front wing is *weak* or absent; they nest in sandy areas and provision their nests with flies.

Bees: Superfamily Apoidea

Pronotum short, collarlike, with *a rounded lobe on each side that does not reach tegulae.* Body usually quite hairy, the body hairs branched or plumose. 1st segment of hind tarsi generally *enlarged and flattened.*

This is a large group, with more than 3300 N. American species; its members are to be found almost everywhere, particularly on flowers. Most species are solitary, nesting in the ground or in various natural cavities; the bumble bees and the Honey Bee are social. The cuckoo bees make no nest but lay their eggs in nests of other bees.

Bees are very valuable insects, largely because of the role they play in the pollination of plants; they are the most important insect pollinators. Insect-pollinated plants include most of our fruits, many vegetables, and important field crops such as clover, cotton, and tobacco.

Most pollen-collecting bees carry the pollen on their hind tibiae; the pollen-carrying surface of the tibia is usually bare and shiny and bordered with long hairs. Pollen sticks to the bee's body hairs when the bee visits a flower and is periodically combed off and placed on the hind tibiae. The 1st segment of the hind tarsi in most pollen-collecting bees is enlarged and flattened and bears a brush of hairs. Some bees do not have the hind legs so modified, and the leafcutting bees (Megachilidae) carry pollen on a brush of hairs on the ventral side of the abdomen.

The groups of bees are distinguished chiefly by characters of wings and tongue; the tongue characters are sometimes difficult to see, because the tongue when not in use is folded up tight against the ventral side of the head. The parts of the tongue in a large carpenter bee (*Xylocopa*) are shown opposite. To facilitate identification, the families of bees may be arranged in 3 groups:

1. Jugal lobe of HW as long as or longer than submedian cell; tongue (galeae and glossa) short; segments of labial palps usually similar and cylindrical; maxillary palps well developed: Colletidae, Andrenidae, and Halictidae.
2. Jugal lobe of HW shorter than submedian cell; tongue (galeae and glossa), labial palps, and maxillary palps as in Group 1: Melittidae (rare bees).
3. Jugal lobe of HW shorter than submedian cell, or lacking; tongue (galeae and glossa) long and usually slender; first 2 segments of labial palps long and flattened; maxillary palps well developed or vestigial: Megachilidae and Apidae.

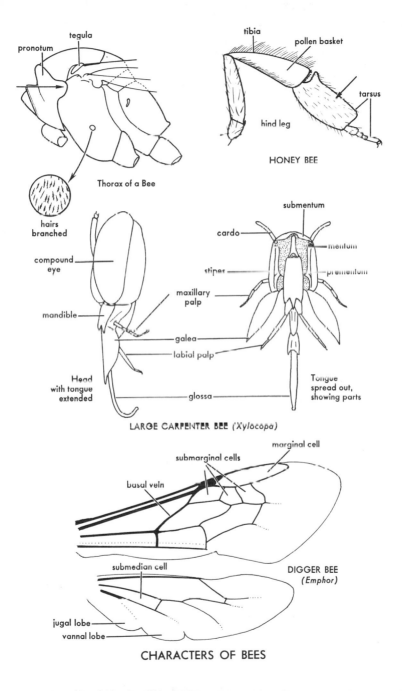

pronotum

tegula

Thorax of a Bee

hairs
branched

tibia

pollen basket

tarsus

hind leg

HONEY BEE

compound
eye

mandible

Head
with tongue
extended

submentum

cardo

mentum

stipes

prem- ntum

maxillary
palp

galea

labial palp

glossa

Tongue
spread out,
showing parts

LARGE CARPENTER BEE (*Xylocopa*)

marginal cell

submarginal cells

basal vein

submedian cell

DIGGER BEE
(*Emphor*)

jugal lobe

vannal lobe

CHARACTERS OF BEES

YELLOW-FACED and PLASTERER BEES   Family Colletidae
Identification: Jugal lobe in HW *longer* than submedian cell.
Basal vein straight or nearly so. *2* (Hylaeinae) *or 3* (Colletinae)
submarginal cells. 1 subantennal suture below each antennal
socket. Glossa short, bilobed or truncate.

Yellow-faced Bees, Subfamily Hylaeinae. Slender black bees,
10 mm. or less, usually with *whitish or yellowish areas* on face.
2 submarginal cells. Body relatively bare. These bees lack a
pollen-collecting apparatus on the hind legs, and resemble small
wasps; they can be recognized as bees by the branched body
hairs. They are quite common, and nest in the ground or in
various natural cavities. Our species belong to the genus
Hylaeus.

Plasterer Bees, Subfamily Colletinae. More robust and hairy
than Hylaeinae. Brownish, the abdomen banded with pale
hairs. *3 submarginal cells.* 2nd recurrent vein somewhat
S-shaped. These bees are much less common than the Hylaeinae.
They nest in the ground and line their galleries with a thin
transparent film (hence the common name).

ANDRENID BEES   Family Andrenidae
Identification: Jugal lobe in HW *longer* than submedian cell.
2 subantennal sutures below each antennal socket. Basal vein
straight or nearly so. 2 or 3 submarginal cells. Glossa short but
pointed. Nest in burrows in the ground, often in colonies,
usually in areas of sparse vegetation.

Subfamily Andreninae. Most are dark brown to brownish
black, 20 mm. or less. Marginal cell *pointed*, its apex on costal
margin of wing. Usually *3 submarginal cells.* This is a large and
widely distributed group, and many species are quite common.
Nests usually consist of a vertical tunnel in the ground, with
lateral tunnels branching off this vertical tunnel.

Subfamily Panurginae. Most are reddish brown, 10 mm. or
less. Marginal cell *truncate. 2 submarginal cells.* Stigma large.
These bees are much less common than the Andreninae but are
widely distributed.

Subfamily Oxaeinae (not illus.). Similar to Panurginae but
with 3 submarginal cells, and stigma very small. This small
group is restricted to the southwestern states.

HALICTID BEES   Family Halictidae   See also Pl. 16
Identification: Similar to Andrenidae but with *only 1 subantennal suture* below each antennal socket, and basal vein
strongly *arched.*

These resemble the andrenids in nesting habits: sometimes
large numbers nest close together, often so close that different
bees may use the same passageway to the outside. Halictids
vary from about 5 to 15 mm.; many are quite small. Most of

them are black or dark-colored, but some are partly or entirely brownish or metallic green. Some of the smaller halictids are attracted to perspiration, and are called sweat bees. The family contains 3 subfamilies: most of our species belong to the Halictinae (3 submarginal cells, 1st longer than 3rd); Nomiinae have

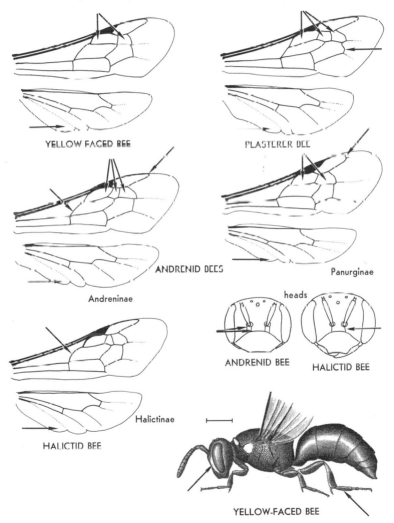

YELLOW FACED BEE

PLASTERER BEE

ANDRENID BEES

Panurginae

Andreninae

heads

ANDRENID BEE

HALICTID BEE

Halictinae

HALICTID BEE

YELLOW-FACED BEE

3 submarginal cells, with 1st about same size as 3rd; and Du-foureinae have only 2 submarginal cells.

MELITTID BEES Family Melittidae **Not illus.**
Identification: By the characters of Group 2 (p. 354).
Melittids are small, dark-colored, and rather rare; their nesting habits are similar to those of the Andrenidae.

LEAFCUTTING BEES Family Megachilidae
Identification: Most are stout-bodied, dark-colored, and 10–20 mm. Jugal lobe of HW *shorter* than submedian cell. FW with *2 nearly equal-sized submarginal cells.* Pollen carried on *underside* of abdomen (except in inquiline species). Subantennal sutures rise on *outer margins* of antennal sockets. First 2 segments of labial palps elongate and flattened. Maxillary palps vestigial. Glossa long and slender.
Many leafcutting bees are very common. They usually nest in the ground or in some natural cavity, with cell partitions of mud, resin, or leaf pulp. A few are inquilines.

DIGGER BEES, CARPENTER BEES, BUMBLE BEES, HONEY BEES, and Others. Family Apidae **See also Pl. 16**
Identification: Jugal lobe in HW *shorter than submedian cell* (rarely absent). Usually 3 submarginal cells. First 2 segments of labial palps elongate and flattened. Maxillary palps well developed or vestigial. Glossa long and slender.
This is a large group that varies in size, appearance, and habits. The family contains 3 subfamilies, Anthophorinae, Xylocopinae, and Apinae. Apinae include bumble bees and the Honey Bee, which are social; bees in the other 2 subfamilies are solitary. Each subfamily is further divided into tribes. Only more common groups in the family are mentioned here.
Digger Bees (Subfamily Anthophorinae, chiefly the tribes Anthophorini, Eucerini, and Emphorini). Robust and hairy, usually brownish, mostly 10–20 mm.; they nest in the ground. These differ from bumble bees and large carpenter bees in having 2nd submarginal cell *shorter than the 1st* (along posterior side); they differ from cuckoo bees in the Anthophorinae and from small carpenter bees in being much more hairy; and differ from the Honey Bee in having eyes bare and in having 2 *apical spurs* on the hind tibiae.
Cuckoo Bees (Subfamily Anthophorinae, chiefly the tribes Nomadini and Epeolini). Do not construct a nest but lay their eggs in nests of other bees. They are relatively bare, and wasp-like in appearance; hind legs *do not have* a pollen-collecting apparatus. Nomadini are reddish or brownish, about 8–10 mm., and have *a very small rounded jugal lobe* in hind wing. Epeolini are larger, usually black, with whitish or yellowish markings.

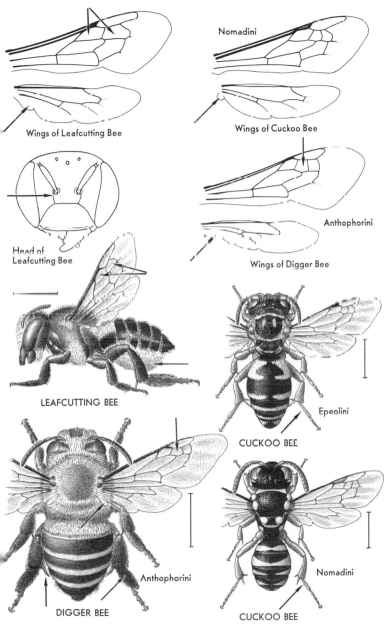

Nomadini

Wings of Leafcutting Bee

Wings of Cuckoo Bee

Anthophorini

Head of
Leafcutting Bee

Wings of Digger Bee

LEAFCUTTING BEE

Epeolini

CUCKOO BEE

DIGGER BEE

Anthophorini

CUCKOO BEE

Nomadini

359

Large Carpenter Bees (*Xylocopa*, Subfamily Xylocopinae); see also Pl. 16. Large, robust, and blackish; resemble bumble bees. They differ from bumble bees in having 2nd submarginal cell *triangular*, dorsal surface of abdomen bare and shining, mandibles immediately below the compound eyes, and in having a *small rounded jugal lobe* in hind wing. Most of them are about 1 in. They nest in cavities excavated in wood (sometimes in buildings).

Small Carpenter Bees (*Ceratina*, Subfamily Xylocopinae). These are *6–10 mm.*, relatively robust, not very hairy, and dark bluish green. Basal vein is *distinctly arched* and they may be confused with halictids, but the *much shorter jugal lobe* in hind wing distinguishes them. Small carpenter bees nest in galleries that they excavate in the pith of stems of various bushes.

Bumble Bees (Subfamily Apinae, Tribe Bombini); see also Pl. 16. Common and well-known insects; robust, hairy, generally 15–25 mm., and black with yellow (rarely orange) markings. 2nd submarginal cell is *more or less rectangular* and about as long as the 1st, dorsal surface of abdomen is hairy, there is a distinct space between base of the compound eye and base of the mandible (most other bees have mandibles attached very close to the eyes), and hind wings *lack a jugal lobe*. Most bumble bees nest in or on the ground, often in a deserted mouse nest. *Psithyrus* species are inquilines, laying their eggs in nests of other bumble bees. Bumble bees are social, and their colonies contain 3 castes: queens, drones (males), and workers (*Psithyrus* has no worker caste). Colonies are generally annual, the queens overwintering and starting new colonies in the spring. Queens are usually much larger than workers and drones. Bees in the genus *Psithyrus* do not collect pollen, and their hind tibiae are rounded, dull, and hairy; other bumble bees (which do collect pollen) have the hind tibiae bare, smooth, and shiny.

Honey Bee (*Apis mellifera* Linn., Subfamily Apinae, Tribe Apini). Only 1 species occurs in N. America, though there are several strains or races that differ slightly in color and other characters. Honey Bees are very common, widely distributed, and well-known insects. They differ from other bees in having the eyes *hairy*, *no apical spurs* on the hind tibiae, and they have a characteristic venation (marginal cell in front wing *narrow and parallel-sided*, 3rd submarginal cell *oblique*). Most Honey Bees nest in man-made hives; escaped swarms usually nest in hollow trees. Colonies contain 3 castes: workers (the most abundant individuals and ones most often seen), drones (a little larger, with eyes meeting dorsally), and the queen (abdomen longer than in workers). Honey Bees are extremely valuable insects, not only because of the honey and beeswax they produce, but because of their pollinating activities; their pollinating services are 15 to 20 times as valuable as their honey and wax.

It is often possible to increase greatly the yields of such crops as orchard fruits and clover seed by introducing hives of Honey Bees into orchards or clover fields when the crop is in bloom. The normal yield of red clover seed, for example (about 1 bushel per acre), can be increased to 4 or more bushels per acre with a dense Honey Bee population in the clover fields.

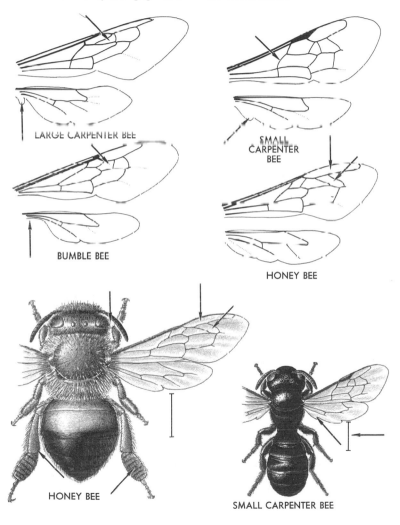

LARGE CARPENTER BEE

SMALL CARPENTER BEE

BUMBLE BEE

HONEY BEE

HONEY BEE

SMALL CARPENTER BEE

Honey Bees have an interesting "language": a worker that discovers a flower with a good flow of nectar can come back to the hive and "tell" the other workers what direction the flower is from the hive, how far away it is, and what kind of flower it is. Information on the direction and distance of the flower from the hive is communicated by means of a peculiar dance performed by the bee inside the hive. Information on the kind of flower involved is communicated by odor of the flower on the body of the bee or in its honey.

Glossary

THE DEFINITIONS given below are specific to use in this book; elsewhere some terms may have additional meanings. Page numbers refer to illustrations except when preceded by *see;* here the references are to definitions given in the text or to text in conjunction with illustrations shown opposite. Terms consisting of two or more words are generally listed under the most significant word.

Abdomen. The hindmost of the 3 main body divisions.

Acrostichal bristles. Two longitudinal rows of bristles along midline of mesonotum (Diptera, p. 287).

Anal. The posterior basal part (e.g., of wing); pertaining to last abdominal segment (which bears the anus). *Anal lobe,* a lobe in posterior part of wing. *Anal loop,* a group of cells in basal part of hind wing (Odonata, p. 73).

Annulate. With ringlike subdivisions (annulate 3rd antennal segment of certain Diptera, p. 273).

Anteapical. Just before the apex.

Antenna (pl., **antennae**). Feelerlike appendages located on head above mouth parts (p. 30). *Antennal club,* the enlarged terminal segments of a clubbed antenna. *Antennal scrobe,* a groove in the beak into which the base of the antenna fits (snout beetles). *Elbowed antennae,* with 1st segment elongated and the remaining segments coming off 1st at an angle (as in ants, p. 345). *Flabellate antennae,* with tonguelike processes on the terminal segments (as in Coleoptera). *Lamellate antennae,* with terminal segments expanded to one side and flattened or platelike.

Antenodal. Between base of wing and nodus (as in Odonata).

Antepygidial. Just in front of last abdominal segment (pygidium).

Anterior. Front; in front of.

Anterolateral. In front and to one side.

Anus. The posterior opening of the alimentary tract.

Apical. At the end, tip, or outermost part.

Arculus. A basal cross vein between R and Cu (Odonata, p. 71).

Areole. *See* Basal areole.

Arista. A large bristle on dorsal side of 3rd antennal segment (Diptera, p. 273). *Aristate,* with an arista.

Atrophied. Rudimentary, reduced in size.

Attenuated. Slender, gradually tapering toward the end.

Axilla (pl., **axillae**). A small sclerite on dorsal side of thorax, usually anterolateral to scutellum (Hymenoptera, p. 325).

Basal. At the base, or point of attachment. *Basad,* toward the base. *Basal areole*, the cell at base of hind wing between Sc and R (Lepidoptera, p. 239).

Beak. The proboscis, or protruding mouth part structures of a sucking insect (p. 32).

Bucca (pl., **buccae**). An area on head below compound eye (Diptera, p. 283).

Calypter. One of 2 small lobes at base of wing on posterior side (Diptera, p. 261).

Carapace. A hard dorsal covering formed by fusion of certain sclerites, in Crustacea.

Carina (pl., **carinae**). A keel or ridge.

Caudal. Pertaining to the tail or posterior part of body.

Cell. A space in the wing partly or completely surrounded by veins. *Accessory cell*, a closed cell in front wing of Lepidoptera, on anterior side of discal cell (p. 221). *Anal cell*, one in anal area of wing (Diptera, p. 261). *Apical cell*, 1 or more cells near wing tip (Hymenoptera, p. 313, *AP*). *Basal anal cell*, an anal cell near base of wing (Plecoptera, p. 95). *Basal cells*, the R and M cells in Diptera (p. 261); cells *MD*, *SMD*, and *L* in hind wing of Hymenoptera (p. 313). *Closed cell*, one bordered by veins and not extending to wing margin. *Discal cell*, one near central or basal part of wing (Diptera, p. 261, and Lepidoptera, p. 221). *Discoidal cell*, one near middle of wing (Hymenoptera, p. 313, *1D, 2D, 3D*). *Lanceolate cell*, one in anal area of wing (Hymenoptera, p. 313, *L*). *Marginal cell*, one bordering front margin near tip of wing (Diptera, p. 261, and Hymenoptera, p. 313, *MC*). *Median cell*, one in basal portion of wing (Hymenoptera, p. 313, *MD*). *Open cell*, one extending to wing margin. *Posterior cells*, those bordering rear edge of wing between R and Cu_2 (Diptera, p. 261). *Submarginal cell*, 1 or more cells just behind marginal cell (Diptera, p. 261, and Hymenoptera, p. 313, *SM*). *Submedian cell*, the cell in basal part of wing just behind median cell (Hymenoptera, p. 313, *SMD*).

Cephalothorax. A body region consisting of head and thoracic segments, in Crustacea and Chelicerata.

Cercus (pl., **cerci**). One of a pair of dorsally located appendages at posterior end of abdomen (p. 30).

Chelicera (pl., **chelicerae**). The anterior, usually fanglike, pair of appendages in arachnids.

Chrysalis (pl., **chrysalids** or **chrysalides**). The pupa of a butterfly (*see* p. 219).

Claval suture. The suture in the front wing of Hemiptera and Homoptera between the clavus and corium (p. 113).

Clavate. Clubbed, or enlarged toward tip.

Clavus. The portion of the front wing in Hemiptera and Homoptera behind the claval suture (p. 113).

Clubbed. With the tip enlarged or swollen.

Clypeus. A sclerite on face between frons and labrum (p. 30).

Cocoon. A case of silk in which the pupa is formed.

Collophore. A small tubular structure on ventral side of 1st abdominal segment, in Collembola.

Compound eye. An eye composed of many individual elements, each of which is marked externally by a facet; the facets are usually more or less hexagonal in shape (p. 30, *eye*).

Compressed. Flattened from side to side.

Corium. The basal, usually thickened, part of front wing (Hemiptera, p. 113).

Cornicle. One of a pair of elongate processes located dorsally near apex of abdomen (aphids, p. 135).

Costa. A longitudinal vein, usually forming front margin of wing (p. 35). *Costal area of wing,* that part just behind front margin. *Costal break,* a point on costa where the vein appears broken or weakened. *Costal margin of wing,* the front margin.

Coxa. The basal leg segment (p. 34). *Closed coxal cavities* (front, Coleoptera), bounded posteriorly by a prothoracic sclerite. *Open coxal cavities* (front, Coleoptera), bounded posteriorly by a mesothoracic sclerite.

Crochets. Tiny hooks on the prolegs of a caterpillar.

Cross vein. A vein connecting adjacent longitudinal veins. *Antenodal cross veins,* those just behind front edge of wing between base and nodus, extending from C to R (Odonata, p. 71). *Discal cross vein,* one just behind discal cell (Diptera, p. 261). *Humeral cross vein,* one in basal part of wing between C and Sc (p. 35, *h*). *Medial cross vein,* one connecting 2 branches of M (p. 35, *m*). *Mediocubital cross vein,* one connecting M and Cu (p. 35, *m-cu*). *Radial cross vein,* one just behind R_1 (p. 35, *r*). *Sectorial cross vein,* one connecting 2 branches of Rs (p. 35, *s*).

Cubital intercalaries. Longitudinal veins in distal part of wing between Cu_1 and Cu_2 (Ephemeroptera, p. 67).

Cubitus. The longitudinal vein just behind M (p. 35).

Cuneus. A more or less triangular apical piece of the corium, set off from rest of corium (Hemiptera, p. 113).

Deflexed. Bent downward.

Dentate. Toothed, or with toothlike projections.

Denticulate. With tiny toothlike projections.

Depressed. Flattened (from top to bottom).

Distal. Near or pertaining to that part farthest from body. *Distad,* toward the end farthest from body or base.

Dorsal. Pertaining to the back or upper side; top or uppermost. *Dorsad,* toward the top or back.

Dorsocentral bristles. A longitudinal row of bristles on mesonotum, just toward side from acrostichal bristles (Diptera, p. 287).

Dorsolateral. Above and to one side.

Ecdysis (pl., **ecdyses**). Molting, or shedding of exoskeleton.

Ectoparasite. A parasite living on the outside of its host.

Elytron (pl., **elytra**). A thickened, horny, or leathery front wing, in Coleoptera, Dermaptera, and some Homoptera.

Emarginate. Notched.

Emergence. The process of the adult leaving the pupal case or the last nymphal skin.

Endoparasite. A parasite living inside its host.

Entire. With a smooth outline, without teeth or notches.

Epimeron (pl., **epimera**). The thoracic sclerite just behind a pleural suture (p. 34).

Epiproct. A process or appendage just above anus, appearing to rise from 10th abdominal segment (p. 30).

Episternum (pl., **episterna**). The thoracic sclerite just in front of a pleural suture (p. 34).

Epistomal suture. The suture on face between frons and clypeus (p. 31).

Exoskeleton. A skeleton, or supporting structure, on outside of body.

Eye cap. An enlarged basal segment of the antenna that overhangs or caps the compound eye (Microlepidoptera, p. 253, head of Lyonetiidae).

Femur (pl., **femora**). The leg segment between trochanter and tibia (p. 34).

File. A filelike ridge on ventral side of front wing, part of sound-producing mechanism in certain Orthoptera (p. 77).

Filiform. Hairlike or threadlike.

Fontanelle. *See* p. 89.

Frenulum. A bristle or group of bristles at humeral angle of hind wing (Lepidoptera, p. 221).

Frons. An area on face, between frontal and epistomal sutures and including median ocellus (p. 30). *Frontal suture*, one of 2 sutures on each side of frons (p. 30); a suture in the form of an inverted U or V, with its apex just above base of antennae (Diptera, p. 283). *Fronto-orbital bristles*, bristles on upper front part of head next to compound eyes (Diptera, p. 283).

Furcula. The forked "tail" of springtails (*see* p. 62).

Gena (pl., **genae**). The area on head below compound eye (p. 30). *Genal comb*, a row of strong spines on lower front border of head (Siphonaptera, p. 309).

Genitalia. *See* p. 36.

Glabrous. Smooth, without hairs.

Globose or globular. Spherical or nearly so.

Glossa (pl., **glossae**). One of a pair of lobes at apex of labium, between the paraglossae (bees, p. 355).

Grub. *See* scarabaeiform larvae, p. 39.

Gula. A sclerite on underside of head in Coleoptera. *Gular sutures*, longitudinal sutures on underside of head, bordering gula.

Haltere. A small knobbed structure on each side of metathorax, representing the hind wings (Diptera, p. 261).

Homonyms. *See* p. 45.

Honeydew. *See* p. 130.

Humeral. Pertaining to the shoulder; pertaining to front basal part of the wing. *Humeral angle*, the basal front angle of the wing. *Humeral bristles*, bristles on humeral callus (Diptera, p. 287). *Humeral callus*, a rounded area on outer front portions of thoracic notum (Diptera, p. 287). *Humeral vein*, a branch of Sc extending into humeral angle of wing (p. 225).

Hyaline. Transparent and colorless, like glass.

Hypermetamorphosis. *See* p. 41.

Hyperparasite. A parasite whose host is another parasite.

Hypopharynx. A median mouth-part structure just in front of labium.

Hypopleuron (pl., **hypopleura**). A sclerite on thorax just above hind coxae, in Diptera (p. 287). *Hypopleural bristles*, a row of bristles on hypopleuron (p. 287).

Inquiline. An animal that lives in nest of another species.

Instar. The stage of an insect between successive molts.

Integument. The outer covering of the body.

Interstitial. Of trochanters: distal margin of trochanter at a right angle to long axis of leg. Of 2 veins: the ends of the veins meeting.

Intra-alar bristles. A row of bristles on mesonotum above wing bases (Diptera, p. 287).

Jugal lobe. A lobe at base of hind wing, on rear side (Hymenoptera, p. 313, *jl*).

Jugum. A fingerlike lobe at base of front wing on rear side (Hepialidae, p. 259).

Labium. The lower lip, or hindmost mouth-part structure (p. 30).

Labrum. The upper lip, situated in front of mandibles and below clypeus (p. 30).

Lacinia (pl., **laciniae**). The jawlike inner lobe of a maxilla (p. 30).

Lanceolate (wings). Narrow and tapering to a point at tip.

Larva (pl., **larvae**). The immature stage, between egg and pupa of an insect with complete metamorphosis (*see* pp. 38–41).

Lateral. On or pertaining to the right or left side. *Laterad*, toward the side, away from the midline.

Maggot. A legless larva without a well-developed head.

Mandible. A jaw (p. 30).

Margined. With a sharp lateral edge.

Maxilla (pl., **maxillae**). One of the paired mouth parts just behind the mandibles (p. 30).

Media. The longitudinal vein between R and Cu (p. 35).

Membrane. The part of wing surface between veins; thin tip part of the wing (Hemiptera, p. 113). *Membranous*, thin and more or less transparent (wings); thin and not hardened (body wall).

Mes-, meso-. Prefixes for names of mesothoracic structures (*see* p. 33).

Mesal. On or near the midline of the body. *Mesad*, toward the midline of the body.

Met-, meta-. Prefixes for names of metathoracic structures (*see* p. 33).

Millimeter. 0.001 meter, or 0.03937 in. (about 1/25 in.).

Nasutus (pl., **nasuti**). A termite caste having the head narrowed in front into a snout (p. 91).

Nodus. A strong cross vein near middle of front border of wing (Odonata, p. 71).

Notopleuron (pl., **notopleura**). An area on dorsolateral surface of thorax, at end of transverse suture, in Diptera (p. 287). *Notopleural bristles*, a group of bristles on notopleuron (p. 287).

Notum (pl., **nota**). The dorsal surface of a thoracic segment.

Nymph. Young of an insect with simple metamorphosis.

Occiput. The upper surface of head between occipital and postoccipital sutures (p. 30). *Occipital suture*, a suture on hind part of head, between vertex and occiput dorsally and between genae and postgenae laterally (p. 30).

Ocellus (pl., **ocelli**). A simple eye (p. 30). *Ocellar bristles*, a pair of bristles near ocelli (Diptera, p. 283). *Ocellar triangle*, a slightly raised triangular area in which the ocelli are located (Diptera, p. 283).

Oral. Pertaining to the mouth. *Oral vibrissae*, a pair of stout bristles at lower edge of face (Diptera, p. 283).

Orbital plate. An area on head next to compound eye (Diptera, p. 283).

Oviparous. Egg-laying.

Oviposit. To lay eggs.

Ovipositor. The egg-laying apparatus (p. 30).

Paedogenesis. *See* p. 150.

Palp. A feelerlike structure borne by the maxillae or labium (p. 30).

Paraproct. One of a pair of lobes located below and on each side of the anus (p. 30).

Parapsidal sutures. A pair of longitudinal sutures on mesonotum (Hymenoptera, p. 327).

Parasite. An animal that lives in or on the body of another animal (its host), at least during part of its life cycle.

Parthenogenesis. *See* p. 38.

Patella. A leg segment between the femur and tibia (chelicerate arthropods).

Pectinate. Bearing processes like the teeth of a comb.

Pedipalps. The pair of appendages (usually feelerlike) just behind the chelicerae (in chelicerate arthropods).

Petiole. A stalk or stem; basal stalk of abdomen in Hymenoptera. *Petiolate*, attached by a stalk or stem. *Petiolate abdomen*, with the basal segment slender and cylindrical, as in Anacharitinae (p. 333).

Pleural. Pertaining to lateral areas of body.

Plumose. Feathery, or bearing many long hairs.

Polyembryony. *See* p. 38.

Posterior. Hind or rear.

Posthumeral bristle. A bristle on outer front surface of thorax, just behind humeral callus (Diptera, p. 287).

Postscutellum. An area just below or behind scutellum (Diptera, p. 287).

Postvertical bristles. A pair of bristles behind ocellar triangle (Diptera, p. 283).

Preapical. Located just before the apex.

Predaceous. Feeding on other animals that are usually smaller or less powerful.

Presutural bristles. Bristles just in front of lateral ends of transverse suture (Diptera, p. 287).

Pro-. A prefix for names of prothoracic structures (*see* p. 33).

Proboscis. The beak (which *see*).

Pronotal comb. A row of strong spines on rear margin of pronotum (Siphonaptera, p. 309).

Propleural bristles. Bristles on propleuron (Diptera, p. 287).

Propodeum. A dorsal area on thorax behind metanotum, actually the 1st abdominal segment, in apocritous Hymenoptera (p. 319).

Proximal. Near body, or the base of an appendage. *Proximad*, toward the base, or the portion nearest body.

Pteropleuron (pl., **pteropleura**). An area on side of thorax, just below wing base (Diptera, p. 287). *Pteropleural bristles*, a group of bristles on pteropleuron (p. 287).

Ptilinum. *See* p. 282.

Pubescent. Covered with short fine hairs.

Pulvillus (pl., **pulvilli**). A pad or lobe beneath each tarsal claw (as in Diptera).

Puncture. A tiny pit or depression.

Pupa (pl., **pupae**). The stage between larva and adult in insects with complete metamorphosis (*see* p. 39).

Pygidium. The last dorsal segment of the abdomen.

Quadrate. Four-sided; square or rectangular.

Radial sector. The posterior of the 2 main branches of the radius (p. 35, *Rs*).

Radius. The longitudinal vein between Sc and M (p. 35).

Recurved. Curved upward or backward.

Reticulate. With a network of ridges or lines.

Scape. The basal segment of an antenna.

Scapula (pl., **scapulae**). An area on mesonotum just toward side from parapsidal suture (Hymenoptera).

Sclerite. A hardened body wall plate, usually bordered by sutures or membranous areas. *Sclerotized*, hardened.

Scraper. The sharpened angle of front wing of a cricket or long-horned grasshopper, a part of the sound-producing mechanism (p. 77).

Scutellum. A dorsal thoracic sclerite (p. 34); in Coleoptera, Hemiptera, and Homoptera the mesoscutellum, a more or less triangular sclerite behind pronotum.

Segment. A subdivision of the body or an appendage, between joints or articulations.

Sensoria (sing., **sensorium**). *See* p. 111.

Sessile. Attached, and not capable of moving from place to place; attached without a stem (petiole).

Seta (pl., **setae**). A bristle. *Setaceous*, bristlelike. *Setate*, with bristles.

Sigmoid. S-shaped.

Simple. Unmodified; not forked, toothed, branched, or divided.

Species. *See* p. 43.

Spindle-shaped. Elongate, cylindrical, tapering at ends.

Spiracle. An external opening of the tracheal system (p. 30). *Spiracular bristle*, one adjacent to a spiracle (Diptera).

Spur. A spine that is usually movable.

Stalked. With a stalk or stemlike base; (of veins) fused together at base.

Sternopleuron (pl., **sternopleura**). A sclerite on side of thorax, just above middle leg (Diptera, p. 287). *Sternopleural bristles*, bristles on sternopleuron (p. 287).

Stigma. A dark spot formed by a thickening of the wing membrane, located in distal part of wing along front edge (Odonata, p. 71, and Hymenoptera, p. 313, *st*).

Striate. With narrow grooves or suturelike lines.

Stridulate. To produce a noise by rubbing 2 surfaces or structures together.

Style. A slender elongate process at apex of antennae (as in snipe flies, p. 275).

GLOSSARY

Stylus (pl., styli). A short, slender, fingerlike process.

Subantennal suture. A suture on face extending downward from base of antenna.

Subapical. Located just before apex.

Subcosta. The longitudinal vein between C and R (p. 35).

Subgenal suture. A horizontal suture on head below gena (p. 30).

Subgenital plate. A platelike structure underlying the genitalia.

Subimago. *See* p. 66.

Subspecies. *See* pp. 43–44.

Supra-alar bristles. A group of bristles on mesonotum just above wing base (Diptera, p. 287).

Suture. A linelike groove in the body wall.

Synonyms. *See* p. 45.

Tarsus (pl., tarsi). The part of the leg beyond the tibia, usually consisting of 2–5 subdivisions (p. 34). *Tarsal claw*, a claw at apex of tarsus.

Tegmen (pl., tegmina). The thickened front wing of an orthopteran.

Tegula (pl., tegulae). A scalelike structure overlying base of front wing (Hymenoptera, p. 319).

Tergum (pl., terga). The dorsal surface of an abdominal segment.

Terminal. At the end; at posterior end (of abdomen).

Thorax. The body region behind head which bears legs and wings.

Tibia (pl., tibiae). The leg segment between femur and tarsus (p. 34). *Tibial spur*, a large spur or spine on tibia, usually at apex of tibia.

Trachea (pl., tracheae). *See* p. 36.

Transverse. Across, or at right angles to longitudinal axis. *Transverse suture*, a suture across mesonotum (Diptera, p. 287).

Triangle. A triangular cell or group of cells in central basal part of wing (Odonata, pp. 71 and 73).

Trochanter. The small leg segment between coxa and femur (p. 34).

Trochantin. A small sclerite in thoracic wall adjacent to base of coxa (as in Coleoptera).

Truncate. Cut off square at end.

Tuberculate. With small rounded protuberances.

Vannal lobe. A lobe in anal area of hind wing, just before end of anal vein (Hymenoptera, p. 313, *vl*).

Vein. A thickened line in wing. *Accessory vein*, the hindmost vein in anal area of front wing (Hymenoptera, p. 313, *ac*). *Anal veins*, longitudinal veins behind Cu (p. 35). *Basal vein*, a more or less transverse vein near middle of front wing (Hymenoptera, p. 313, *bv*). *Brace vein*, a slanting cross vein behind basal end of stigma (Odonata, p. 71). *Humeral vein*, a branch of Sc extending into humeral angle of wing (as in Neuroptera and Lepidoptera). *Intercostal vein*, a longitudinal vein in costal cell (Hymenoptera,

p. 313, *ic*). *Marginal vein*, one on or just inside wing margin; the vein forming posterior side of marginal cell (Hymenoptera, (p. 313, *mv*). *Recurrent vein*, 1 of 2 transverse veins just behind cubital vein (Hymenoptera, p. 313, *rv*). *Spurious vein*, a veinlike thickening between R and M (Syrphidae, p. 281). *Subdiscal* (or *subdiscoidal*) *vein*, the vein along rear side of 3rd discoidal cell (Hymenoptera, p. 313, *sd*). *Submarginal vein*, one just behind front margin of wing (as in Chalcidoidea). *Transverse costal vein*, a cross vein in costal cell (Hymenoptera, p. 313, *tc*). *Transverse cubital vein*, a cross vein between marginal and cubital veins (Hymenoptera, p. 313, *tcb*). *Transverse median vein*, a cross vein between median or discoidal and anal veins (Hymenoptera, p. 313, *tm*).

Ventral. Lower or underneath; pertaining to the underside. *Ventrad,* toward the underside, downward.

Vertex. Top of head, between compound eyes and in front of occipital suture (p. 30).

Vestigial. Small, poorly developed, nonfunctional.

Viviparous. Giving birth to live young, not egg-laying.

References

THE FOLLOWING LIST is designed for those who seek informaion beyond that given in this book. Although not intended to be complete, it includes some of the more important references under each heading, and the bibliography in each publication will lead to additional literature.

GENERAL

Borror, Donald J., and Dwight M. DeLong. 1970 (3rd ed.). An introduction to the study of insects. New York: Holt, Rinehart and Winston.

Brues, Charles T., Axel L. Melander, and Frank M. Carpenter, 1954. Classification of insects. Bull. Mus. Comp. Zool., Harvard Univ., 73·vi + 1–917.

Jaques, Harry E. 1947 (2nd ed.). How to know the insects. Dubuque, Ia.: Wm. C. Brown Co.

Lutz, Frank E. 1935. Field book of insects. New York: Putnam.

Swain, Ralph B. 1948. The insect guide. New York: Doubleday.

SPECIFIC GROUPS

ARTHROPODS OTHER THAN INSECTS

Baker, Edward W., John H. Camin, Frederick Cunliffe, Tyler A. Woolley, and Conrad E. Yunker. 1958. Guide to the families of mites. Contrib. No. 3, Inst. Acarology, Univ. Maryland.

Comstock, John H., and Willis J. Gertsch. 1940. The spider book. New York: Doubleday.

Kaston, Benjamin J., and Elizabeth Kaston. 1953. How to know the spiders. Dubuque, Ia.: Wm. C. Brown Co.

Miner, Roy W. 1950. Field book of seashore life. New York: Putnam.

(*See also* references under "Aquatic Insects," below.)

AQUATIC INSECTS

Morgan, Ann H. 1930. Field book of ponds and streams. New York: Putnam.

Pennak, Robert W. 1953. Fresh-water invertebrates of the United States. New York: Ronald Press.

Usinger, Robert L. (ed.). 1956. Aquatic insects of California, with keys to North American genera and California species. Berkeley: Univ. California Press.

PROTURA, THYSANURA, AND COLLEMBOLA

Maynard, Elliott A. 1951. A monograph of the Collembola or springtail insects of New York State. Ithaca, N.Y.: Comstock.

Scott, Harold G. 1961. Collembola: pictorial keys to the nearctic genera. Annals Entomol. Soc. Amer., 54:104–113.

Tuxen, S. L. 1964. The Protura. A revision of the species of the world with keys for determination. Paris: Hermann.

EPHEMEROPTERA

Burks, Barnard D. 1953. The mayflies, or Ephemeroptera, of Illinois. Bull. Ill. Natural Hist. Surv., 26: 1–216.

Needham, James G., Jay R. Traver, and Tin-Chi Hsu. 1935. The biology of mayflies. Ithaca, N.Y.: Comstock.

(*See also* references under "Aquatic Insects," p. 373.)

ODONATA

Corbet, Philip S. 1963. A biology of dragonflies. Chicago: Quadrangle Books.

Needham, James G., and Hortense B. Heywood. 1929. A handbook of the dragonflies of North America. Springfield, Ill.: Charles C. Thomas.

Needham, James G., and Minter J. Westfall, Jr. 1955. A manual of the dragonflies of North America (Anisoptera). Los Angeles: Univ. California Press.

Walker, Edmund M. 1953–58. The Odonata of Canada and Alaska. Vol. 1 (1953): General; The Zygoptera — damselflies. Vol. 2 (1958): The Anisoptera — 4 families. Toronto: Univ. Toronto Press.

(*See also* references under "Aquatic Insects," p. 373.)

ORTHOPTERA AND DERMAPTERA

Alexander, Richard D., and Donald J. Borror. 1956. The songs of insects. Boston: Houghton Mifflin (since 1966). A 12-inch LP in the Sounds of Nature series.

Blatchley, Willis S. 1920. Orthoptera of northeastern America. Indianapolis, Ind.: Nature Publishing Co.

Hebard, Morgan. 1934. The Dermaptera and Orthoptera of Illinois. Bull. Ill. Natural Hist. Surv., 20:iv + 125–179.

ISOPTERA

Banks, Nathan, and Thomas E. Snyder. 1920. A revision of the nearctic termites. Bull. No. 108, U.S. Natl. Museum.

Snyder, Thomas E. 1954. Order Isoptera — the termites of the United States and Canada. New York: Natl. Pest Control Assn.

———. 1965. Our native termites. Smiths. Inst. Rept. for 1964, pp. 497–506.

PLECOPTERA

Frison, Theodore H. 1935. The stoneflies, or Plecoptera, of Illinois. Bull. Ill. Natural Hist. Surv., 20:281–471.

———. 1942. Studies of North American Plecoptera, with special reference to the fauna of Illinois. Bull. Ill. Natural Hist. Surv., 22:231–355.

Needham, James G., and Peter W. Claassen. 1925. A monograph of the Plecoptera or stoneflies of America north of Mexico. Publ. No. 2, Thomas Say Foundation.

(*See also* references under "Aquatic Insects," p. 373.)

EMBIOPTERA, ZORAPTERA, AND PSOCOPTERA

Gurney, Ashley B. 1938. A synopsis of the order Zoraptera, with notes on the biology of *Zorotypus hubbardi* Caudell. Proc. Entomol. Soc. Wash., 40:57–87.

Pearman, J. V. 1936. The taxonomy of the Psocoptera; preliminary sketch. Proc. Roy. Entomol. Soc. London, Ser. B, 5:58–62.

Ross, Edward S. 1944. A revision of the Embioptera, or web-spinners, of the New World. Proc. U.S. Natl. Museum, 94:401–504.

MALLOPHAGA AND ANOPLURA

Ewing, Henry E. 1929. A manual of external parasites. Springfield, Ill.: Charles C. Thomas.

Ferris, Gordon F. 1951. The sucking lice. Mem. No. 1, Pacific Coast Entomol. Soc.

THYSANOPTERA

Stannard, Lewis J., Jr. 1957. The phylogeny and classification of the North American genera of the suborder Tubulifera (Thysanoptera). Ill. Biol. Monog. No. 25.

———. 1968. The thrips, or Thysanoptera, of Illinois. Bull. Ill. Natural Hist. Surv., 29·vi + 215–552.

HEMIPTERA

Blatchley, Willis S. 1926. Heteroptera or true bugs of eastern North America, with special reference to the faunas of Indiana and Florida. Indianapolis, Ind.: Nature Publishing Co.

(*See also* references under "Aquatic Insects," p. 373.)

HOMOPTERA

DeLong, Dwight M. 1948. The leafhoppers, or Cicadellidae, of Illinois (Eurymelinae — Balcluthinae). Bull. Ill. Natural Hist. Surv., 24:91–376.

Ferris, Gordon F. 1937–53 (6 v.). Atlas of the scale insects of North America. Stanford Univ., Calif.: Stanford Univ. Press.

Hottes, Frederick C., and Theodore H. Frison. 1931. The plant lice, or Aphididae, of Illinois. Bull. Ill. Natural Hist. Surv., 19:121–447.

Metcalf, Zeno P. 1923. Fulgoridae of eastern North America. Jour. Elisha Mitchell Sci. Soc., 38:139–230.

NEUROPTERA

Carpenter, Frank M. 1940. A revision of nearctic Hemerobiidae, Berothidae, Sisyridae, Polystoechotidae, and Dilaridae (Neuroptera). Proc. Amer. Acad. Arts Sci., 74:193–280.

Parfin, Sophy I., and Ashley B. Gurney. 1956. The spongillaflies, with special reference to those of the western hemisphere (Sisyridae, Neuroptera). Proc. U.S. Natl. Museum, 105:421–529.

(*See also* references under "Aquatic Insects," p. 373.)

COLEOPTERA AND STREPSIPTERA

Arnett, Ross H., Jr. 1968. The beetles of the United States (a manual for identification). Ann Arbor, Mich.: The Amer. Entomol. Inst.

Blatchley, Willis S. 1910. An illustrated and descriptive catalogue of the Coleoptera or beetles (exclusive of the Rhynchophora) known to occur in Indiana. Indianapolis, Ind.: Nature Publishing Co.

——, and Charles W. Leng. 1916. Rhynchophora or weevils of northeastern North America. Indianapolis, Ind.: Nature Publishing Co.

Bohart, Richard M. 1941. A revision of the Strepsiptera with special reference to the species of North America. Calif. Univ. Pub. Entomol., 7:91–160.

Dillon, Elizabeth S., and Lawrence S. Dillon. 1961. A manual of common beetles of eastern North America. Evanston, Ill.: Row, Peterson & Co.

Edwards, J. Gordon. 1949. Coleoptera or beetles east of the Great Plains. Ann Arbor, Mich.: Edwards.

Jaques, Harry E. 1951. How to know the beetles. Dubuque, Ia.: Wm. C. Brown Co.

(*See also* references under "Aquatic Insects," p. 373.)

MECOPTERA AND TRICHOPTERA

Carpenter, Frank M. 1931. Revision of nearctic Mecoptera. Bull. Mus. Comp. Zool., Harvard Univ., 72:205–277.

Ross, Herbert H. 1944. The caddis flies or Trichoptera of Illinois. Bull. Ill. Natural Hist. Surv. 23:1–326.

LEPIDOPTERA

Forbes, William T. M. 1923–54. Lepidoptera of New York and neighboring states. Cornell Univ. Agric. Expt. Sta. Mem. 68 (1923), 274 (1948), and 329 (1954).

Holland, William J. 1949. The butterfly book. New York: Doubleday.

——. 1968. The moth book. New York: Dover.

Klots, Alexander B. 1951. A field guide to the butterflies. Boston: Houghton Mifflin.

DIPTERA

Curran, Charles H. 1965 (2nd rev. ed.). The families and genera of North American Diptera. Woodhaven, N.Y.: Henry Tripp.

Stone, Alan, et al. 1965. A catalogue of the Diptera of America north of Mexico. U.S.D.A. Agric. Handbook No. 276.

(*See also* references under "Aquatic Insects," p. 373.)

SIPHONAPTERA

Ewing, Henry E., and Irving Fox. 1943. The fleas of North America. U.S.D.A. Misc. Pub. No. 500.

Holland, George P. 1949. The Siphonaptera of Canada. Tech. Bull. No. 70, Canada Dept. Agric.

HYMENOPTERA

Muesebeck, Carl F. W., et al. 1951. Hymenoptera of America north of Mexico; synoptic catalogue. U.S.D.A. Agric. Monog. No. 2. First Supplement, 1958; Second Supplement, 1967, by Karl V. Krombein et al.

Index

THIS INDEX includes the names of insects and other animals and of insect-borne diseases mentioned in this *Field Guide;* the location of accounts of various subjects may be found in the Contents, and in some cases in the Glossary. Numbers in **boldface** refer to illustrations. Where two or more page references to the text occur, the asterisk after a number indicates the main text description. The index also includes synonyms, other spellings, and groups sometimes recognized as distinct but not mentioned in this book.

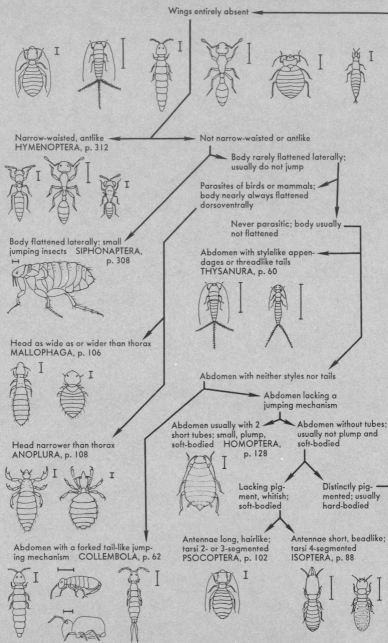

PICTURED KEY TO THE PRINCIPAL

Wings entirely absent

Narrow-waisted, antlike
HYMENOPTERA, p. 312

Not narrow-waisted or antlike

Body rarely flattened laterally;
usually do not jump

Parasites of birds or mammals;
body nearly always flattened
dorsoventrally

Never parasitic; body usually
not flattened

Body flattened laterally; small
jumping insects   SIPHONAPTERA,
p. 308

Abdomen with stylelike appendages or threadlike tails
THYSANURA, p. 60

Head as wide as or wider than thorax
MALLOPHAGA, p. 106

Abdomen with neither styles nor tails

Abdomen lacking a
jumping mechanism

Abdomen usually with 2
short tubes; small, plump,
soft-bodied   HOMOPTERA,
p. 128

Abdomen without tubes;
usually not plump and
soft-bodied

Head narrower than thorax
ANOPLURA, p. 108

Lacking pig-
ment, whitish;
soft-bodied

Distinctly pig-
mented; usually
hard-bodied

Abdomen with a forked tail-like jumping mechanism   COLLEMBOLA, p. 62

Antennae long, hairlike;
tarsi 2- or 3-segmented
PSOCOPTERA, p. 102

Antennae short, beadlike;
tarsi 4-segmented
ISOPTERA, p. 88